The ESG and Sustainability Deskbook for Business

A Guide to Policy, Regulation, and Practice

Kristyn Noeth

The ESG and Sustainability Deskbook for Business: A Guide to Policy, Regulation, and Practice

Kristyn Noeth
Los Angeles, CA, USA

ISBN-13 (pbk): 979-8-8688-0260-7   ISBN-13 (electronic): 979-8-8688-0261-4
https://doi.org/10.1007/979-8-8688-0261-4

Copyright © 2024 by Kristyn Noeth

This work is subject to copyright. All rights are reserved by the Publisher, whether the whole or part of the material is concerned, specifically the rights of translation, reprinting, reuse of illustrations, recitation, broadcasting, reproduction on microfilms or in any other physical way, and transmission or information storage and retrieval, electronic adaptation, computer software, or by similar or dissimilar methodology now known or hereafter developed.

Trademarked names, logos, and images may appear in this book. Rather than use a trademark symbol with every occurrence of a trademarked name, logo, or image we use the names, logos, and images only in an editorial fashion and to the benefit of the trademark owner, with no intention of infringement of the trademark.

The use in this publication of trade names, trademarks, service marks, and similar terms, even if they are not identified as such, is not to be taken as an expression of opinion as to whether or not they are subject to proprietary rights.

While the advice and information in this book are believed to be true and accurate at the date of publication, neither the authors nor the editors nor the publisher can accept any legal responsibility for any errors or omissions that may be made. The publisher makes no warranty, express or implied, with respect to the material contained herein. The content of this book is for informational purposes, and it does not constitute legal or professional advice or an endorsement of any company, product, or service.

Managing Director, Apress Media LLC: Welmoed Spahr
Acquisitions Editor: Shiva Ramachandran
Development Editor: James Markham
Project Manager: Jessica Vakili

Cover designed by eStudioCalamar

Distributed to the book trade worldwide by Apress Media, LLC, 1 New York Plaza, New York, NY 10004, U.S.A. Phone 1-800-SPRINGER, fax (201) 348-4505, e-mail orders-ny@springer-sbm.com, or visit www.springeronline.com. Apress Media, LLC is a California LLC and the sole member (owner) is Springer Science + Business Media Finance Inc (SSBM Finance Inc). SSBM Finance Inc is a **Delaware** corporation.

For information on translations, please e-mail booktranslations@springernature.com; for reprint, paperback, or audio rights, please e-mail bookpermissions@springernature.com.

Apress titles may be purchased in bulk for academic, corporate, or promotional use. eBook versions and licenses are also available for most titles. For more information, reference our Print and eBook Bulk Sales web page at http://www.apress.com/bulk-sales.

Any source code or other supplementary material referenced by the author in this book is available to readers on GitHub (https://github.com/Apress). For more detailed information, please visit https://www.apress.com/gp/services/source-code.

If disposing of this product, please recycle the paper

To Timothy Beatley (Ph.D., Teresa Heinz Professor of Sustainable Communities at the University of Virginia) and Edith Brown Weiss (J.D., Ph.D., University Professor Emeritus at the Georgetown University Law Center)

Table of Contents

About the Author xvii

About the Technical Reviewer xix

Acknowledgments xxi

Endorsements xxiii

Foreword xxvii

Preface xxxi

Introduction xxxv

Chapter 1: Understanding ESG: Factors, Foundations, and Differentiators 1

1.1 Evolution of ESG 3

1.2 Description of ESG Factors 5

1.3 Corporate Sustainability Movement 7

    Milestones in Corporate Sustainability 8

1.4 Emergence of Corporate Social Responsibility 11

    Role of CSR in Shaping Corporate Behavior 12

    Impact of Corporate Social Responsibility on Businesses ... 13

1.5 Untangling the Terminology: ESG, CSR, and Sustainability 14

    Intersection of ESG and CSR 15

    Overlap of ESG and Sustainability 16

TABLE OF CONTENTS

    ESG, CSR, and Sustainability: Complementary Approaches 16
  1.6 Business Case for ESG: Long-Term Value Creation 17
    ESG is Sound Business Strategy 21

Chapter 2: The Current ESG Outlook 25

  2.1 Recent and Heightened Focus on ESG.. 25
  2.2 Relevance of ESG Initiatives in the Current Business Climate 27
    Influence of ESG on Business Planning 28
    Market and Economic Impacts of ESG 29
    Growing Attention and Expectations from Stakeholders 31
  2.3 Evolving Legal and Regulatory Landscape 33
    Ripple Effects of a Regulatory Vacuum for Businesses 34
    Business Preparation for Regulatory Requirements..................................... 35
  2.4 Coalescing of the Standards Framework 38
    Global Push for Standardized ESG Reporting 38
  2.5 Collision Course on Climate 39
  2.6 Demands for Transparency and Data Integrity... 43

Chapter 3: The International Legal and Policy Foundations of ESG 47

  3.1 Business and Human Rights 48
    Fundamentals of Human Rights 48
    Origins of International Human Rights Law... 48
    Human Rights Laws and Business Practices 55
  3.2 Climate and Natural Resources.. 57
    Guiding Principles of International Environmental Law............................... 57
    Emergence of International Cooperation 57
  3.3 Sustainable Development 58
    Brundtland Commission (1983) 58
    Brundtland Report: Our Common Future (1987) 58

TABLE OF CONTENTS

Defining Sustainability and Sustainable Development 59
Impact and the Origins of Sustainability 59
3.4 Key Global Treaties, Agreements, and Summits 60
Stockholm Conference (1972) 60
The Rio Earth Summit: A Pioneering Conference on Environment
and Development (1992) 61
United Nations Convention on the Law of the Sea (1994) and
High Seas Treaty (2023) 70
Kyoto Protocol (1997) 74
Paris Agreement (2015) 77
COP28 in Dubai (2023) 80
3.5 United Nations Sustainable Development Goals (SDGs) 85
Introduction to the Sustainable Development Goals 85
Key Terms and Concepts 87
The Seventeen SDGs 87
Progress Toward Targets 89
3.6 ESG Integration with the International Principles and SDGs 90
Investor Criteria 90
Key Insights for ESG Integration 93
Challenges in Achieving the SDGs 96

Chapter 4: A Deep Dive into the "E" in ESG 97

4.1 Global Tipping Point 99
4.2 World Economic Forum (WEF) Global Risks Report 100
Global Risks Report 2024 101
Top Global Risks 2024 102
Global Risks Report 2023 104
Top Global Risks 2023 106
Takeaways from the Past Two Annual Global Risks Reports 107

TABLE OF CONTENTS

4.3 Climate Policy and the Intergovernmental Panel on Climate Change (IPCC) 108
Structure and Organization of the IPCC 109
Reporting Cycles 110
Influence on Climate Policy 111

4.4 Greenhouse Gases (GHGs) 112
List of Greenhouse Gases 113
Primary Sources and Impacts of Greenhouse Gases 115
Regulation of Greenhouse Gases 117

4.5 Top Greenhouse Gas (GHG)-Emitting Sectors and Countries 117
Major Contributing Sectors 118
Top Emitting Countries 120
Emissions Data Analysis Informs GHG Reduction Approaches 121

4.6 Carbon Markets 123
Understanding Carbon Markets: Compliance Market and Voluntary Market Models 123
Development of Global and Regional Carbon Markets 124
Challenges Facing Expansion of Carbon Markets 127

4.7 Pollution and Waste 129
Sustainability Trends in Packaging 130

4.8 Biodiversity and Natural Capital Depletion 131
Defining Biodiversity 131
Current State of Biodiversity and Natural Capital 132
Expansion of Natural Capital and Biodiversity Funds 135

4.9 Responsible Sourcing 139
Establishing a Responsible Sourcing Strategy 139

4.10 Water Use and Stewardship 141
Water Stewardship as a Business Imperative 142

TABLE OF CONTENTS

 4.11 Circular Economy 144
  Environmental Justice 147
  Future of the Circular Economy 148

Chapter 5: A Deep Dive into the "S" in ESG 151
 5.1 Social Factors and Their Impacts... 151
 5.2 Human Capital and Employee Welfare 153
  Evolution of Human Capital Management in the Era of ESG....... 153
 5.3 Diversity, Equity, and Inclusion 155
 5.4 Gender Equality and Equity 159
  Gender Pay Gap 160
 5.5 Human Rights and Labor Practices.. 163
 5.6 Supply Chain Transparency.. 165
  Maintaining a Transparent Supply Chain 167
 5.7 Consumer Protection, Product Safety, and Quality..................... 168
 5.8 Philanthropy, Responsibility, and Volunteering........................... 170
 5.9 Community Relations 173

Chapter 6: A Deep Dive into the "G" in ESG 177
 6.1 Governance Factors and Their Impacts....................................... 177
 6.2 Transparent and Accountable Governance Practices.................. 180
 6.3 Management Responsibilities and Transparency........................ 183
  Executive Compensation 184
 6.4 Board Duties and Oversight 187
 6.5 Board Composition and Capabilities 189
  Board Diversity 191
  Board Committee Structure 193
  Board Performance 197

TABLE OF CONTENTS

6.6 Business Ethics and Values.. 198
6.7 Compliance, Risk, and Audit... 200
6.8 Stakeholder Engagement... 202
6.9 Shareholder Activism, Proxy Voting, and Asset Management 204
6.10 Greenwashing 208

Chapter 7: The Global ESG Regulatory Paradigm.......................... 213

7.1 Development of the ESG Regulatory Environment..................................... 213
    Building Upon Existing Environmental, Social, and Governance Regulations.. 215
    ESG Taxonomy 220

7.2 Key Global Regulatory Initiatives 223
    Comprehensive Sustainability Regulation in the European Union............... 231
    Sustainability Disclosure Requirements in the United Kingdom 245
    Stock Exchange and Securities Rules Requiring Sustainability Disclosures 249
    Climate-Focused Regulation, Legislation, Exchange Rules, and Tariffs 251
    Executive Compensation 270
    Human Capital Management 271
    Supply Chain and Human Rights Regulation 274
    Circular Economy and Greenwashing Regulation.................................. 276
    Government and Stock Exchange Board Diversity Rules 282
    Regulatory Efforts to Address the Gender Pay Gap 288
    Regulatory and Legislative Actions on Antitrust 290

7.3 Distinguishing Regulation, Legislation, and Executive Action 292
7.4 Concept of Materiality... 293
    Financial Materiality 293
    Non-financial Materiality and Double Materiality 294

TABLE OF CONTENTS

7.5 Disclosure and Reporting Models 295
7.6 Rise of Enforcement, Litigation, and Shareholder Challenges 297
    Regulatory Enforcement 297
    Climate Litigation 301
    Greenwashing Litigation 304
    Shareholder Litigation 306
7.7 Navigating Regulations and Emerging ESG Risk 308

Chapter 8: The ESG Standards and Frameworks 311

8.1 Progression of Standards and Frameworks 311
8.2 Examination of Key Standards and Frameworks 317
    International Sustainability Standards Board (ISSB) 317
    ISSB's Inaugural Standards: IFRS S1 and IFRS S2 318
    Global Reporting Initiative (GRI) 322
    CDP (Formerly Carbon Disclosure Project) 323
    Taskforce on Nature-Related Financial Disclosures (TNFD) 324
8.3 Carbon Accounting and the GHG Protocol 327
    Carbon Accounting 327
    Greenhouse Gas (GHG) Protocol 328
8.4 Incorporation of Standards into Legislation and Regulation 329
8.5 Implications for Compliance and Risk Management 330

Chapter 9: The ESG Ratings Providers and Indices 333

9.1 Role of ESG Ratings 333
9.2 ESG Ratings Providers 336
9.3 ESG Indices and Relevancy in the Market 339
9.4 Criticism of Ratings Providers 340
9.5 Potential Regulation and Oversight of Ratings Providers 342

xi

TABLE OF CONTENTS

Chapter 10: The Role of the Corporation..................................349
10.1 Focus on Emerging ESG Trends and Risks.................................349
10.2 Fiduciary Duties and ESG..351
Fundamentals of the Fiduciary Duties.................................351
Duty of Care...352
Duty of Loyalty...352
Duty of Good Faith...353
Business Judgment Rule..353
Fiduciary Duties and ESG Application..............................354
10.3 Rise of Corporate Purpose and the Aligned Corporation.............354
10.4 Alternative Corporate Structures..356
Public Benefit Corporations (PBCs)..................................356
Distinguishing PBCs from B Corps....................................358
10.5 Stakeholder Engagement..361
10.6 ESG Strategy, Program Development, and Reporting................362
10.7 ESG in Mergers and Acquisitions...367
10.8 Creating Green Jobs and Reskilling the Workforce...................368
Leading Sustainability Within the Corporation.....................370
10.9 Continuous Innovation..372
Blue Ocean Strategy...373
Blue Ocean Economy..374

Chapter 11: The Role of the Financial Sector..............................379
11.1 Rise in ESG-Aligned Investing..379
11.2 Growth in Sustainable Finance...380
11.3 Managing Climate Risk as a Strategic Priority.........................384
11.4 Advancing the New Economy and the Clean Energy Transition.....385

TABLE OF CONTENTS

Chapter 12: The Role of International Bodies and UN Programs 391

12.1 Driving International Agreement 392

12.2 Integration of the UN Principles for Responsible Investment (PRI) 393

12.3 Business Engagement with the UN Global Compact (Global Compact) 396

12.4 United Nations Environment Assembly (UNEA) and Multilateralism 397

12.5 Group of 7 (G7) and Group of 20 (G20) Weathervanes 398

    G7 and Global Sustainability 399

    G20 and the Green Growth Agenda 401

12.6 Global Reach of the Organisation for Economic Co-operation and Development (OECD) 402

12.7 International Union for Conservation of Nature (IUCN): A Convening of Governments and Civil Society 406

12.8 International Organization of Securities Commissions Oversight (IOSCO) 408

Chapter 13: The Role of Business Interest Groups 411

13.1 Convening the World Economic Forum (WEF) 411

13.2 World Business Council for Sustainable Development (WBCSD) and the SDGs 413

13.3 Business for Social Responsibility (BSR) as an Early Mover 414

13.4 Business Roundtable (BRT) and the "Statement on the Purpose of a Corporation" 415

13.5 The Ceres Influence 416

13.6 We Mean Business Coalition and Climate Action 418

13.7 State of Play for Trade Associations and Business Groups 419

Chapter 14: The Role of NGOs and Third-Party Mechanisms 423

14.1 Multipurpose Role of the Nongovernmental Organization (NGO) 423

14.2 Leading the Way on Science-Based Targets 426

14.3 Organizational Advocacy Platforms 429

14.4 Independent Role of Certifying Organizations 430

TABLE OF CONTENTS

Chapter 15: The Role of the Philanthropic Sector 437

15.1 Philanthropic Impact and Influence 437

15.2 Mission Shift to Achieve Impact 438

15.3 Increased Level of Funding and Grants Activity 440

15.4 Innovative Financing Vehicles 442

15.5 Private Foundations Making their Mark in Climate Philanthropy 448

15.6 Public Charities Programming for Impact and Alignment
with the SDGs 451

Chapter 16: The Pathways to Decarbonizing Key Sectors 453

16.1 Decarbonizing the Energy Sector 455

    Decarbonization Challenges and Potential Solutions 457

    Role of Innovation and Technology 459

16.2 Decarbonizing the Transportation Sector 460

    Road Transportation: Advancing Technology 460

    Aviation and Maritime Transport: Energy Density Matters 463

    Road Ahead 464

16.3 Decarbonizing the Agricultural Sector 465

    GHG Emissions in Agriculture 466

    Pathways to Decarbonization 467

16.4 Decarbonizing the General Manufacturing Sector 469

    Challenges to Decarbonization 470

    Pathways to Decarbonization 471

16.5 Decarbonizing the Building Industry 472

    Green Buildings and Sustainable Urban Planning 473

    Continuous Innovation in Green Design and ESG Business Strategy 476

    Challenges to and the Future of Decarbonization in the
Building Sector 477

TABLE OF CONTENTS

16.6 Decarbonizing the Fashion Industry 477

    Raw Material Production and Processing.. 478

    Manufacturing and Distribution 479

    Consumer Use and Disposal 479

    Challenges in Decarbonizing the Fashion Sector 479

    Key Steps to Decarbonize the Fashion Sector.. 481

Index..**485**

About the Author

Kristyn Noeth is an accomplished advisor and speaker who collaborates with leaders and organizations worldwide as the Founder of Verde Impact, a purpose-driven advisory and consulting firm. Her current portfolio includes advising a climate tech company nominated for The Earthshot Prize, working as the sustainability advisor to an award-winning Bloomberg TV show, and partnering with changemakers and companies in all growth stages to advance their environmental, social, and governance goals toward a more equitable and sustainable future. She is frequently interviewed and writes and presents on a range of global change topics.

Kristyn previously held in-house executive and legal roles in global corporations, law firms, charitable organizations, and the White House. Those positions include Chief Sustainability Officer, Chief Counsel for Sustainability, Head of Government Relations and Public Policy, and Chief Compliance Officer for Charitable Foundations and Corporate Social Responsibility at market-leading companies, including Nestlé, Hess, and USAA. She also worked as an attorney at Weil, Gotshal & Manges and other top law firms. Kristyn began her career as a Presidential Management Fellow in the White House, working on COP3 and the Kyoto Protocol, and has subsequently been engaged in the primary global policy, legislative, and regulatory initiatives on climate and sustainability.

ABOUT THE AUTHOR

With a keen interest in the power of philanthropy and nonprofit programs to advance social good, Kristyn has been a member of the Board of Directors of Voss Foundation, Alzheimer's Association, CITYarts, and MAG America. She has also served as pro bono counsel to many nonprofits throughout the course of her career.

Kristyn holds a J.D. from Georgetown Law and a B.U.E.P. and an M.U.E.P. from the University of Virginia (UVA). Her thesis was entitled, "Implementing Market Mechanisms to Address Environmental and Social Cost," and she was a US Environmental Protection Agency NNEMS Fellow at UVA. Kristyn lives in Los Angeles and teaches the "Global Business Practices in Sustainability" course at UCLA Extension.

Contact information and more is available at www.verdeimpact.com.

About the Technical Reviewer

As the Founder and Managing Partner of Faust Global Partners, **Jennifer Faust** has over 20 years of experience in global investment and business advisory, specializing in ESG, sustainability, and impact investing. She also provides strategic counsel in business development, investor relations, and capital raising, primarily focusing on private equity, venture capital, and development finance. As an International Advisor at the UN's International Trade Centre, Jennifer supports initiatives that promote sustainable agricultural value chains and focus on inclusive and innovative finance and agricultural investment in the African, Caribbean, and Pacific states. She serves on the board of the Aruna Project, advises Catalyst Investment Management and Bankers without Boundaries, and has previous experience with the US Development Finance Corporation and 57 Stars. Jennifer is dedicated to fostering social and environmental change through innovative, collaborative efforts.

Acknowledgments

Writing this book has been a full circle experience. It brought me back to my introduction to sustainability and the complexities of our global environmental, social, and governance challenges more than 30 years ago. For that, I have Professor Tim Beatley and Professor Edith Brown Weiss to thank, which is why this book is dedicated to them.

I was fortunate to be an undergraduate student when Professor Beatley launched the sustainability coursework at the University of Virginia, and it opened a new window into the world for me and others. Professor Beatley is a pioneer in the sustainable communities field, and, subsequently, as my graduate advisor, he encouraged me to undertake the first sustainability plan for the campus as an independent study. Then-dean and author of *Cradle to Cradle: Remaking the Way We Make Things*, William McDonough, was also incredibly supportive of my graduate work.

Studying with Professor Weiss at Georgetown Law expanded those foundations with international environmental law and policy courses and seminars. Professor Weiss has served in prominent posts at the United Nations, the World Bank, and other global organizations. She also gave me my first opportunity to become a published author all those years ago as a chapter author in the book, *Reconciling Environment and Trade*.

My sincere appreciation to the colleagues who each lead by example and have helped shape my perspective throughout my career, including Ted Tsekerides, Holly Loiseau, Steve Reiss, Annemargaret Connolly, Sal Romanello, David Singh, Bill Brownell, Wally Martinez, Steve Solow, Tim Goodell, Charlie Broll, Bob Long, Trey Sadiq, Brian Nurse, and Susan Gordon.

ACKNOWLEDGMENTS

My deepest gratitude to Melissa J. Kopolow for writing the foreword, to Jennifer Faust for serving as the reviewer, and to Charlie Broll, Olly Purnell, Eric Fins, Rebekah Jefferis, Susan Koehler, and Anna Andreis for providing advance reviews of the book.

A heartfelt thank you to all the family members, friends, and colleagues who were delighted to learn that this book was coming to fruition. Your support means the world to me.

I wish to acknowledge Shiva Ramachandran, the Senior Editor at Apress, who greenlit and guided the production of this book; her colleagues Jim Markham, Jessica Vakili, and Shonmirin P.A.; and the entire Apress and Springer Nature team.

Endorsements

"In a fast-moving ESG landscape touching many areas, Kristyn Noeth's *The ESG and Sustainability Deskbook for Business* provides the detailed yet practical guidance many senior executives are in need of as they plan out how best to approach their corporate ESG strategy. While the stakeholders are passionate and varied, and the enacted and proposed laws and regulations are not yet by any means settled, this book is high on facts and insights about the various ESG and sustainability issues businesses must consider as they plan ahead. As a lawyer, I recall the days where *The Bluebook* and *Black's Law Dictionary* were always within arm's reach; I envision this book to be a well-earmarked and highlighted reference kept close by for many an ESG professional and their senior executive colleagues looking to understand the myriad of issues associated with ESG and sustainability."

  Charlie Broll, Chief Legal Officer, Tropicana Brands Group

"With the forensic analysis of a lawyer, Kristyn leaves nothing on the table in her quest to make the world of ESG so much easier to navigate and know. Wherever you're leading a business, whichever continent, country or virtual space you inhabit—this book can be your 'go to' guide. It's easy to digest, thoughtful in tone... and, unlike power outlets, sockets and chargers, it's universally relevant and compatible."

  Olly Purnell, Founder and Managing Partner, Q5 Partners

ENDORSEMENTS

"From comprehensive policies to regulatory structures, sustainability efforts are here to stay and *The ESG and Sustainability Deskbook for Business* primer on what ESG means, what it does, and why it matters is a must-read for investors, business leaders, academics, and practitioners alike. While details across jurisdictions and international borders may vary and change, especially as it relates to climate risk disclosures, the need for uniform, transparent, and data driven decision-making will not. The requirement to understand how we arrived at this moment in history and where we go from here is critical, and for that, Kristyn has us covered."

Eric Fins, Vice President, Grove Climate Group

Former Deputy Staff Director,
US House Select Committee on the Climate Crisis

"Today's Founders need to be the creative architects of business, designing with intention, purpose, and sustainability from the start. Their ability to design fluid and flexible systems for innovation as a daily practice is critical for building forever companies, and this book sets the context and the challenge to do just that."

Rebekah Jefferis, Founder & Co-CEO, FNDR

"Kristyn Noeth's *The ESG and Sustainability Deskbook for Business* deconstructs ESG's complexities into understandable and actionable guidance. As a comprehensive resource it's vital for board of directors' oversight, strategic business leaders implementing sustainability, and students seeking a 360-degree overview to accelerate global impact."

Susan Koehler, Founder, Growth4Future

ENDORSEMENTS

"In an era of increased complexity, rapid change, and looming risks, *The ESG and Sustainability Deskbook for Business* from Kristyn Noeth is a necessary and grounding reading. It gives perspectives on how the sustainability and ESG principles took shape and provides an exhaustive overview of regulations, principles, and terminology that surround the work of all professionals in the space. At such a critical moment for action, when companies accelerate not just pledges and commitments, but tangible action plans, maintaining a solid foothold in the facts and frameworks that led us here is essential."

Anna Andreis, EVP, Sustainability
and Social impact, MSL Group

Foreword

Madeleine Albright, the first female Secretary of State of the United States, once said, "One of the issues I kept saying to my students is you have to learn to interrupt. When you raise your hand at a meeting, by the time they get to you, the point is not germane. So, the bottom line is active listening. If you are going to interrupt, you look for opportunities. You have to know what you're talking about." In an era where businesses are increasingly scrutinized not only for their financial performance but also for their impact related to environment, social, and governance practices, Kristyn Noeth's *The ESG and Sustainability Deskbook for Business* interrupts with gusto and arrives as an indispensable guide, illuminating the intricate tapestry of international agreements, regulatory frameworks, and evolving practices that shape our collective journey toward a more sustainable future. Where oh where was this book when I was starting my career in this space?

Written in a refreshingly straightforward and conversational tone, this Deskbook invites you on a comprehensive exploration of the ESG landscape, from its historical origins to its contemporary applications in business and the markets. Through insightful discussions, topical briefings, and substantive analyses, it provides a 360-degree perspective on the organizations, frameworks, and stakeholders driving the evolution of sustainability practices worldwide.

From the foundational principles elucidated in the early chapters to the nuanced examinations of ESG factors and the roles of various stakeholders, each section of the Deskbook serves as a vital piece in the

FOREWORD

mosaic of understanding. It navigates seamlessly through the labyrinth of international treaties, policy developments, and regulatory paradigms, offering readers a roadmap to navigate the complexities of the ESG terrain.

What sets this Deskbook apart is Noeth's ability to demystify complex concepts and distill them into actionable insights. Whether exploring the differentiation between ESG, sustainability, and corporate social responsibility or delving into the intricacies of corporate governance and investor engagement, Noeth seamlessly guides readers through the maze of information with clarity and precision.

As ESG and sustainability practices transition from voluntary initiatives to regulated imperatives, the Deskbook serves as a timely resource for corporations and investors navigating this evolving landscape. With mandatory reporting requirements on an array of ESG metrics, from greenhouse gas emissions to supply chain transparency, the need for practical guidance has never been greater. This Deskbook transcends the realm of corporate boardrooms, delving into the collaborative efforts of international bodies, business interest groups, NGOs, and philanthropic organizations in fostering a collective commitment to sustainability and responsible stewardship of our planet.

The ESG and Sustainability Deskbook for Business rises to the occasion, threading the needle between theory and practice, offering best practices, case studies, and actionable takeaways to empower organizations, investors, and even students alike. In an era defined by unprecedented challenges and opportunities, this Deskbook stands as an indispensable companion for those committed to advancing sustainability and responsible business practices.

FOREWORD

The Deskbook is as practical as it is inspiring. It provides the knowledge from which individuals and organizations can make genuine and sustainable change. It should be required reading for sustainability courses, as part of an on-boarding "gift" for new ESG/sustainability/CSR hires, and a Boardroom "must read." The Deskbook is exactly what I would have used to help me when I was hired to turn Secretary Albright's vision for a sustainability practice into reality. As she often said, it was our job to help companies see that they could do well by doing good.

Melissa J. Kopolow
March 19, 2024
Washington, DC

Preface

I can trace my early interest in how people interact with the natural and built environments back to Philadelphia. I was twelve years old when I started competing at a serious level in the sport of swimming. That pursuit would later take me to the national and collegiate levels, but back then I was just a kid curious about the world around her and traveling across the Philly area to train and compete at different facilities for various swim meets. The urban infrastructure of Philly in the mid-1980s was aging (actually, crumbling) and disproportionately affecting so many of its communities. I observed that first-hand as we'd drive through the different neighborhoods, and the disparities were a regular topic of conversation among the area swimmers.

Live Aid was held that same year with two concerts cast simultaneously at Wembley Stadium in London and JFK Stadium in Philly. Riveted, I watched the broadcast on our family television with the standard rabbit ears antenna. Although a member of the MTV generation, I was rarely allowed to watch TV, so I was excited to see so many of my favorite artists of the time perform. Tom Petty, Queen, Tina Turner, Eric Clapton, David Bowie, U2, and so many more. When we learned that Phil Collins was performing at Wembley and then taking a supersonic plane called the Concorde to perform in Philly, it blew my young mind and heightened my anticipation.

But what struck me and stayed with me most from Live Aid was its humanitarian purpose. I recall sitting on our living room floor and seeing the images of the devastating famine in Ethiopia, the Sudan, and other regions in Africa and feeling the magnitude of the suffering. It had a profound effect. That so many artists and organizers had come together for

PREFACE

Live Aid showed me the power of people and platforms to do good in the world. That message was clear to my impressionable 12-year-old self, and it has always been a guiding light, particularly in my service to charitable organizations. I am now advising a nonprofit partnering with Earth Aid, which is the global reboot and expansion of Live Aid set to launch in 2025. Full circle, indeed.

I speak with so many colleagues in the sustainability space as well as other leaders and founders who are building with purpose, and most of them have origin stories too. The desire to be effective stewards and to solve our global challenges is strong. I also talk to early career professionals and students regularly who want to make an impact. We want to believe the work we are doing is meaningful. It's important to recognize that since the spotlight trained on corporate sustainability and ESG in recent years can be demoralizing as criticism comes from all directions and tells us that our efforts are either falling short or going too far or somehow incongruously both. Pretty much every one of us feels the pressure of the moment and knows that we are in the most pivotal decade to reach the global climate and sustainable development goals. Constructive engagement is paramount given the scale of the interconnected environmental, social, and governance challenges we face.

I am reminded of a time a decade ago when I met one of my heroes. I had the honor of serving on the Board of Directors at Voss Foundation, a charitable organization that partnered with local communities to build water, sanitation, and hygiene (WASH) projects in sub-Saharan Africa. The foundation had been formed by then-CEO of Voss Water, Knut Brundtland, with the support and involvement of his father, Arne Olav Brundtland, and mother, Gro Harlem Brundtland. As we'll cover in Chapter 3, Dr. Gro Harlem Brundtland is the architect of modern sustainability. Dr. Brundtland was the Prime Minister of Norway and later became Director-General of the World Health Organization (WHO) and a member of The Elders, the renowned group of world leaders convened by Nelson Mandela. She chaired the World Commission on Environment and Development (WCED), commonly known as the Brundtland Commission. The Brundtland Commission coined and poignantly defined the term "sustainable development" as, in pertinent part, "meeting the needs of the present without compromising the ability of future generations to meet their own needs." It is still the universally accepted definition of sustainability today.

Voss Foundation held a Women Helping Women luncheon with Jewel as the headliner and a wonderful turnout at Donna Karan's Urban Zen studio in New York. There were many luminaries in the room. I was on the lookout for Dr. Brundtland, as I knew she was coming from Norway for the event. We held a bazaar featuring artisans from the communities in which the foundation supported WASH projects prior to the luncheon, and, as I was milling about greeting guests, I turned and saw Dr. Brundtland approaching me. As she reached me, she pointed to the necklaces I was wearing and then to hers. We had both layered the same two cast brass pendants—a talon pendant and a circle pendant—from the Kenyan artisan collective, Ziya. We smiled and introduced ourselves and had a lovely conversation.

The shared connection over our matching pendants was unexpected and so refreshing. I reflect on that moment warmly as I realized then that our heroes are people, too. We are all just people. Everyone that I have met who has been working in sustainability for some time cares deeply about making a difference. Sometimes it's tough to feel like we are putting any wins on the board given the scope of what we are confronting, so we all seek those moments of connection and mutual support. I think that's one reason why attendance at the annual COP meetings and Climate Week has risen astronomically in recent years, and why Climate Week was famously referred to as "Burning Man for Climate Geeks," in an article by Pulitzer Prize-winning journalist Cara Buckley, who sits on the Climate Desk at *The New York Times*.

Collaboration and concerted effort are the keys to our collective future. We have big problems to solve and it's going to take all of us to come together and advance solutions. This book is part of my contribution.

PREFACE

Through years of training and toil, I have learned how to best break down complex concepts, policies, and regulations into cogent steps for strategy implementation and operational integration. I have also worked across every industry and engaged with stakeholders around the world—elected officials, regulators, investors, scientists, philanthropists, NGOs, activists, consumers, and business and community groups—that have provided a 360-degree view and informed my perspective.

I love doing the work. I have always loved doing the work. I have been writing, speaking, and teaching on the subject for decades and I believe it is vitally important to share knowledge and build community. That's why when I never had a good answer to the question posed by so many asking what they could read that would provide a comprehensive overview of ESG and sustainability in business, I decided to write this book.

My goal in writing this book was to capture much of what I've assimilated in the field over the years and to synthesize those insights in an informative manner. I purposefully wrote a deskbook as a resource for readers to meet you wherever you are on your sustainability journey. I hope this book serves you well and that you enjoy reading it as much as I enjoyed writing it.

Kristyn Noeth
March 17, 2024
Los Angeles, CA

Introduction

This book is a practical guide and reference tool meant for a wide audience. Companies and their boards of directors, public policy leaders, sustainability professionals, founders, investors, philanthropists, and students wanting to advance their knowledge and familiarity with the core principles, policies, practices, and players in the global marketplace will benefit from this book. Readers will gain an understanding of the progression of ESG and sustainability and how related considerations increasingly drive business, policy, and economic decisions to inform an advanced take on the multi-stakeholder profile in business and the markets.

We have entered the Age of Accountability. Stakeholders are demanding transparency and accountability, and the pace and record of regulatory activity worldwide on ESG and sustainability is remarkable. Governments have stepped up their oversight efforts. We've seen all manner of legislative, regulatory, and executive action as well as new stock exchange rules on financial- and non-financial-related climate risk and sustainability disclosure, supply chain, human rights, greenwashing, board diversity, ratings, investment fund labels, and other environmental, social, and governance subjects. We've also seen the advancement of national green taxonomies and sovereigns in the sustainable finance market.

We are also in a Race to the Top. The advent of disregard for and scaling back of global environmental policy in the past was commonly referred to as the "Race to the Bottom" (borrowing a socioeconomic phrase). Companies were thought to do the minimum to comply with regulations, particularly in curbing pollution and waste generated from their manufacturing activities. We've entered a new phase in which many market-leading companies view ESG and sustainability as a differentiator.

INTRODUCTION

A demonstrated commitment to transparent, ethical, and sustainable business practices showing measurable outcomes is not only reflective of a responsible company, but it also brings a competitive advantage. The expectation of stakeholders—regulators, investors, shareholders, consumers, employees, and the community—across the board is driving change. The risks (e.g., physical risk, legal risk, and reputational risk) and opportunities (e.g., financial performance, access to capital, and social license to operate) have become evident and ESG is unlocking new potential. A sound ESG strategy is a long-term value creator for companies.

Recently, there has also been unprecedented public investment which has spurred exponential private investment to advance the new economy with a focus on clean and green technologies, decarbonization and related greenhouse gas reduction pathways, and natural capital opportunities linked to biodiversity protection. Historical impediments to advancement, particularly in the hard-to-abate sectors, included a lack of investment and available financing, slow regulatory approvals and permitting processes, difficulties transitioning a world built for conventional sources, and hurdles to get to market and scale. Those obstacles are now being removed from the runway.

The topics of ESG and sustainability are multifaceted and require some navigation. This book provides a comprehensive overview of the current state of play as well as the emerging trends, evolving regulatory landscape, cross-sector collaborations, and market outlook for investment and innovation. You will get a good sense of the book's organization and information flow from the Table of Contents. We all absorb information differently, so this book is structured so you could either read it from start to finish as the material in each chapter progressively builds or flip directly to any given chapter to review a specific topic. Each chapter closes with takeaways, and there are best practices, case studies, and infographics supporting key points throughout the book.

CHAPTER 1

Understanding ESG: Factors, Foundations, and Differentiators

WHAT IS ESG?

ESG is a framework used to assess a company's business practices, operations, and performance on a variety of interconnected environmental, social, and corporate governance factors that have the potential to impact the company's ability to execute business strategy and to create long-term value.

Environmental, Social, and Governance (ESG) is a multifaceted concept that has gained widespread recognition and implementation in the business world. ESG criteria focus on how a company impacts the natural environment (Environmental), how it relates to its employees, suppliers, customers, investors, shareholders, and communities (Social), and how the company is managed (Governance). The three central pillars are considered a critical guide to assessing risk and to measuring the sustainability and societal impact of a company. A circular design like Figure 1-1 is typically used to depict the interrelated aspects of ESG and how business performance on ESG is viewed as a whole.

CHAPTER 1   UNDERSTANDING ESG: FACTORS, FOUNDATIONS, AND DIFFERENTIATORS

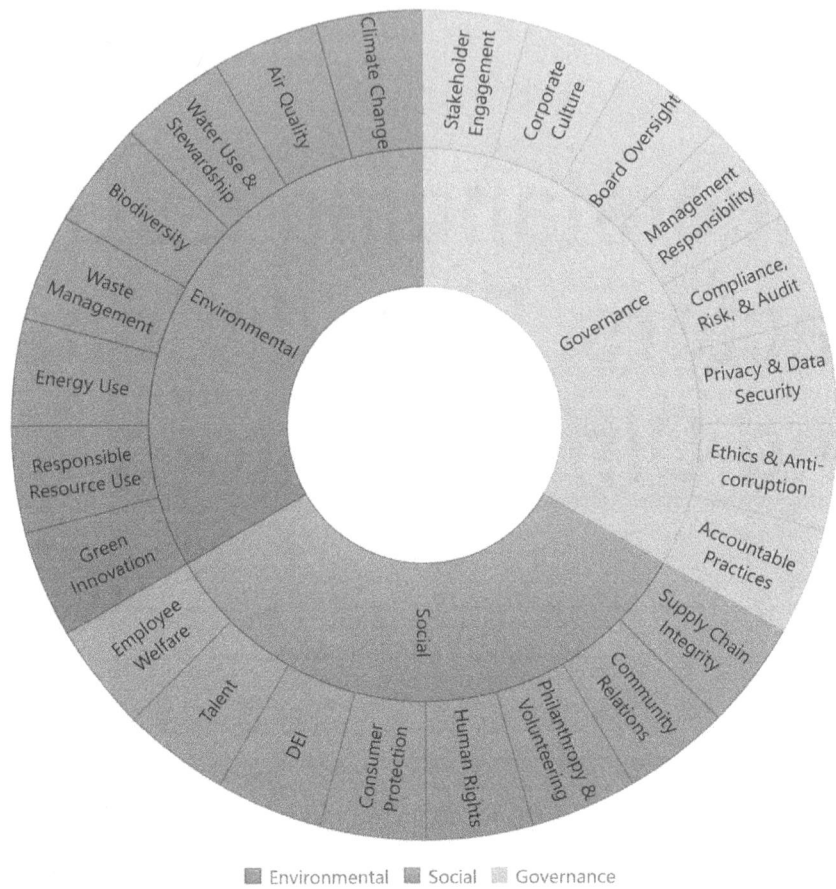

Figure 1-1. Environmental, Social, and Governance (ESG)

The term **ESG** first appeared in the *Who Cares Wins: Connecting Financial Markets to a Changing World* report by the United Nations Global Compact in 2004, which included "[r]ecommendations by the financial industry to better integrate environmental, social and governance issues in analysis, asset management and securities brokerage."[1] Twenty financial

---

[1] United Nations Global Compact, *Who Cares Wins: Connecting Financial Markets to a Changing World* (2004) (www.unepfi.org/fileadmin/events/2004/stocks/who_cares_wins_global_compact_2004.pdf).

institutions, including Calvert Group, Morgan Stanley, Goldman Sachs, AXA, HSBC, BNP Paribas, and UBS, with a combined USD$6 trillion in assets under management (AUM), collaborated on and endorsed the *Who Cares Wins* report. The report stated that the endorsing institutions were "convinced that a better consideration of environmental, social and governance factors will ultimately contribute to stronger and more resilient investment markets, as well as contribute to the sustainable development of societies."[2]

The report correlated performance on ESG issues with increased shareholder value and noted the associated benefits of proper risk management, anticipated regulatory action, and access to new markets.[3] At its core, ESG involves using non-financial information as inputs into financial decisions. Companies are scored by ratings providers on metrics around environmental, social, and governance factors. The original use of ESG was as a tool to screen and assess investment risk, and it has since expanded to a broader application used by many stakeholders to gauge company performance and business responsibility. The terms ESG and sustainability are often used interchangeably, and that usage is addressed later in this chapter.

1.1 Evolution of ESG

ESG has evolved significantly over the years. From its initial stages as an offshoot of socially responsible investing (SRI), it grew into a fully-fledged investment philosophy adopted by many leading global investors and asset managers. SRI initially involved avoiding investments in companies whose practices conflict with an investor's ethical considerations or run afoul of contemporary public policy. Over time, this evolved into an approach that incorporated the consideration of ESG issues into investment decision-making.

---

[2] Id. at ii.
[3] Id. at i.

CHAPTER 1   UNDERSTANDING ESG: FACTORS, FOUNDATIONS, AND DIFFERENTIATORS

Today, ESG has moved beyond just screening out potentially harmful investments. It now also involves actively seeking out companies that demonstrate strong ESG performance, with the belief that these companies will be better long-term investments. ESG has now become an umbrella term encompassing sustainability and social responsibility. It is used somewhat interchangeably with corporate responsibility and sustainability, although those are distinct terms and with differing functions.

The concept of the "triple bottom line" of people, planet, and profit for a business was developed by John Elkington in 1994.[4] The triple bottom line was initially an accounting concept to value complete internal and external costs, including ecological, social, and economic cost. It is part of the corporate lexicon and represents alignment with and transparency around the full cost of doing business.

Around the same time, in 1997, Business Roundtable (BRT), a leading nonprofit business organization headquartered in Washington, D.C., issued its first "Statement on the Purpose of a Corporation." The statement was updated in 2019 and signed by nearly 200 CEOs, including chief executives from Apple, General Motors, BlackRock, JPMorgan Chase, Bank of America, Amazon, and Vista Equity Partners. It sent a strong message that "shareholder value is no longer everything."[5] The statement focuses

---

[4] John Elkington, *Cannibals with Forks*. Oxford: Capstone (1997). See also Jesse Finfrock, *Q&A: John Elkington.* Mother Jones (2008). See also The Economist. Triple Bottom Line, November 17, 2009 (www.economist.com/news/2009/11/17/triple-bottom-line).

[5] David Gelles and David Yaffe-Bellany, *Shareholder Value Is No Longer Everything, Top C.E.O.s Say*, The New York Times (August 19, 2019) (www.nytimes.com/2019/08/19/business/business-roundtable-ceos-corporations.html).

CHAPTER 1   UNDERSTANDING ESG: FACTORS, FOUNDATIONS, AND DIFFERENTIATORS

on commitments to all stakeholders, including to generate "long-term value for shareholders" and "good jobs, a strong and sustainable economy, innovation, a healthy environment and economic opportunity for all."[6]

We've seen the advancement from shareholder capitalism[7] to stakeholder capitalism. ESG is largely focused on stakeholders. The term "stakeholders" as used in the field generally refers to customers, employees, shareholders, investors, regulators, communities, and often society at large. Recent discussion among corporate leadership, boards of directors, and institutional investors has centered around the role of ESG and how best to incorporate it into corporate and investment decision-making. Companies must effectively manage both short- and long-term risks while considering stakeholder viewpoints.

1.2 Description of ESG Factors

There is no uniform definition of ESG factors; however, there are a set of core criteria that are consistently applied by ratings providers while accounting for variabilities across industries. We'll discuss ESG ratings more fully in Chapter 9. Table 1-1 provides a commonly identified set of environmental, social, and governance topics.

---

[6] Business Roundtable, *Statement on the Purpose of a Corporation* (2022) (https://opportunity.businessroundtable.org/ourcommitment/).
[7] Milton Friedman, *A Friedman doctrine—The Social Responsibility of Business is to Increase its Profits*, The New York Times (September 13, 1970) (www.nytimes.com/1970/09/13/archives/a-friedman-doctrine-the-social-responsibility-of-business-is-to.html).

CHAPTER 1   UNDERSTANDING ESG: FACTORS, FOUNDATIONS, AND DIFFERENTIATORS

Table 1-1. *ESG Pillars and Factors*

Environmental	Social	Governance
Climate Change	Employee Welfare, Health & Safety	Accountable Policies, Procedures, & Practices
Air Quality & Emissions	Talent Development & Retention	Ethics & Anti-corruption
Water Use & Stewardship	Diversity, Equity, & Inclusion	Privacy & Data Security
Biodiversity & Natural Resources	Consumer Protection and Product Safety & Quality	Compliance, Risk, & Audit
Waste Management	Human Rights & Labor Practices	Management Responsibilities & Transparency
Energy Use	Philanthropy, Responsibility, & Volunteering	Board Oversight & Diversity
Responsible Resource Use	Community Relations	Corporate Culture
Green Innovation	Supply Chain Integrity	Stakeholder Engagement

ESG offers a set of defined and quantifiable criteria used to evaluate a company's performance in these three pillars. It also provides a framework by which a company can assess material risk, identify challenges and opportunities, and integrate environmental, social, and governance metrics into its operations.

ESG is a lens that offers a comparable view of a company's societal and environmental impact. We'll take a deep dive into each of the substantive environmental, social, and governance issues in Chapters 4–6, respectively. We'll next look at the precursors to ESG—sustainability and social responsibility—and differentiate the three terms.

1.3 Corporate Sustainability Movement

WHAT IS SUSTAINABILITY?

Sustainability "meets the needs of the present without compromising the ability of future generations to meet their own needs."

-Our Common Future (1987)[8]

The practice of corporate sustainability dates back nearly 50 years and coincided with the modern environmental movement and the advent of environmental law. Concerns about pollution and resource depletion prompted companies to adopt responsible business practices. Rooted in compliance with air, water, and natural resources law, many companies began to take a strategic approach to environmental stewardship.

There was a growing realization that corporations needed to take responsibility for their impact on the environment and society. Sustainability encompasses the interconnectedness of environmental, economic, and social issues. Companies looked at the long-term consequences of their actions and started to implement measures to minimize their negative effects. This marked a shift in mindset and laid the foundations for further development of sustainability efforts.

---

[8] *Report of the World Commission on Environment and Development: Our Common Future* (1987) (https://sustainabledevelopment.un.org/content/documents/5987our-common-future.pdf).

CHAPTER 1   UNDERSTANDING ESG: FACTORS, FOUNDATIONS, AND DIFFERENTIATORS

Milestones in Corporate Sustainability

Two important developments helped guide and advance corporate sustainability. They set the stage for standards setting, corporate reporting, and business engagement with the United Nations (UN) on sustainability. Summaries of those two developments are as follows, with an in-depth discussion provided in Chapter 13.

1. **Global Reporting Initiative (GRI):** GRI is an international and independent standards organization that developed one of the foundational frameworks for companies to report sustainability performance, enabling stakeholders to assess environmental, social, and economic impacts, in 1997.

2. **United Nations Global Compact (Global Compact):** The Global Compact is an initiative established by the UN in 2000 that calls upon corporations to align business operations and strategy with ten principles in the areas of human rights, labor, the environment, and anti-corruption (Table 1-2). The principles are derived from the Universal Declaration of Human Rights, the International Labour Organization's Declaration on Fundamental Principles and Rights at Work, the Rio Declaration on Environment and Development, and the United Nations Convention Against Corruption.[9]

---

[9] United Nations Global Compact, *The Ten Principles of the Global Compact* (https://unglobalcompact.org/what-is-gc/mission/principles).

Table 1-2. *The Ten Principles of the UN Global Compact*[10]

Human Rights
Principle 1: Businesses should support and respect the protection of internationally proclaimed human rights; and
Principle 2: make sure that they are not complicit in human rights abuses.
Labor
Principle 3: Businesses should uphold the freedom of association and the effective recognition of the right to collective bargaining;
Principle 4: the elimination of all forms of forced and compulsory labour;
Principle 5: the effective abolition of child labour; and
Principle 6: the elimination of discrimination in respect of employment and occupation.
Environment
Principle 7: Businesses should support a precautionary approach to environmental challenges;
Principle 8: undertake initiatives to promote greater environmental responsibility; and
Principle 9: encourage the development and diffusion of environmentally friendly technologies.
Anti-Corruption
Principle 10: Businesses should work against corruption in all its forms, including extortion and bribery.

Those platforms enabled a shift in corporate sustainability from a primary focus on facility and footprint operations to a broader perspective on corporate values and stakeholder relationships. Stakeholders began to voice their opinions and exercise their purchasing and investing power to align with environmentally and socially responsible decisions.

---

[10] Id.

Companies began to realize the business benefits of sustainability. Sustainable practices can lead to cost savings and increased competitiveness, as well as a social license to operate. Sustainability was also seen to drive innovation and new market opportunities by way of consumer preference for eco-friendly and ethically sourced products. A strong corporate purpose also aids in attracting talent and employee retention.[11]

A challenge for early and mid-growth corporate sustainability teams was receiving the resources and internal support to develop and implement comprehensive programs. Often, sustainability was seen as a cost center rather than a profit center and took a back seat within companies. That perception has largely changed over the intervening years, depending on the industry and company profile. We'll discuss corporate sustainability programs in Chapter 10.

CASE STUDY: BROOKS

Brooks is a US-based running apparel company. The company imbedded its sustainability ethos into its product life cycle. Its corporate sustainability report is titled *Running Responsibly*.

The company's public priorities are: (1) *Diversity, equity, and inclusion*; (2) *Responsible sourcing*; (3) *Climate action*; (4) *Sustainable consumption*; and (5) *Community impact.* Brooks provides transparency on the elements of and progress against those commitments and has gone the step further to map their strategy to the UN Sustainable Development Goals (SDGs) (more on the importance of charting corporate goals to the SDGs in Chapter 3).[12]

---

[11] *Economic uncertainty puts pressure on sustainable behavior change*, Deloitte Insights (July 25, 2023) (www2.deloitte.com/us/en/insights/environmental-social-governance/sustainable-consumer-behaviors.html).

[12] Brooks *Running Responsibly* (www.brooksrunning.com/en_us/meet-brooks/running-responsibly/).

1.4 Emergence of Corporate Social Responsibility

> **WHAT IS CSR?**
>
> ***Corporate Social Responsibility*** refers to voluntary actions taken by a business to improve its social and environmental impact and that demonstrate the importance of ethical corporate behavior.

Corporate Social Responsibility (CSR) refers to a company's voluntary actions aimed at improving its societal and environmental impact beyond what is legally required. It is an organizational philosophy that stresses the importance of behaving ethically and contributing to societal development. The concept of CSR suggests that corporations have an obligation to consider the interests of their stakeholders in all aspects of their operations. This could involve a range of activities, such as conducting fair labor practices, supporting community partnerships, and practicing ethical marketing.

CSR is not a one-size-fits-all concept. Different companies interpret and implement CSR in unique ways, depending on factors such as their size, industry, and geographic location. At its core, CSR is about companies making a conscious effort to do good and to minimize harm within the company, within the community, and within society.

CSR initiatives may encompass outreach to and support of the community with philanthropic donations, community service programs, employee volunteer efforts, and any other actions that contribute positively to society. Unlike ESG, CSR is more of an internally defined and self-regulating framework, primarily used by companies to set and carry out objectives for responsible corporate citizenship practices. The success of CSR programs is a more qualitative measure and can be challenging to define, as implementation varies from company to company.

The charge of corporate responsibility is for companies to take responsibility for their impact on society. This can be anything from creating employment opportunities to investing in nonprofit organizations and community projects. It's all about companies recognizing that their behavior can impact a wide range of people and places.

The core principles of CSR are

1. Ethical Behavior

2. Promoting Sustainability

3. Transparency

4. Fostering Stakeholder Engagement

5. Accountability

Role of CSR in Shaping Corporate Behavior

In the early days of capitalism, businesses were primarily focused on profit maximization. However, as industries grew and their impact on society and the environment became more apparent, the idea of corporate responsibility began to take shape.

The origins can be traced back to the early industrialists of the 20th century, who funded philanthropic causes and established charitable organizations. The concept was further formalized in 1953, with the publication of *Social Responsibilities of the Businessman*, by Howard R. Bowen, which advocated for corporate ethics and social responsibility.[13] Much has changed since then, including the makeup of the workforce and the associated nomenclature.

Modern CSR both shapes and shows corporate behavior. It encourages companies to ensure their operations are safe and consider the welfare of employees and citizens. It also promotes concepts of fairness and

---

[13] Howard R. Bowen, *Social Responsibilities of the Businessman*, University of Iowa Press (republished 2013) (www.jstor.org/stable/j.ctt20q1w8f).

stewardship, thereby fostering a positive relationship between businesses and society. CSR initiatives can also influence corporate behavior by encouraging businesses to engage in philanthropic activities, such as making grants to charities or creating social programs that benefit the community.

Impact of Corporate Social Responsibility on Businesses

The impact of CSR on businesses can be significant. CSR initiatives can help companies improve their reputation, increase customer loyalty, and attract and retain employees. CSR can also help businesses differentiate themselves from their competitors, thereby gaining a competitive advantage. CSR can also build trust with stakeholders. By behaving ethically, respecting stakeholder interests, and being accountable and transparent, companies can foster better relationships with their stakeholders.

Many companies struggle to fully integrate CSR into their management decisions, which can limit the effectiveness of their efforts and hamper their ability to measure impact. Companies may also face criticism of their CSR programs with assertions of window-dressing or that the efforts don't go far enough to address more fundamental problems. The most effective CSR programs are ones that stay true to their purpose and adjust to keep up with emerging social issues and stakeholder expectations.

CHAPTER 1   UNDERSTANDING ESG: FACTORS, FOUNDATIONS, AND DIFFERENTIATORS

> **CASE STUDY: PATAGONIA**
>
> **Patagonia** is a US-based outdoor clothing company. Patagonia is known for its strong commitment to sustainability and its responsibility as a corporate citizen. The company uses recycled materials in its products, encourages customers to repair and recycle their gear, and donates a portion of its profits to environmental causes.
>
> Through its CSR efforts, Patagonia has not only reduced its environmental footprint but also built a strong brand and loyal customer base. In 2022, the company took its efforts a step further and converted its corporate legal structure to a charitable trust, declaring to the world: *"Earth is now our only shareholder."*[14]

1.5 Untangling the Terminology: ESG, CSR, and Sustainability

In the corporate world, the acronyms **ESG** (Environmental, Social, and Governance) and **CSR** (Corporate Social Responsibility) and the term **sustainability** have gained significant prominence. Although related, these concepts have distinct meanings, implications, and applications. To comprehend the paradigm, it's key to understand these three concepts and their interrelationships.

---

[14] *Patagonia Ownership* (www.patagonia.com/ownership/).

Intersection of ESG and CSR

While ESG, CSR, and sustainability are distinct concepts, they are not mutually exclusive. In fact, they often work together in the pursuit of sustainable and responsible business practices. While ESG is fundamentally external facing, CSR is primarily internally focused as it is grounded in company culture and values. The initiatives that form a company's CSR strategy often contribute to its ESG performance.

CSR relates primarily to the "S" (Social) pillar of ESG, which measures a company's impact on and commitment to making a positive impact on society. A company's CSR initiatives can be charted to many of the social factors of ESG reporting. ESG can lend credibility to specific aspects of CSR as social metrics.

Integrating CSR into corporate practices requires a focus on a company's culture, its impact on people and society, and its relationships with the local community. Developing initiatives that align with core competencies and the issues that matter to customers, employees, and communities is a strong foundation. A continuous improvement loop is necessary for the success of a CSR program, and that involves:

1. Collecting Insights from Stakeholders
2. Setting Measurable Goals
3. Developing a Roadmap
4. Reporting on Progress

Overlap of ESG and Sustainability

Integrating ESG and sustainability practices can yield numerous benefits in terms of reputation, brand recognition, customer loyalty, talent recruitment and retention, and long-term value. Both ESG and sustainability mitigate risk and encourage responsible business practices. Sustainability is grounded in systems and largely in how a company's operations, such as facilities and sourcing, impact the environment.

While related concepts, ESG and sustainability differ in scope and focus. ESG primarily focuses on measuring a company's performance in relation to specific environmental, social, and governance factors. It provides a framework for evaluating corporate practices, risks, and opportunities. Sustainability encompasses a holistic approach to business operations and a broader set of considerations beyond ESG metrics.

That said, the terms sustainability and ESG are often used interchangeably in the corporate sector. This is primarily due to ease of reference and avoidance of confusion. For the purposes of this book, we'll do the same while remaining steadfast that there are distinctions between the terms. It's about understanding how these concepts can complement each other to create a more sustainable and responsible business.

ESG, CSR, and Sustainability: Complementary Approaches

In an era marked by climate change, civil unrest, and evolving regulatory landscapes, ESG, CSR, and sustainability will continue to play crucial roles in shaping corporate strategies. They are no longer business optional—they are business imperatives.

The future will likely see a greater integration of ESG, sustainability, and CSR, with companies increasingly recognizing the value of a fulsome approach to sustainability and social impact. As investors, consumers, and regulators demand greater transparency and accountability, businesses that embrace these principles will be better positioned for success.

At the end of the day, companies apply different labels to their efforts, but the goal is the same: to build businesses that are not only profitable but also responsible, ethical, sustainable, and resilient. By integrating these practices into business strategy, companies can drive growth, enhance brand reputation, and create value for all stakeholders. It's about doing good—and doing well.

1.6 Business Case for ESG: Long-Term Value Creation

The value of ESG is multidimensional and extends beyond mere financial returns. ESG metrics have multifaceted applications for risk management and for the investment community. Companies with strong ESG performance are often better at mitigating risk, which can, in turn, lead to more stable returns. Along with financial performance and operational performance, ESG performance relates to how a company is measured, evaluated, and valued (Figure 1-2).

CHAPTER 1   UNDERSTANDING ESG: FACTORS, FOUNDATIONS, AND DIFFERENTIATORS

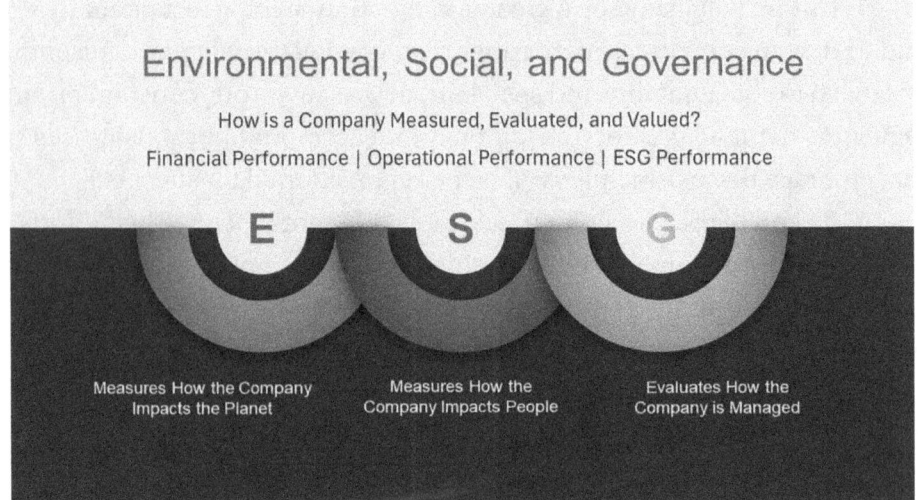

Figure 1-2. How is a Company Measured, Evaluated, and Valued?

ESG criteria can also help investors identify potential investment opportunities. Companies that lead in ESG performance may be better positioned for future challenges and could represent attractive investment opportunities. ESG offers a value alignment for investment portfolios. This alignment can contribute to investor satisfaction and potentially lead to greater investor engagement.

Myriad studies and surveys have found a positive correlation between ESG and corporate performance. The following are sources evidencing the continued expansion and value of ESG:

- Bloomberg Intelligence and Bloomberg New Economy conducted a survey of 250 C-suite executives and 250 investors globally.[15] Key findings include:

---

[15] *Bloomberg Intelligence Survey Finds Investors and C-Suite Embrace ESG, Despite Concerns*, Bloomberg (November 7, 2023) (www.bloomberg.com/company/press/bioomberg-intelligence-survey-finds-investors-and-c-suite-embrace-esg-despite-concerns/).

CHAPTER 1   UNDERSTANDING ESG: FACTORS, FOUNDATIONS, AND DIFFERENTIATORS

- ○ ESG remains a backbone of financial markets and corporate strategy, driven by global regulation, consumer demands, and competitive ambitions.
- ○ 84% of executives say ESG helps deliver a more robust corporate strategy and 85% of investors reported that ESG leads to better returns, resilient portfolios and enhanced fundamental analysis.
- ○ 85% of investors plan to boost ESG investment over the next five years despite geopolitical risks and macro-economic factors.[16]

- Bain & Company worked with EcoVadis to assess how ESG activities impact the companies for which EcoVadis is a sustainability ratings provider. They reviewed 100,000 companies (95% of which are privately held) and found "that positive ESG outcomes are a trait of successful companies and that sustainability measures [including in supply chain, renewable energy, employee satisfaction, and DEI] correlate with better financial performance."[17]

- IBM's Institute for Business Value, in collaboration with Oxford Economics, studied the linkage between sustainability and business value. They surveyed 5,000 C-suite executives across 22 industries in 22 countries

---

[16] Id.

[17] Axel Seemann, Sylvain Guyoton, Anna Bianchi, and Jacqueline Han, *Do ESG Efforts Create Value?* Bain & Company and EcoVadis (April 17, 2023) (www.bain.com/insights/do-esg-efforts-create-value/).

and found that companies that embed sustainability throughout their operations are 52% more likely to outperform peer companies with an average 16% higher revenue growth rate. The complete findings are contained in the 2024 report, *Beyond checking the box: How to create business value with embedded sustainability*.[18]

- Rockefeller Asset Management and the NYU Stern Center for Sustainable Business collaborated to evaluate the link between ESG and financial performance as studied in more than 1,000 research papers published from 2015 to 2020. They "found a positive relationship between ESG and financial performance for 58% of the 'corporate' studies focused on operational metrics such as ROE, ROA, or stock price with 13% showing neutral impact, 21% mixed results (the same study finding a positive, neutral or negative results) and only 8% showing a negative relationship."[19]

- According to the oft-cited 2019 McKinsey *Five Ways That ESG Creates Value* white paper, which was supported by independent research, "A strong ESG proposition correlates with higher equity returns,

---

[18] *Beyond checking the box: How to create business value with embedded sustainability*, IBM Institute for Business Value (February 2024) (www.ibm.com/thought-leadership/institute-business-value/en-us/report/sustainability-business-value).

[19] Tensie Whelan, Ulrich Atz, Tracy Van Holt, and Casey Clark, CFA, *ESG and Financial Performance: Uncovering the Relationship by Aggregating Evidence from 1,000 Plus Studies Published between 2015 – 2020* (Rockefeller Asset Management and NYU Center for Sustainable Business) (www.stern.nyu.edu/sites/default/files/assets/documents/NYU-RAM_ESG-Paper_2021%20Rev_0.pdf).

from both a tilt and momentum perspective. Better performance in ESG also corresponds with a reduction in downside risk, as evidenced, among other ways, by lower loan and credit default swap spreads and higher credit ratings."[20]

ESG is Sound Business Strategy

ESG has emerged as a requisite for contemporary business strategy. Environmental, social, and governance considerations are not just the ethical or moral aspects of business anymore; they have substantial implications for a company's financial performance, risk management, and brand reputation. We'll consider the business case for ESG and its purpose in value creation and provide some key insights.

ESG is no longer a peripheral aspect of business. It's now at the heart of corporate strategies, and for a good reason. There are compelling reasons why ESG matters, and they can be broadly categorized under three main themes: financial performance, risk mitigation, and brand perception. The leading benefits of ESG include

- **Financial performance:** The correlation between strong ESG performance and financial performance is increasingly evident. Companies that excel in their ESG practices often outperform their peers in the long term, both in stock performance and profitability. The profit and growth correlation is a clear indication that companies are wise to integrate ESG into their core business strategy.

---

[20] *Five ways that ESG creates value*, McKinsey Quarterly (November 14, 2019) (www.mckinsey.com/capabilities/strategy-and-corporate-finance/our-insights/five-ways-that-esg-creates-value).

CHAPTER 1   UNDERSTANDING ESG: FACTORS, FOUNDATIONS, AND DIFFERENTIATORS

- **Access to capital:** Private and public companies at all growth stages increase their access to capital with strong ESG performance. For instance, many emerging companies with a strong purpose and ESG record appeal to ESG-focused and socially conscious funds. Public companies have been able to attract major institutional investors, particularly with funds that have mandatory ESG requirements.

- **Risk mitigation:** ESG factors are essential in identifying and managing risks that could detrimentally affect a company's profitability and survival. Such risks could be related to climate change, unethical practices, or poor governance structures. Companies with robust ESG programs and reporting practices can anticipate, manage, and mitigate these risks more effectively as their management teams and boards make more informed and responsive decisions.

- **Social license and reputation:** Consumers, employees, investors, and regulators are increasingly conscious of a company's ESG performance. As a result, companies that prioritize ESG are likely to have a better reputation, attract and retain top talent, and enjoy customer loyalty. Demonstration of and transparency around ESG commitments serves to mitigate reputational risk.

- **Company valuation for mergers and acquisitions (M&A) activity:** The potential of ESG to create value is immense. ESG has also become a differentiator in the mergers and acquisitions (M&A) due diligence process and a harbinger of deal risk. A KPMG survey of over

200 ESG practitioners including corporate investors, financial investors, and M&A debt providers found that more than half had cancelled deals based on poor ESG due diligence findings.[21]

- **Competitive advantage:** Strong ESG performance can differentiate a company from its competitors. It can enhance its reputation, make its products or services more appealing to consumers, and make it a more attractive employer. ESG can create a deeper understanding of stakeholder needs, as well as better aligned stakeholder relationships.

- **New opportunities and innovation:** ESG can help companies identify and seize new market opportunities. Companies can develop and adapt products and services to meet emerging consumer demand and the needs of civil society. ESG challenges can spur innovation. Businesses that take their environmental and social responsibilities seriously are more likely to come up with novel solutions to problems and to make their operations more transparent and accountable.

---

[21] KPMG, *2023 ESG Due Diligence Study* (https://kpmg.com/us/en/articles/2023/esg-due-diligence/us-esg-due-diligence-study.html).

CHAPTER 1   UNDERSTANDING ESG: FACTORS, FOUNDATIONS, AND DIFFERENTIATORS

CHAPTER TAKEAWAYS

- ESG has become the new norm for businesses committed to sustainable and responsible practices
- ESG is a driver of innovation and long-term value creation
- Environmental criteria consider a company's impact on the environment, such as its carbon footprint, natural resource use, and water management strategies
- Social criteria evaluate a company's societal impact, including its labor practices, community engagement, and diversity, equity, and inclusion initiatives
- Governance criteria scrutinize a company's management practices, board oversight, ethical standards, and stakeholder relations

CHAPTER 2

The Current ESG Outlook

In today's rapidly changing business landscape, ESG issues are pivotal in shaping the future of corporations. From reducing the carbon footprint to promoting diversity on boards of directors, companies that prioritize ESG are not only aligning with societal values but also positioning for long-term success. We will explore the importance of ESG for corporations, the benefits and opportunities it brings, the increasing focus from investors, its influence on business planning, and the market and economic impacts.

2.1 Recent and Heightened Focus on ESG

It has been widely reported that ESG assets are expected to exceed USD$50 trillion by 2025—representing one-third of total global AUM.[1] In 2023, sustainable funds returned to the long-term trend of outperforming their peers.[2] The allocation of capital in the private markets to ESG investments

---

[1] Bloomberg Intelligence. *ESG assets may hit $53 trillion by 2025, a third of global AUM* (February 23, 2021) (www.bloomberg.com/professional/blog/esg-assets-may-hit-53-trillion-by-2025-a-third-of-global-aum/).

[2] *Sustainable Funds Outperformed Peers in 2023*, Morgan Stanley Institute for Sustainable Investing (February 29, 2024) (www.morganstanley.com/ideas/sustainable-funds-performance-2023-full-year).

CHAPTER 2   THE CURRENT ESG OUTLOOK

sends a strong signal. ESG has significant implications for the broader economy. It reflects a new era where investors seek out companies that contribute positively to society and the environment. By steering investment toward companies with strong ESG performance, investors can encourage all companies to adopt more sustainable business practices. In turn, companies with strong ESG performance are often better equipped to handle economic downturns, which can lead to a more stable economy.

Layered upon market activity is the flurry of ESG regulation in recent years. We've seen executive, legislative, and regulatory action across the globe on climate disclosure and reporting, human capital, human rights, and related topics. The nascent regulatory regimes require businesses to get up to speed on new requirements, to onboard qualified talent, to align their corporate governance, and to update their risk management, compliance, legal, and audit programs with ESG requirements.

At the same time, the debate on ESG has heated up politically in the United States.[3] This is not new, as there has always been concerted political opposition to climate, sustainability, and social impact initiatives—it's just now on a bigger scale with broader media coverage as awareness of and engagement in these issues has increased.

In a turn, Larry Fink, the Chairman and Chief Executive Officer (CEO) of BlackRock, and a frontrunner in stakeholder capitalism, publicly shied away from the continued use of the moniker "ESG" as it had been used as a lightning rod for detractors.[4] But the consensus is

---

[3] Michael Copley, *Republican attacks on ESG aren't stopping companies in red states from going green*, NPR (June 27, 2023) (www.npr.org/2023/06/27/1183794068/republican-esg-attacks-nucor-steel-climate-change).

[4] Isla Binnie, *BlackRock's Fink says he's stopped using 'weaponised' term ESG*, Reuters (June 26, 2023) (www.reuters.com/business/environment/blackrocks-fink-says-hes-stopped-using-weaponised-term-esg-2023-06-26/).

that ESG is here for the long haul. Investors say that ESG terminology is now mainstream, and the term shouldn't be replaced despite the backlash.[5]

2.2 Relevance of ESG Initiatives in the Current Business Climate

ESG has become one of the top issues at the corporate leadership and board levels. The COVID-19 pandemic, climate change, social justice, human rights, and other current global issues have heightened awareness and demand for business responsibility and sustainability. These challenges have underscored the need for the corporate sector to address ESG issues more urgently than ever before.

Investors, customers, employees, regulators, and other stakeholders are increasingly demanding that companies demonstrate their commitment to ESG. They expect businesses to operate in a socially and environmentally responsible manner and to contribute positively to society. By aligning with stakeholder expectations, companies can enhance their reputation, attract investment, and foster long-term loyalty.

ESG initiatives can unlock new opportunities for business growth and innovation. By investing in sustainable practices and technologies, companies can reduce costs, enhance operational efficiency, and tap into emerging markets. For example, companies that prioritize renewable energy can benefit from government incentives, cost savings, and a competitive edge in a rapidly evolving energy landscape.

ESG-focused companies have a competitive advantage when it comes to accessing capital. Investors are increasingly incorporating ESG criteria into decision analysis, and companies that prioritize ESG are more likely

---

[5] Tim Quinson, *'ESG' Is Too Important to Ax*, Investors Say, Bloomberg (November 7, 2023) (www.bloomberg.com/news/articles/2023-11-08/investors-ignore-us-attacks-as-esg-judged-too-important-to-ax).

to attract capital. By demonstrating a commitment to ESG, companies can tap into a larger pool of capital and gain a competitive edge in the market.

Strong ESG practices also guide companies to identify and mitigate potential risks. By proactively addressing environmental, social, and governance issues, companies can minimize legal and reputational risks, avoid regulatory penalties, and enhance their ability to adapt to changing market conditions. ESG also enables companies to build resilience, ensuring their long-term viability in the face of evolving challenges.

Influence of ESG on Business Planning

ESG considerations are increasingly integrated into business planning processes, influencing strategic decision-making and shaping the future direction of companies. ESG influences various aspects of business planning, including

- **Setting Long-Term Goals and Targets**

    ESG considerations are fundamental in setting long-term goals and targets for companies. By aligning business strategies with ESG principles, companies can define their purpose, articulate their values, and establish a clear vision for the future. Setting ambitious ESG targets helps companies drive continuous improvement, track progress, and demonstrate their commitment to sustainability.

- **Enhancing Risk Management and Resilience**

    ESG considerations are essential for effective risk management and resilience-building. By identifying and addressing ESG-related risks, companies can enhance their ability to navigate uncertainties and disruptions. For example, companies that prioritize supply chain transparency and resilience are better equipped to mitigate risks associated with disruptions, such as those caused by natural disasters or social unrest.

- **Strengthening Stakeholder Engagement and Relationships**

    ESG considerations foster stronger engagement and can build trust, loyalty, and positive reputation. Effective stakeholder engagement is vital for understanding expectations, aligning strategies, and fostering mutually beneficial relationships.

- **Driving Innovation and Adaptation**

    ESG considerations drive innovation and adaptation by encouraging companies to explore new business models, technologies, and practices. By embracing sustainable innovation, companies can develop products and services that meet evolving customer demands, address societal challenges, and create positive environmental and social impacts. ESG-driven innovation fosters creativity, resilience, and a competitive edge in the market.

Market and Economic Impacts of ESG

ESG considerations have significant market and economic impacts, shaping investment trends, influencing consumer behavior, and driving economic transformation. Let's delve into the market and economic impacts of ESG.

CHAPTER 2   THE CURRENT ESG OUTLOOK

- **Attracting Sustainable Investments**

    ESG-focused companies have a competitive advantage in attracting sustainable investments. Investors are increasingly incorporating ESG criteria into their decision-making process, seeking companies that demonstrate a commitment to sustainability and responsible business practices. By prioritizing ESG, companies can access a larger pool of capital and benefit from the growing demand for sustainable investments.

- **Influencing Consumer Behavior and Preferences**

    Consumer behavior is shifting toward greater sustainability and conscious consumption. Increasingly, customers are seeking products and services from companies that align with their values and contribute positively to society and the environment. By adopting ESG practices, companies can gain a competitive edge, enhance brand reputation, and attract a growing base of environmentally and socially conscious consumers.

- **Complying with Regulatory and Legal Changes**

    ESG considerations are driving regulatory and legal changes worldwide. Governments and regulatory bodies are implementing policies and regulations that encourage companies to adopt sustainable practices and mitigate environmental and social risks. By proactively embracing ESG, companies can stay ahead of regulatory requirements, minimize compliance costs, and demonstrate their commitment to responsible business conduct.

- **Fostering Economic Resilience and Sustainable Growth**

    Companies that prioritize ESG are better equipped to navigate economic uncertainties, manage risks, and seize opportunities. By integrating sustainability into their business models, companies can drive innovation, enhance operational efficiency, and contribute to long-term economic growth.

Growing Attention and Expectations from Stakeholders

A common critique of corporate sustainability planning is that there is a disconnect between stated commitments and a credible action plan to reach those goals. Early net-zero pledges announced across the corporate sector were lauded, but most failed to hold up to scrutiny when leadership was pressed for details and a timeline to phase out carbon emissions. Stakeholders expect companies to walk the walk and not just talk the talk.

The clarion call to the financial sector on climate risk as investment risk came from Larry Fink with his *2020 Letter to CEOs: A Fundamental Reshaping of Finance.* He said, "purpose is the engine of long-term profitability" and "[e]very government, company, and shareholder must confront climate change."[6]

Carbon Disclosure Project (CDP) conducted an analysis of more than 13,000 companies that had made public net-zero pledges. That study set accounted for 64% of global market capital (approximately USD$64 trillion). CDP found that only one-third of those companies were developing a carbon transition plan and that just 1% of the companies

---

[6] Larry Fink, *A Fundamental Reshaping of Finance*, BlackRock (2020) (www.blackrock.com/us/individual/larry-fink-ceo-letter).

CHAPTER 2   THE CURRENT ESG OUTLOOK

were providing information across a series of key indicators for climate transition planning.⁷

The rise of shareholder activism on ESG issues coincided with the realization of that gap in credible planning and information disclosure, as it is seen as a tool to effect corporate change. As State Street identified in its *2020 CEO Letter*, ESG is "good business practice and essential to a company's long-term financial performance" as "shareholder value is increasingly being driven by issues such as climate change, labor practices, and consumer product safety."⁸ That letter also provided that "fewer than 25% of the companies we've evaluated have meaningfully identified, incorporated, and disclosed material ESG issues into their strategies."⁹

The number of shareholder resolutions proposed to US companies in 2023 on environmental and social topics—climate change being the most popular—was 337, which represented an increase of 23% over the prior year.¹⁰ Shareholders are exercising their rights to drive change on ESG at the highest levels of the corporation. We'll discuss shareholder activism and proxy voting in more detail in subsequent chapters.

Corporate leadership is attuned to Wall Street and to Main Street. The financial markets outlook tells us that tailwinds on ESG and sustainability are strong. The leading asset managers and institutional investors share

---

⁷ *Just a third of companies that disclosed through CDP in 2021 have climate transition plans*, Carbon Disclosure Project (March 2, 2022) (www.cdp.net/en/articles/companies/just-a-third-of-companies-4002-13-100-that-disclosed-through-cdp-in-2021-have-climate-transition-plans).

⁸ *State Street Global Advisors Letter* (January 28, 2020) (www.ssga.com/library-content/pdfs/insights/CEOs-letter-on-SSGA-2020-proxy-voting-agenda.pdf).

⁹ Id.

¹⁰ Lindsey Stewart, *Proxy Voting Insights: Key ESG Resolutions*, Harvard Law Forum on Corporate Governance (October 4, 2023) (https://corpgov.law.harvard.edu/2023/10/04/proxy-voting-insights-key-esg-resolutions/#:~:text=The%20number%20of%20shareholder%20resolutions,616%2C%20from%20522%20in%202022).

that perspective with individual investors. According to the *Sustainable Signals* report produced by the Morgan Stanley Institute for Sustainable Investing and Morgan Stanley Wealth Management, in January 2024, more than 70% of individual investors believe a company's strong ESG practices can lead to higher returns and more than half of individual investors plan to increase allocations to sustainable investments within the next year.[11]

2.3 Evolving Legal and Regulatory Landscape

As the world becomes increasingly concerned about sustainability, organizations are facing a myriad of new legislative, executive, and regulatory requirements (the terms "regulatory" and "regulations" are commonly used as umbrella terms to describe these requirements) related to ESG and sustainability practices. These regulations vary significantly from one region to another, creating a complex landscape that organizations must navigate in order to remain compliant and competitive.

Confronted with increasing legal and enforcement risk, companies are placing compliance at the forefront of their sustainability strategy. That requires a solid understanding of the nascent regulatory regimes to inform priorities to effectively manage risk and to avoid falling behind the market. We will delve into the global regulatory outlook, summarizing the key ESG, climate, and sustainability requirements, and discuss the business imperatives to adapt to the shifting landscape.

The past few years have been a watershed of global executive, legislative, and regulatory action on ESG issues, particularly climate disclosure and transparency. Significant requirements have been enacted in the United Kingdom, the European Union, China, Singapore, Canada,

---

[11] *Individual Investors' Interest in Sustainability is on the Rise,* Morgan Stanley Institute for Sustainable Investing (January 26, 2024) (www.morganstanley.com/ideas/sustainable-investing-on-the-rise?linkId=356500813).

CHAPTER 2   THE CURRENT ESG OUTLOOK

New Zealand, Japan, Switzerland, Brazil, the United States, and the State of California. The corporate sector and investors have been actively engaged in these developments. We'll discuss each of those legal and regulatory initiatives in detail in Chapter 7.

The new regulations have moved what was largely a voluntary reporting model to a mandatory reporting model for businesses. The trajectory has also gone from regulation solely of financial-related risks to regulation of both financial-related and non-financial related risks. That is a paradigm shift, and one that brings risk and necessitates organizational change. Regulatory disclosure and reporting requirements vary globally and reflect jurisdictional challenges and priorities. This patchwork creates complex issues, particularly for companies operating across borders.

Ripple Effects of a Regulatory Vacuum for Businesses

The ongoing expansion of ESG into the regulatory and legislative arena creates uncertainty for companies subject to compliance and enforcement. That lack of certainty inhibits strategic business planning. Clear expectations and cogent guidance from regulators are needed.

> *As the international community moves forward with their own disclosure rules, slow regulatory movement on our domestic front will hurt U.S. competitiveness.*[12]
>
> —Mindy Lubber, CEO, Ceres

---

[12] *With financial losses from climate change mounting, the SEC must act now*, Ceres (September 14, 2022) (www.ceres.org/news-center/blog/financial-losses-climate-change-mounting-sec-must-act-now).

The US case study is illustrative. The country is lagging other leading economies in adopting second-generation ESG regulations. In 2023, US Securities and Exchange Commission (SEC) Chair Gary Gensler warned that if the SEC did not issue a final climate disclosure rule with substance consistent with EU regulations to meet substituted compliance, then potentially thousands of global US companies with European operations or entities would have to comply with conflicting European and US standards.[13] That came to fruition when the SEC promulgated its final climate disclosure rule on March 6, 2024, as substituted compliance was omitted from the effective rule and covered companies were left to navigate the international regulatory patchwork. We'll discuss that in more detail in Chapter 7.

Business Preparation for Regulatory Requirements

The global regulatory landscape for ESG is complex and rapidly evolving. From the United States to the European Union to Asia Pacific and beyond, regulators are imposing new requirements on organizations to promote sustainability and transparency. As with all evolving regulatory regimes, additional state and national action is expected within the coming years.

The lack of certainty in an evolving regulatory landscape is a through line issue and creates compliance complexities for companies. A KPMG survey of 750 companies globally yielded that only 25% believe they are adequately prepared for the ESG assurance requirements in the new regulatory arena.[14]

---

[13] Mark Segal, *SEC Chair Warns Lack of Climate Rule Exposes Thousands of U.S. Companies to EU Climate Reporting Requirements*, ESG Today (December 7, 2023) (www.esgtoday.com/sec-chair-warns-lack-of-climate-rule-exposes-thousands-of-u-s-companies-to-eu-climate-reporting-requirements/).

[14] *Three-quarters of firms globally are not ready for new ESG rules, KPMG finds,* Reuters (September 26, 2023) (www.reuters.com/sustainability/three-quarters-firms-globally-are-not-ready-new-esg-rules-kpmg-finds-2023-09-26/).

CHAPTER 2   THE CURRENT ESG OUTLOOK

The ESG regulatory landscape may be complex, but it also presents opportunities. By proactively addressing ESG risks, organizations can not only ensure compliance with regulatory requirements but also drive sustainable growth and create long-term value for their stakeholders.

As these regulations and new requirements continue to take shape, organizations will need to stay abreast of developments and adapt their practices accordingly. Corporate planning at the outset is critical to reporting, compliance, and risk management. It is also imperative for internal teams to have direction from leadership so that they can execute their duties. Key steps include

1. **Monitor Compliance and Develop a Strategy in Advance**

    - Track government updates and developments in all jurisdictions where the company is subject to regulation.

    - Develop a strategy that incorporates required standards to adequately prepare and avoid surprises.

    - Establish appropriate internal controls to ensure accuracy and reliability of data sources and reporting.

    - Engage financial, risk, and legal teams at an early stage in the process.

    - Update audit and assurance processes as needed for compliance.

2. **Establish Clear Board of Directors Oversight of ESG Matters**

    - Review existing corporate governance for coverage of ESG issues and ensure it is fit-for-purpose.

    - Ensure the board has requisite expertise and necessary oversight of ESG matters.

    - Align board committees and charters to ESG topics.

3. **Assign a Responsible Corporate Officer**

    - Empower a corporate officer to be responsible for ESG at the company.

    - Assess ESG expertise and talent resources and fill gaps.

    - Establish multidisciplinary teams to serve as spokes with the Responsible ESG Officer as the hub.

4. **Assess Data Collection Infrastructure and Expand the Reporting Program**

    - Implement robust processes to collect, analyze, and report ESG data across the value chain.

    - Review data collection platforms and adapt as needed to meet compliance requirements.

    - Anticipate and plan for ESG-related questionnaires and information disclosure requirements from partners, vendors, and other counterparts.

5. **Continuously Review ESG Metrics and Commitments**

    - ESG metrics and commitments should be reviewed against regulatory requirements and updated as needed.

- Key Performance Indicators (KPIs) should be established for the ESG program and continuously monitored.

- ESG plans should be updated and progress against goals should be shared in a transparent manner with stakeholders.

2.4 Coalescing of the Standards Framework
Global Push for Standardized ESG Reporting

As ESG regulation continues to evolve globally, there is a clear push for standardized ESG reporting. Absent government oversight, independent standards were developed over the course of decades by nonprofit and business organizations. The frameworks collectively made tremendous progress in promoting transparency and accountability in sustainability reporting.

The difficulty in measuring certain ESG factors and the availability and reliability of company-reported data have been a consistent challenge. Early frameworks were limited in application across sectors, and the standards-setting organizations worked to break down those silos and provide sector-specific reporting guidance.

A common refrain was that there were too many frameworks that were overlapping and labor-intensive. The patchwork created a host of problems, and this caused confusion for businesses. Companies were hampered by "conflicting standards, definitions and accounting rules" and "a cacophony of reporting frameworks."[15]

---

[15] Kenza Bryan, *Weak policy holds back corporate climate action, says Sarah Bloom Raskin*, Financial Times (October 26, 2023) (www.ft.com/content/d407d8d5-a1cc-4662-88e4-c47b64a46ecd).

CHAPTER 2   THE CURRENT ESG OUTLOOK

In recent years, there has been a concerted effort to consolidate and align some of the major standards. That has been driven largely by businesses, as well as legislators and regulators who seek to incorporate reporting requirements and mandate the use of specific standards within the language of ESG legislation and regulation. We've seen the merger of the former Sustainability Accounting Standards Board (SASB) and the International Integrated Reporting Council (IIRC) into the Value Reporting Foundation (VRF), which led to the creation of the International Sustainability Standards Board (ISSB) and the subsequent consolidation with and disbanding of the Task Force on Climate-related Financial Disclosures (TCFD).

The next phase of global standards realignment will require the issuance of guidance, continued consultation, and technical assistance to reporting corporations. While the alignment was welcome in the regulated industries, the application of the newly redrawn standards will require process and technology changes for companies. We'll discuss the ongoing developments and specifics of the standards architecture in Chapter 8.

2.5 Collision Course on Climate

> *As long as greenhouse gas concentrations keep rising, we can't expect different outcomes from those seen this year. The temperature will keep rising and so will the impacts of heatwaves and droughts. Reaching net zero as soon as possible is an effective way to manage our climate risks.*[16]
>
>   —Dr. Carlo Buontempo, Director of C3S

---

[16] *Record warm November consolidates 2023 as the warmest year*, Climate Change Service, European Commission (December 7, 2023) (https://climate.copernicus.eu/record-warm-november-consolidates-2023-warmest-year#:~:text=C3S%20Director%2C%20Carlo%20Buontempo%20adds,impacts%20of%20heatwaves%20and%20droughts).

CHAPTER 2   THE CURRENT ESG OUTLOOK

The World Meteorological Organization (WMO) has confirmed that 2023 was the hottest year on record.[17] The average annual global temperate in 2023 was 1.45 ± 0.12°C above pre-industrial levels, which meant we were approaching the Paris Agreement threshold of +1.5°C.[18] We surpassed that barrier in January 2024, with the warmest January on record and the cumulative average for the previous 12 months rising above the 1.5°C mark.[19] An estimated four billion people live in countries where extreme weather and natural disasters related to climate change are increasing in severity and frequency.[20] We've seen increased heat waves and droughts, tidal and storm surges, melting glaciers, and wildfires.

The world's leading climate scientists have warned for decades that increasing greenhouse gas (GHG) concentrations would continue to warm the Earth's atmosphere and alter the climate. The Intergovernmental Panel on Climate Change (IPCC) was established by the United Nations Environment Programme (UNEP) and the World Meteorological Organization (WMO) in 1988. It was endorsed by the UN General Assembly and tasked with preparing a comprehensive review on climate

---

[17] *WMO confirms that 2023 smashes global temperature record*, World Meteorological Organization (January 12, 2023) (https://wmo.int/news/media-centre/wmo-confirms-2023-smashes-global-temperature-record). See also *2023 to be the hottest year on record, EU scientists say*, Reuters (December 6, 2023) (www.reuters.com/business/environment/climate-change-2023-will-be-warmest-year-record-eus-copernicus-2023-12-06/#:~:text=Dec%206%20(Reuters)%20%2D%20European,above%20the%201850%2D1900%20average).

[18] Id.

[19] Kate Abnett and Gloria Dickie, *Climate change drives world to first 12-month spell over 1.5C*, Reuters (February 8, 2024) (www.reuters.com/business/environment/january-was-worlds-warmest-record-eu-scientists-say-2024-02-08/#:~:text=BRUSSELS%2FLONDON%2C%20Feb%208%20(,monitoring%20service%20said%20on%20Thursday).

[20] Jess Shankleman, *COP28: Fossil Fuels Make Climate Deal Draft Text for First Time*, Bloomberg (December 11, 2023) (www.bloomberg.com/news/newsletters/2023-12-11/cop28-fossil-fuels-make-climate-deal-draft-text-for-first-time).

CHAPTER 2   THE CURRENT ESG OUTLOOK

change, specifically recommendations on (1) the state of knowledge on the science, (2) the social and economic impacts, and (3) potential response strategies for inclusion in a future, to-be-established international convention.

The IPCC issued its first report in 1990, which served as the basis for the United Nations Framework Convention on Climate Change (UNFCCC), and it has issued six major assessment reports since its inception. The recently released IPCC Synthesis Report of the Sixth Assessment underscores that immediate action is imperative.[21] Meeting the Paris Agreement climate goals will require a *43% reduction from 2019 baseline GHG emissions* by 2030. Human activity has already warmed the planet markedly since 1880, with most of the warming occurring since 1975.[22] In fact, a "long-term warming trend can be seen as the height of each month increases over time, the result of human activities releasing greenhouse gases like carbon dioxide into the atmosphere."[23]

The level of planetary warming has caused unprecedented changes—some of which are permanent and cannot be remediated by adaptation measures. While popular attacks on the existence of climate change have lessened in the intervening decades given the proven meteorological changes and the devastating impacts on the natural world, the challenge has shapeshifted. We must reverse course within a "rapidly closing window to secure a livable future."[24] This is the most critical decade for climate solutions.

---

[21] *Synthesis Report of the Sixth Assessment Report,* IPCC (2023) (www.ipcc.ch/ar6-syr/).
[22] *Earth Observatory,* National Aeronautics and Space Administration (NASA) (https://earthobservatory.nasa.gov/world-of-change/global-temperatures).
[23] Katy Mesermann, *NASA September 2023 Temperature Data Shows Continued Record Warming,* NASA (October 13, 2023) www.nasa.gov/image-article/nasa-september-2023-temperature-data-shows-continued-record-warming/).
[24] Id.

CHAPTER 2   THE CURRENT ESG OUTLOOK

The impacts are already being observed and felt by humans including with respect to health and welfare, food systems, land use, and property damage. Some industries are more impacted by others, including insurance, energy, agriculture, and consumer packaged goods. Climate change has been categorized as a "threat multiplier" given its tangible systems impact and destabilizing influence on peace and security.[25]

The financial cost of climate change is astounding. A study published in the journal *Nature Communications* found that "the climate change-attributed costs of extreme weather over 2000–2019 are estimated to be US$ 2.86 trillion, or an average of US$ 143 billion per year."[26] Private and public investment in climate solutions have reached new heights; however, it is estimated that the United States alone needs USD$2.5 trillion in the short term (2021-2031)[27] and USD$125 trillion is the global total needed by 2050 to meet decarbonization goals consistent with the Paris Agreement.[28]

Informed by these impacts and costs, businesses with foresight across all sectors are conducting climate risk modeling and transition planning. The recent groundswell on climate and ESG has galvanized many of our public and private sector leaders to action. Governments, nongovernmental

---

[25] *Climate change recognized as 'threat multiplier,' U.N. Security Council Debates impact on peace*, UN News (www.un.org/peacebuilding/fr/news/climate-change-recognized-%E2%80%98threat-multiplier%E2%80%99-un-security-council-debates-its-impact-peace).

[26] Rebecca Newman and Ilan Noy, *The global costs of extreme weather that are attributable to climate change*, Nature Communications (2023) (www.nature.com/articles/s41467-023-41888-1#Sec8).

[27] E. Larson, C. Greig, J. Jenkins, E. Mayfield, A. Pascale, C. Zhang, J. Drossman, R. Williams, S. Pacala, R. Socolow, EJ Baik, R. Birdsey, R. Duke, R. Jones, B. Haley, E. Leslie, K. Paustian, and A. Swan, *Net-Zero America: Potential Pathways, Infrastructure, and Impacts*, Princeton University, Princeton, NJ (December 15, 2020) (https://netzeroamerica.princeton.edu/?explorer=year&state=national&table=2020&limit=200).

[28] United Nations High-Level Climate Action Champions, *What's the cost of net zero?* United Nations (November 3, 2021) (https://climatechampions.unfccc.int/whats-the-cost-of-net-zero-2/).

organizations (NGOs), investors, and corporations have begun collaborating at an accelerated pace for this unprecedented planetary change.[29]

Present levels of investment may not be enough to avert the crisis. UN Secretary-General António Guterres poignantly referred to the latest IPCC report as "an atlas of human suffering and a damning indictment of failed climate leadership."[30] We'll discuss climate change, international climate policy, and ESG integration in more detail in Chapter 3.

2.6 Demands for Transparency and Data Integrity

The progression from a voluntary model for ESG and sustainability reporting to a mandatory reporting model as governments have entered the stage puts pressure on the corporate sector. Two primary areas for scrutiny in the new structure are (1) transparency around disclosures and reporting, and (2) integrity of baseline ESG data.

According to the Governance & Accountability Institute's *2022 Sustainability Reporting in Focus*, 96% of S&P 500 and 81% of Russell 1000 companies publish ESG or sustainability reports.[31] The research also found that the incidence of corporate sustainability reporting has grown most significantly for mid-cap companies from 2020 to 2022.[32]

---

[29] Amy Myers Jaffe, *This Was the Year Investors and Businesses Put Big Bets on Climate*, Wall Street Journal (December 13, 2021) (www.wsj.com/articles/investors-climate-2021-11638372735).

[30] Brad Plumer and Raymond Zhong, *Climate Change is Harming the Planet Faster than We Can Adapt, U.N. Warns*, The New York Times (February 28, 2022) (www.nytimes.com/2022/02/28/climate/climate-change-ipcc-report.html).

[31] *2022 Sustainability Reporting in Focus*, Governance & Accountability Institute, Inc. (2022) www.ga-institute.com/research/ga-research-directory/sustainability-reporting-trends/2022-sustainability-reporting-in-focus.html).

[32] Id.

CHAPTER 2 THE CURRENT ESG OUTLOOK

The preparation and publication of a corporate ESG report is seen as a best practice. The content and quality of those reports vary widely. Many market-leading, publicly traded companies have been preparing and releasing sustainability reports that align to the generally accepted standards frameworks for upwards of 25 years. At the other end of the spectrum lie corporate sustainability reports that are principles-based with little supporting data or corporate governance. The discrepancy is often related to the stable of resources available to larger companies, and that inconsistency correlates with lower investor confidence in the reported data.

Regardless of the sophistication level of a company's reporting program, the data integrity, preparation, and review process will be scrutinized for accuracy given the heightened risk of enforcement and liability with the new regulations and laws. The choice of what data points to share publicly has been taken out of the hands of businesses as new regulations mandate data verification and specific disclosures. This has trained a new lens on data integrity and collection methods and companies are looking toward technological innovation in the field of artificial intelligence (AI) to manage complexity across the supply chain, deliver actionable insights, and transform sustainability programs including reporting and compliance efforts.

Implementation of the new disclosure requirements will yield more consistent reporting over time. That, in turn, will make benchmarking across markets and sectors more accessible and reliable. Investors and other stakeholders have been calling for data consistency, particularly as claims of greenwashing rise.[33] We'll talk more about disclosure risks and greenwashing implications and challenges in Chapter 6.

---

[33] Lily Turnbull and Kim Knickle, *Strategic Focus: Delivering Investor-Grade ESG Data*, Verdantix (November 10, 2023) (www.verdantix.com/report/strategic-focus-delivering-investor-grade-esg-data).

CHAPTER 2   THE CURRENT ESG OUTLOOK

CHAPTER TAKEAWAYS

- ESG is correlated with a range of business benefits:
  - Competitive advantage
  - Investment attraction
  - Enhanced financial performance
  - Increased long-term value
  - Customer loyalty
  - Operational sustainability
- There are major shifts happening in the ESG arena:
  - Global regulatory developments
  - Global standards alignment
  - Growing expectations from stakeholders
  - Demands for transparency and data integrity
- Those shifts are driven by the climate tipping point, which is exacerbating a wide range of interconnected impacts:
  - Business should prepare for the new ESG reporting, standards, and disclosures paradigm
  - Businesses must develop climate risk strategies
  - Businesses should prepare for enhanced transparency requirements

CHAPTER 3

The International Legal and Policy Foundations of ESG

International environmental law has a profound influence on public policy worldwide. It has successfully addressed many serious environmental and social issues and continues to play a significant part in shaping national policies and legal frameworks. It is also impeded by the need for stricter enforcement and broader cooperation and the ongoing challenge of productively addressing the disparity between developed and developing nations, particularly in areas such as climate impacts and environmental justice.

To understand ESG and sustainability, it is important to look to the foundations of environmental and social law on both the national and international scales. Familiarity with the origins, key treaties, agreements, and influence of international law in shaping public policy on human rights, climate, and natural resources provides an informed perspective of the underlying environmental, social, and governance factors as well as the basis for the evolving regulation on those topics.

CHAPTER 3   THE INTERNATIONAL LEGAL AND POLICY FOUNDATIONS OF ESG

3.1 Business and Human Rights
Fundamentals of Human Rights

Human rights are universal principles that outline the standards of human behavior acceptable to the global community. These rights are inherent to all individuals, regardless of nationality, sex, ethnicity, religion, language, or any other distinguishing factor. They are founded on the values of respect, equality, and dignity and are enshrined in international law and policy. The international human rights movement has been instrumental in defining these rights and ensuring their protection and enforcement.

Origins of International Human Rights Law

The adoption of the Universal Declaration of Human Rights (UDHR) by the United Nations General Assembly on December 10, 1948, marked a significant milestone in the history of human rights.[1] The core tenets of human rights law are rooted in civil, political, economic, social, and cultural issues.

The landmark Universal Declaration of Human Rights (UDHR) consists of 30 articles, each documenting distinct rights that are inherent to all individuals (Table 3-1). It continues to be a guiding force in addressing injustices and conflicts and in the quest for universal enjoyment of human rights.

Numerous international human rights treaties and instruments have been adopted since 1945, further developing the body of international human rights law and policy. These declarations, principles, and guidelines provide mechanisms for rights protection. The corpus of human

---

[1] United Nations. *Universal Declaration of Human Rights,* Paris, December 10, 1948 (www.un.org/en/about-us/universal-declaration-of-human-rights).

rights international law and policy is considered *customary international law*, a term derived from general practice, such as norms and customs, that is accepted as law and exists independent of treaty law. Many nations have also adopted constitutions and specific legislation that formally protect basic human rights. International human rights law imposes obligations that nation-states are bound to respect. The United Nations Human Rights Council is responsible for the enforcement of international human rights law.

Table 3-1. *Articles of the Universal Declaration of Human Rights*[2]

Article 1

All human beings are born free and equal in dignity and rights. They are endowed with reason and conscience and should act towards one another in a spirit of brotherhood.

Article 2

Everyone is entitled to all the rights and freedoms set forth in this Declaration, without distinction of any kind, such as race, colour, sex, language, religion, political or other opinion, national or social origin, property, birth or other status. Furthermore, no distinction shall be made on the basis of the political, jurisdictional or international status of the country or territory to which a person belongs, whether it be independent, trust, non-self-governing or under any other limitation of sovereignty.

Article 3

Everyone has the right to life, liberty and security of person.

Article 4

No one shall be held in slavery or servitude; slavery and the slave trade shall be prohibited in all their forms.

(continued)

---

[2] Id.

Table 3-1. (*continued*)

Article 5

No one shall be subjected to torture or to cruel, inhuman or degrading treatment or punishment.

Article 6

Everyone has the right to recognition everywhere as a person before the law.

Article 7

All are equal before the law and are entitled without any discrimination to equal protection of the law. All are entitled to equal protection against any discrimination in violation of this Declaration and against any incitement to such discrimination.

Article 8

Everyone has the right to an effective remedy by the competent national tribunals for acts violating the fundamental rights granted him by the constitution or by law.

Article 9

No one shall be subjected to arbitrary arrest, detention or exile.

Article 10

Everyone is entitled in full equality to a fair and public hearing by an independent and impartial tribunal, in the determination of his rights and obligations and of any criminal charge against him.

Article 11

1. Everyone charged with a penal offence has the right to be presumed innocent until proved guilty according to law in a public trial at which he has had all the guarantees necessary for his defence.
2. No one shall be held guilty of any penal offence on account of any act or omission which did not constitute a penal offence, under national or international law, at the time when it was committed. Nor shall a heavier penalty be imposed than the one that was applicable at the time the penal offence was committed.

(*continued*)

Table 3-1. (*continued*)

Article 12

No one shall be subjected to arbitrary interference with his privacy, family, home or correspondence, nor to attacks upon his honour and reputation. Everyone has the right to the protection of the law against such interference or attacks.

Article 13

1. Everyone has the right to freedom of movement and residence within the borders of each state.
2. Everyone has the right to leave any country, including his own, and to return to his country.

Article 14

1. Everyone has the right to seek and to enjoy in other countries asylum from persecution.
2. This right may not be invoked in the case of prosecutions genuinely arising from non-political crimes or from acts contrary to the purposes and principles of the United Nations.

Article 15

1. Everyone has the right to a nationality.
2. No one shall be arbitrarily deprived of his nationality nor denied the right to change his nationality.

Article 16

1. Men and women of full age, without any limitation due to race, nationality or religion, have the right to marry and to found a family. They are entitled to equal rights as to marriage, during marriage and at its dissolution.
2. Marriage shall be entered into only with the free and full consent of the intending spouses.
3. The family is the natural and fundamental group unit of society and is entitled to protection by society and the State.

(*continued*)

Table 3-1. (*continued*)

Article 17

    1. Everyone has the right to own property alone as well as in association with others.

    2. No one shall be arbitrarily deprived of his property.

Article 18

Everyone has the right to freedom of thought, conscience and religion; this right includes freedom to change his religion or belief, and freedom, either alone or in community with others and in public or private, to manifest his religion or belief in teaching, practice, worship and observance.

Article 19

Everyone has the right to freedom of opinion and expression; this right includes freedom to hold opinions without interference and to seek, receive and impart information and ideas through any media and regardless of frontiers.

Article 20

    1. Everyone has the right to freedom of peaceful assembly and association.

    2. No one may be compelled to belong to an association.

Article 21

    1. Everyone has the right to take part in the government of his country, directly or through freely chosen representatives.

    2. Everyone has the right of equal access to public service in his country.

    3. The will of the people shall be the basis of the authority of government; this will shall be expressed in periodic and genuine elections which shall be by universal and equal suffrage and shall be held by secret vote or by equivalent free voting procedures.

(*continued*)

Table 3-1. (*continued*)

Article 22

Everyone, as a member of society, has the right to social security and is entitled to realization, through national effort and international co-operation and in accordance with the organization and resources of each State, of the economic, social and cultural rights indispensable for his dignity and the free development of his personality.

Article 23

1. Everyone has the right to work, to free choice of employment, to just and favourable conditions of work and to protection against unemployment.
2. Everyone, without any discrimination, has the right to equal pay for equal work.
3. Everyone who works has the right to just and favourable remuneration ensuring for himself and his family an existence worthy of human dignity, and supplemented, if necessary, by other means of social protection.
4. Everyone has the right to form and to join trade unions for the protection of his interests.

Article 24

Everyone has the right to rest and leisure, including reasonable limitation of working hours and periodic holidays with pay.

Article 25

1. Everyone has the right to a standard of living adequate for the health and well-being of himself and of his family, including food, clothing, housing and medical care and necessary social services, and the right to security in the event of unemployment, sickness, disability, widowhood, old age or other lack of livelihood in circumstances beyond his control.
2. Motherhood and childhood are entitled to special care and assistance. All children, whether born in or out of wedlock, shall enjoy the same social protection.

(*continued*)

Table 3-1. (*continued*)

Article 26
1. Everyone has the right to education. Education shall be free, at least in the elementary and fundamental stages. Elementary education shall be compulsory. Technical and professional education shall be made generally available and higher education shall be equally accessible to all on the basis of merit.
2. Education shall be directed to the full development of the human personality and to the strengthening of respect for human rights and fundamental freedoms. It shall promote understanding, tolerance and friendship among all nations, racial or religious groups, and shall further the activities of the United Nations for the maintenance of peace.
3. Parents have a prior right to choose the kind of education that shall be given to their children.

Article 27
1. Everyone has the right freely to participate in the cultural life of the community, to enjoy the arts and to share in scientific advancement and its benefits.
2. Everyone has the right to the protection of the moral and material interests resulting from any scientific, literary or artistic production of which he is the author.

Article 28
Everyone is entitled to a social and international order in which the rights and freedoms set forth in this Declaration can be fully realized.

(*continued*)

Table 3-1. (*continued*)

Article 29
1. Everyone has duties to the community in which alone the free and full development of his personality is possible.
2. In the exercise of his rights and freedoms, everyone shall be subject only to such limitations as are determined by law solely for the purpose of securing due recognition and respect for the rights and freedoms of others and of meeting the just requirements of morality, public order and the general welfare in a democratic society.
3. These rights and freedoms may in no case be exercised contrary to the purposes and principles of the United Nations.

Article 30
Nothing in this Declaration may be interpreted as implying for any State, group or person any right to engage in any activity or to perform any act aimed at the destruction of any of the rights and freedoms set forth herein.

Human Rights Laws and Business Practices

The principles of human rights are not only applicable to states and individuals; they are also requisite for businesses. Respect for human rights is a fundamentally expected standard for businesses.

Companies are expected to incorporate human rights considerations into their business practices and corporate governance. This includes

1. Conducting robust due diligence across the supply chain to identify, prevent, and mitigate human rights impacts
2. Integrating international human rights standards into corporate reporting frameworks

CHAPTER 3   THE INTERNATIONAL LEGAL AND POLICY FOUNDATIONS OF ESG

3. Taking accountability for preventing and remediating business-related human rights abuses across the value chain

4. Developing comprehensive supplier codes of conduct

5. Collaborating with certified NGOs and third-party assurance providers to assess programs

6. Engaging in consultation with potentially affected groups and other stakeholders

A number of global ESG standards and regulations cover human rights. These include standards developed by the United Nations, such as the UN Guiding Principles on Business and Human Rights,[3] which provide a framework for businesses to respect human rights. Recent sustainability and ESG regulations, such as the European Union's Non-Financial Reporting Directive (NFRD)[4] and the proposed Corporate Sustainability Reporting Due Diligence Directive (CSDDD or CS3D),[5] require businesses to disclose social pillar information, including that related to employee matters, human rights, and anti-bribery and corruption.

Many well-known and otherwise respected companies have committed human rights violations and run afoul of public trust. A common scenario found in those examples is supplier or subcontractor malfeasance that went

---

[3] United Nations. *Guiding Principles on Business and Human Rights*, New York and Geneva (2011) (www.ohchr.org/documents/publications/guidingprinciplesbusinesshr_en.pdf).

[4] European Parliament. *Non-financial Reporting Directive* (2014) (www.europarl.europa.eu/RegData/etudes/BRIE/2021/654213/EPRS_BRI(2021)654213_EN.pdf).

[5] *Corporate sustainability due diligence*, European Commission (https://commission.europa.eu/business-economy-euro/doing-business-eu/corporate-sustainability-due-diligence_en).

undetected. By fully integrating human rights into strategies and operations, businesses can not only minimize risk but also contribute to a more equitable and sustainable world. Achieving truly responsible and rights-respecting business conduct requires the collective efforts of businesses, investors, policymakers, and civil society. This involves strengthening ESG practices, adopting and implementing human rights commitments across the value chain, and gauging and mitigating risks to people and the planet.

3.2 Climate and Natural Resources

Guiding Principles of International Environmental Law

Nations have adopted a set of basic principles that have become the cornerstone of international environmental law. These principles include state sovereignty, common concern, the duty not to cause harm, and the precautionary principle.

The origins of international environmental law can be traced back centuries to when nations began passing laws to protect human health from transboundary environmental pollution. The scale and complexity of global environmental challenges necessitated an expansion from national to international law. International agreement and cooperation became necessary for an effective response to environmental challenges.

Emergence of International Cooperation

The need for a multinational solution to environmental issues became evident in the mid-20th century with the growing evidence of global environmental stresses. This led to the advent of significant international conferences and summits on the environment. These convenings were attended by representatives from various UN member nations and emphasized the need for an international approach to environmental challenges.

3.3 Sustainable Development
Brundtland Commission (1983)

The World Commission on Environment and Development (WCED), commonly known as the Brundtland Commission, was vital in defining and popularizing the concept of sustainable development in the latter part of the 20th century. The commission's final report, "Our Common Future," also known as the Brundtland Report, has had a significant impact on how the world perceives and addresses sustainability today.[6]

The Brundtland Commission was established by the United Nations in 1983. An international group of environmental scientists, government officials, and civil servants were convened, and the commission was named after its chair, Dr. Gro Harlem Brundtland, who was then Prime Minister of Norway and later became Director-General of the World Health Organization (WHO) and a member of The Elders.

The commission was tasked with addressing growing concerns about the accelerating ecological deterioration and the consequences of that decline for economic and social development. Pressing issues of the time included ozone depletion, global warming, and other environmental challenges preventing improvement to the standard of living for the world population.

Brundtland Report: Our Common Future (1987)

The Commission's work culminated in the publication of *Our Common Future*, in 1987. The report highlighted the interconnectedness of the environment and the economy and emphasized that they should not be considered in isolation from each other. The Commission explored the interconnections between social equity, economic growth, and environmental problems.

---

[6] United Nations. *Our Common Future* (https://sustainabledevelopment.un.org/content/documents/5987our-common-future.pdf).

The Commission introduced the term "sustainable development" and defined it as "development that meets the needs of the present without compromising the ability of future generations to meet their own needs."[7] It was a watershed report—the Brundtland Report represents the origins of sustainability and a turning point in the perspective on the interplay of environmental, social, and economic issues.

Defining Sustainability and Sustainable Development

The Brundtland Report's definition of sustainable development has shaped the way sustainability is understood and applied today. It stressed that the pursuit of economic growth must not come at the expense of ecological health and that the needs of the global population cannot be overlooked. The definition given to sustainable development had two key components: (1) the recognition of essential human needs for survival, particularly of the world population living in poverty; and (2) the concept of natural resource limitations and the ability of the environment to meet both present and future needs.

Impact and the Origins of Sustainability

The Brundtland Report was a turning point in the global environmental debate. It shifted the perspective from a narrow understanding of environmental conservation to a broader, more integrated approach to sustainable development. It highlighted that economic development, social development, and environmental protection were interconnected and mutually dependent.

---

[7] Id.

The report's impact is evident in several international environmental agreements and policies that followed, including the 1992 Earth Summit in Rio de Janeiro, the Kyoto Protocol, and the Paris Agreement. It also led to the creation of the United Nations Commission on Sustainable Development.

The Brundtland Commission and its report shaped our understanding of and approach to sustainability and sustainable development. It advocated for a global partnership and cooperation to manage the planet's resources sustainably. While nearly four decades old, the report's principles and vision continue to guide global sustainability efforts and policies. The report's definition of sustainable development has become the most popular and enduring definition of *sustainability*.

3.4 Key Global Treaties, Agreements, and Summits

Stockholm Conference (1972)

The United Nations took a significant step in 1972 by convening the first major international conference on the environment, the UN Conference on the Human Environment, in Stockholm.[8] This conference highlighted emerging global environmental challenges and gave environmental issues a foothold in the international community. The Stockholm Conference also led to the creation of the United Nations Environment Programme (UNEP), which serves as a convener for environmental cooperation and a principal global environmental organization.[9]

---

[8] United Nations Conference on the Human Environment, June 5–16, 1972, Stockholm (www.un.org/en/conferences/environment/stockholm1972).

[9] United Nations Environment Programme (www.unep.org/).

CHAPTER 3   THE INTERNATIONAL LEGAL AND POLICY FOUNDATIONS OF ESG

Stockholm set the global stage for international agreement negotiations that formed the corpus of international environmental law. Global leaders now regularly meet on specific subject matter, including hazardous substances, climate change, biodiversity, and oceans.

The Rio Earth Summit: A Pioneering Conference on Environment and Development (1992)

Following on the work of the Brundtland Commission, the UN gathered for a major summit to steer sustainable development in 1992. The international community came together in Rio de Janeiro, Brazil, to participate in a landmark conference known as the United Nations Conference on Environment and Development (UNCED), also widely referred to as the Rio Conference or the Earth Summit.[10] This conference marked a turning point in the global approach to environmental concerns and developmental challenges.

The Earth Summit initiated a widespread dialogue on key issues such as sustainable development, climate change, and biodiversity. It was also a key next step, following on the work of the Brundtland Commission on sustainable development. It was prolific in its outcomes and remains one of the most notable summits for sustainability and the environment.

Genesis of the Earth Summit

The roots of the Earth Summit can be traced back to Stockholm. Two decades after the Stockholm Conference, the Earth Summit was organized to revisit and expand upon the environmental and developmental issues discussed in Stockholm. The goal of the Earth Summit was not just to

---

[10] *United Nations Conference on Environment and Development*, Rio de Janeiro, Brazil, June 3-14, 1992 (www.un.org/en/conferences/environment/rio1992).

CHAPTER 3   THE INTERNATIONAL LEGAL AND POLICY FOUNDATIONS OF ESG

create awareness but also to produce a broad, actionable agenda for international cooperation on environmental and development issues in the 21st century.

Participants and Objectives

The conference brought together an unprecedented number of participants, including political leaders, diplomats, scientists, NGOs, and media representatives from 179 countries. The primary objective of the Earth Summit was to produce a comprehensive plan for sustainable development, taking into account economic, social, and environmental factors. The conference aimed to highlight the interdependence of these factors and the need for a balanced approach to sustainability in consideration of human life and the planet.

Affirming the Concept of Sustainable Development

One of the major outcomes of the Earth Summit was the universal acceptance of the concept of sustainable development as proposed by the Brundtland Commission. The conference concluded that sustainable development, an integrated approach balancing economic, social, and environmental dimensions, was an achievable goal for all nations, regardless of their level of development. This revolutionized the way societies viewed development, sparking a global debate on how to ensure sustainability for future generations.

Earth Summit's Achievements

The Earth Summit led to significant achievements in the realm of international environmental policy. It was in Rio that the UN member states had the foresight to view the interconnected nature of environmental and social issues, to establish guiding principles on sustainable development, to establish a framework for action on climate change, and to formalize protections for biodiversity, deserts, and forests.

CHAPTER 3   THE INTERNATIONAL LEGAL AND POLICY FOUNDATIONS OF ESG

The specific outcomes of the Earth Summit that have since shaped international policy include

1. **Rio Declaration on Environment and Development (Rio Declaration)**[11]

   The Rio Declaration on Environment and Development was a significant outcome of the Earth Summit. It established 27 principles to guide nations toward more environmentally sustainable patterns of development, which are often referred to as the Rio Principles. The declaration was adopted without negotiation, owing to fears that further debate could jeopardize the consensus.

2. **United Nations Framework Convention on Climate Change (UNFCCC)**[12]

   One of the milestone achievements of the Earth Summit was the United Nations Framework Convention on Climate Change (UNFCCC). The UNFCCC was established by the 154 nations convened at the Earth Summit as a response to mounting scientific evidence pointing toward human-induced climate change. It recognized climate change as *a common concern of humankind* and set out a framework for global action.

---

[11] United Nations. *Report of the United Nations Conference on Environment and Development,* Rio de Janeiro, June 3–14, 1992(www.un.org/en/development/desa/population/migration/generalassembly/docs/globalcompact/A_CONF.151_26_Vol.I_Declaration.pdf).

[12] United Nations. *What is the United Nations Framework Convention on Climate Change?* (https://unfccc.int/process-and-meetings/what-is-the-united-nations-framework-convention-on-climate-change).

The convention established a global framework for efforts to tackle the challenge of climate change, focusing on the reduction of greenhouse gas (GHG) emissions. The primary objective given to the UNFCCC was to stabilize atmospheric concentrations of greenhouse gases to prevent harmful interference with the Earth systems. The convention was predicated on the principle of common but differentiated responsibilities (CBDR), recognizing that while all nations have a task to do in combatting climate change, developed nations bear a greater responsibility due to their historically higher levels of GHGs.

3. **Convention on Biological Diversity (CBD)**[13]

   The Earth Summit also resulted in the Convention on Biological Diversity (CBD). This convention aimed at conserving the world's diverse plant and animal species, prescribing steps for their protection and sustainable use.

   The 15th Conference of the Parties (COP15) to the United Nations Convention on Biodiversity was held in Montreal in 2022. The conference served as a biennial convening of global governments to set ten-year goals on conservation and sustainable resource use and to carve a path to mitigate biodiversity loss. The convention's 195 signatory nations agreed to adopt the Global Biodiversity Framework (GBF) and to protect and restore a minimum of 30% of the

---

[13] United Nations. Convention on Biological Diversity (www.cbd.int/rio/).

planet's land and water by 2030, and a commitment was made to pay an estimated $USD30 billion annually to economically disadvantaged nations through a new biodiversity fund.[14]

The "30x30" plan is the most significant biodiversity commitment in history, yet many experts continue to sound the alarm that it may fall short. The ten-year target set at COP10 in 2010 to halve natural habitat loss and ensure 17% of the Earth's surface were nature reserves by 2020 has not been met.

4. **United Nations Convention to Combat Desertification (UNCCD)**[15]

The UNCCD remains the only legally binding framework established to address desertification and the effects of drought.

5. **Forest Principles**[16]

The Earth Summit also led to the creation of a non-legally binding statement of principles for forest management, known as the Forest Principles. These principles provided a global consensus on the management, conservation, and sustainable development of all types of forests.

---

[14] *COP15 ends with landmark biodiversity agreement*, United Nations Environment Programme (December 20, 2022) (www.unep.org/news-and-stories/story/cop15-ends-landmark-biodiversity-agreement).

[15] United Nations. *United Nations Convention to Combat Desertification*, Rio de Janeiro, June 3–14, 1992 (www.unccd.int/).

[16] United Nations. *Forest Principles—Report of the United Nations Conference on Environment and Development*, Rio de Janeiro, June 3–14, 1992 (https://digitallibrary.un.org/record/170821?ln=en).

CHAPTER 3   THE INTERNATIONAL LEGAL AND POLICY FOUNDATIONS OF ESG

6. **Agenda 21**[17]

    Agenda 21, another major outcome of the Earth Summit, was a comprehensive blueprint for sustainable development in the 21st century. This non-binding and voluntarily implemented action plan offered strategies for addressing environmental issues and promoting sustainable development at local, national, regional, and global levels. The UN outlined Agenda 21 as an integrated plan across the global, national, and local levels of the United Nations System to strengthen participation in addressing human impacts on the environment. It was a precursor to the Sustainable Development Goals.

7. **Global Environment Facility (GEF)**[18]

    The Global Environment Facility (GEF) is a multilateral environmental fund. It was originally established in advance of Rio and co-piloted by the United Nations Development Programme (UNDP), the United Nations Environment Programme (UNEP), and the World Bank. During the Earth Summit, the GEF was restructured as a permanent and independent organization, and it became the financial mechanism for both the U.N. Convention on Biological Diversity and the U.N. Framework Convention on Climate Change.

---

[17] United Nations. *Agenda 21*, Rio de Janeiro, June 3–14, 1992 (https://sustainabledevelopment.un.org/content/documents/Agenda21.pdf).
[18] *Who We Are*, Global Environment Facility (www.thegef.org/who-we-are).

CHAPTER 3   THE INTERNATIONAL LEGAL AND POLICY FOUNDATIONS OF ESG

The GEF's mission is to assist developing countries and economies in transition in their efforts to achieve the objectives set by international environmental conventions and agreements. Its mission and funding are pivotal in addressing critical issues such as biodiversity loss, climate change, and pollution, aiming to foster a sustainable global environment. The GEF is the largest multilateral environmental fund, and it recently announced a plan to establish a biodiversity fund with an allocated USD$1.1 billion in funding.[19]

Legacy of the Earth Summit

The Earth Summit concluded with the establishment of the Commission on Sustainable Development to ensure effective implementation of the agreements at the local, national, regional, and international levels.

Despite its many accomplishments, the Earth Summit faced criticism for the perceived lack of enforcement of its agreements, particularly in areas such as poverty eradication and environmental protection. However, the conference undeniably paved the way for future international environmental conferences and negotiations, marking a significant step in the global endeavor toward sustainable development.

The convening in Rio represented a monumental time in the history of international environmental policymaking. The principles, conventions, and action plans conceived during that conference continue to guide global efforts toward sustainable development and environmental preservation. The legacy of the Earth Summit serves as a potent reminder of the power of international cooperation in addressing the world's most pressing environmental and developmental challenges.

---

[19] *GEF Council approves $1.1 billion and sets plans for biodiversity fund*, Global Environment Facility (February 9, 2024) (www.thegef.org/newsroom/news/gef-council-approves-1-1-billion-and-sets-plans-biodiversity-fund).

Addressing Climate Change

The UNFCCC framework is designed to facilitate international cooperation in addressing climate change. It establishes mechanisms for data collection, knowledge building, and clear communication among the parties involved.

The parties to the convention are organized into groups based on their economic capabilities and historical greenhouse gas emission levels. These groups include Annex I countries, which are mainly developed countries, and non-Annex I countries, which are primarily developing countries. The concept of equity and proportionate responsibility has always been a tenet of the framework.

The convention addresses the related need for technology transfer and financial support from developed to developing countries. This is instrumental as developing countries transition to a low-carbon economy and adapt to the impacts of climate change.

UNFCCC Programs and Mechanisms

The UNFCCC has established various programs and mechanisms to help countries fulfill their commitments under the convention. One notable program is the Green Climate Fund (GCF),[20] which was established to assist developing countries in their efforts to mitigate climate change and adapt to its impacts. The GCF provides financial resources to developing countries to help them implement projects and programs that contribute to their goals under the UNFCCC.

Another important mechanism under the UNFCCC is the Clean Development Mechanism (CDM). The CDM allows developed countries to invest in emission reduction projects in developing countries and earn carbon credits, which they can use to meet their own emission reduction targets.

---

[20] Green Climate Fund, UNEP (`www.unep.org/about-un-environment/funding-and-partnerships/green-climate-fund`).

Implementation of the UNFCCC

The implementation of the UNFCCC is guided by the decisions of the Conference of the Parties (COP), the decision-making body of the convention. The COP meets annually to assess progress in dealing with climate change and to negotiate and adopt decisions that further the implementation of the convention.

The UNFCCC Secretariat, based in Bonn, Germany, coordinates the implementation of the convention. The Secretariat organizes the annual COP meetings, prepares reports on the implementation of the convention, and provides technical and logistical support to the parties.

Key Milestones for UNFCCC

The UNFCCC has overseen several significant milestones in the global response to climate change. In 1997, the Kyoto Protocol was adopted. It introduced legally binding emission reduction targets for developed countries. This marked a significant step forward in the global effort to combat climate change.

In 2015, the Paris Agreement was adopted, marking another major milestone in global climate action. The Paris Agreement set to limit global warming to well below 2°C and to pursue efforts to limit the temperature increase to 1.5° C.[21]

Engagement of Civil Society in the UNFCCC

The UNFCCC recognizes the importance of engaging civil society in the global response to climate change. Civil Society Observers, including NGOs, indigenous persons, and other stakeholders, formally participate in

---

[21] United Nations. *Paris Agreement*, Paris, December 12, 2015 (https://treaties.un.org/Pages/ViewDetails.aspx?src=TREATY&mtdsg_no=XXVII-7-d&chapter=27&clang=_en).

CHAPTER 3   THE INTERNATIONAL LEGAL AND POLICY FOUNDATIONS OF ESG

the UNFCCC process and contribute to its decision-making. The aim was for inclusive discussion and representation of all populations impacted by climate change.

Despite its accomplishments, the UNFCCC process has been subject to criticism. Some critics argue that the process has been too slow and has failed to achieve significant reductions in global greenhouse gas emissions. Others point to the lack of enforcement mechanisms for countries that fail to meet their commitments.

Impact and Legacy of the UNFCCC

The UNFCCC has a primary position in shaping the global response to climate change. It has provided a platform for international cooperation and negotiation on climate change, leading to landmark agreements such as the Kyoto Protocol and Paris Agreement.

Despite the challenges and criticisms, the UNFCCC remains the primary intergovernmental instrument in the global effort to combat climate change. As the world continues to grapple with the impacts of climate change, the work of the UNFCCC continues to be of vital importance. It has provided a framework for cooperation, established key principles and mechanisms, and facilitated important agreements. Its function in shaping the global response to climate change remains instrumental.

United Nations Convention on the Law of the Sea (1994) and High Seas Treaty (2023)

United Nations Convention on the Law of the Sea (UNCLOS)

The United Nations Convention on the Law of the Sea (UNCLOS) was a milestone in international maritime law. It provides a legal framework for marine and maritime activities, which largely focused on resource use and international peace and security.²² The concept of freedom of the seas had long been established, which provides that all waters beyond national boundaries are international waters and open to all nations.

With an increasing desire of nations to extend national claims to include mineral resources, protect fish stocks, and enforce pollution controls, it became clear that a more comprehensive set of regulations was required. The third United Nations Conference on the Law of the Sea, which took place between 1973 and 1982, was the major outcome of the conference and came into force in 1994, replacing earlier treaties. ²³

One of the most significant provisions of UNCLOS pertains to the limit of territorial waters and the establishment of Exclusive Economic Zones (EEZs). According to the convention, a coastal state has sovereignty over its territorial sea. The territorial sea extends up to 12 nautical miles from the baseline, and beyond that baseline and up to 200 nautical miles from the shoreline lies the Exclusive Economic Zone. ²⁴ Within this zone, the coastal nation has sole exploitation rights over all natural resources. UNCLOS also established the International Seabed Authority (ISA), a specialized agency of the United Nations tasked with overseeing mineral-related activities on the seabed beyond any state's territorial waters or EEZs.²⁵

---

²² *United Nations Convention on the Law of the Sea*, Montego Bay, Jamaica, November 16, 1994 (www.un.org/depts/los/convention_agreements/texts/unclos/unclos_e.pdf).

²³ *United Nations Convention on the Law of the Sea of 10 December 1982 Overview and full text*, United Nations Division for Ocean Affairs and the Law of the Sea (www.un.org/depts/los/convention_agreements/convention_overview_convention.htm#:~:text=by%20%22*%22.-,The%20United%20Nations%20Convention%20on%20the%20Law%20of%20the%20Sea,the%20oceans%20and%20their%20resources).

²⁴ *United Nations Convention on the Law of the Sea*, Montego Bay, Jamaica, November 16, 1994 (www.un.org/depts/los/convention_agreements/texts/unclos/unclos_e.pdf).

²⁵ Id. See also International Seabed Authority (www.isa.org.jm/).

CHAPTER 3   THE INTERNATIONAL LEGAL AND POLICY FOUNDATIONS OF ESG

The convention placed an obligation on all signatory nations to protect and preserve the marine environment. It set out a general legal framework for all states to collaborate in the conservation and sustainable use of oceans and their resources. The impacts of UNCLOS have been significant and wide-ranging. It has been instrumental in preventing and resolving disputes related to maritime boundaries and has facilitated peaceful and cooperative use of the world's oceans. UNCLOS promoted the conservation of marine resources, contributing to the achievement of United Nations Sustainable Development Goal 14, which focuses on the conservation and sustainable use of oceans.

Despite its many achievements, UNCLOS has been subject to several controversies and challenges. A key area of contention has been around the provisions relating to the seabed and ocean floor beyond national jurisdiction. Despite efforts to resolve such concerns, some countries, notably the United States, have yet to ratify the convention.

United Nations High Seas Treaty

In 2023, the United Nations adopted a monumental accord on the conservation and sustainable use of marine biodiversity of the ocean areas beyond national jurisdiction, which is said to cover over two-thirds of the planet's oceans.[26] This agreement is commonly known as the United Nations High Seas Treaty with the formal name given as the agreement on Biodiversity Beyond National Jurisdiction (BBNJ).[27]

---

[26] *Historic agreement adopted at the U.N. for conservation and sustainable use of biodiversity in over two-thirds of the ocean*, United Nations (June 19, 2023) (www.un.org/sustainabledevelopment/blog/2023/06/press-release-historic-agreement-adopted-at-the-un-for-conservation-and-sustainable-use-of-biodiversity-in-over-two-thirds-of-the-ocean/).

[27] *Draft agreement under the United Nations Convention on the Law of the Sea on the conservation and sustainable use of marine biological diversity of areas beyond national jurisdiction*, United Nations General Assembly, New York, NY June 12, 2023 (https://undocs.org/A/CONF.232/2023/L.3).

A key feature of the High Seas Treaty is establishing a fair and equitable sharing of benefits arising from activities related to marine genetic resources in the high seas. The treaty embodies technical elements for area-based management of marine resources and environmental impact assessment (EIA), and capacity building for marine genetic resources and the transfer of marine technology.[28] The treaty also requires the formation of various committees and working groups to be inclusive and diverse and includes provisions to incorporate indigenous voices and those of small island and archipelago communities.[29]

As UNCLOS did not encompass a comprehensive legal framework for biodiversity in the high seas, there had been a long-running push for a treaty to fill the gaps and to elevate biodiversity conservation. The ocean plays an integral role in sustaining life on our planet, and serves as a massive carbon sink, mitigating the impacts of climate change. It is estimated that over 91% of marine species remain undiscovered and that approximately 80% of our ocean is uncharted and unexplored.[30]

The High Seas Treaty provides a pathway for more effective marine conservation, which signifies a major step toward global ocean management and marine conservation. It also addresses inequality issues that have often been overlooked in international agreements. The treaty is expected to enable species and habitat preservation and support the marine and coastal conversation targets of the 30x30 plan on biodiversity, which is discussed in this chapter.

---

[28] Id.

[29] Id.

[30] *How Many Species Live in the Ocean?* National Ocean Service, National Oceanic and Atmospheric Administration (https://oceanservice.noaa.gov/facts/ocean-species.html#:~:text=The%20number%20of%20species%20that,as%20possible%20about%20ocean%20life).

CHAPTER 3   THE INTERNATIONAL LEGAL AND POLICY FOUNDATIONS OF ESG

Kyoto Protocol (1997)

The Kyoto Protocol stands as an iconic symbol of global efforts to combat climate change, both for its international collaboration and for the use of green finance instruments, which were a cornerstone in the development of carbon trading markets. The carbon trading markets are essentially environmental commodities markets for nontangible energy credits.

This international treaty was conceived during the COP3 meeting held in Kyoto in 1996 to curb the escalating levels of greenhouse gas emissions and mitigate the adverse effects of global warming. A meticulous analysis of the Kyoto Protocol offers a rich understanding of its inception, objectives, achievements, challenges, and broader influence on the climate change narrative.

Background on the Kyoto Protocol

The Kyoto Protocol tracks back to the creation of the UNFCCC at the 1992 Earth Summit in Rio. The primary objective outlined for the UNFCCC was to stabilize atmospheric concentrations of greenhouse gases to prevent harmful interference with the climate system.

This mandate culminated in the Kyoto Protocol, which was adopted following COP3 on December 11, 1997. The Protocol was a significant stride toward legally binding commitments for developed nations to reduce GHG emissions. The ratification process was complex and lengthy. The protocol entered into force on February 16, 2005, after ratification by the minimum required states (at least 55 states), accounting for a baseline of 55% of total carbon dioxide emissions from developed countries (1990 levels). Currently, there are 192 nations party to the Kyoto Protocol.[31]

---

[31] United Nations Climate Change. *What is the Kyoto Protocol?* (https://unfccc.int/kyoto_protocol#:~:text=The%20Kyoto%20Protocol%20was%20adopted,Parties%20to%20the%20Kyoto%20Protocol).

Objectives and Key Provisions of the Kyoto Protocol

The Kyoto Protocol pursued a two-pronged objective. First, it aimed to reduce the global emissions of six key greenhouse gases, including carbon dioxide (CO_2), methane (CH_4), and nitrous oxide (N_2O). Second, the protocol sought to harness market mechanisms to achieve these reductions cost-effectively.

The Kyoto Protocol imposed binding emissions reduction targets on developed nations, recognized as Annex I Parties. These countries committed to reducing their combined emissions by an average of 5.2% below 1990 levels during the first commitment period from 2008 to 2012. Setting internationally binding emissions targets facilitated carbon emissions trading and related financial instruments. To facilitate cost-effective achievement of these targets, the Protocol introduced three innovative *flexibility mechanisms*: International Emissions Trading, Joint Implementation (JI), and the Clean Development Mechanism (CDM).

1. **International Emission Trading:** This mechanism established the global carbon market as it allowed nations with surplus emissions allowances to sell them to countries exceeding reduction targets.

2. **Joint Implementation:** This mechanism enabled Annex I countries to earn emission reduction credits for investing in emission-reducing projects in other Annex I nations.

3. **Clean Development Mechanism:** This mechanism encouraged Annex I countries to invest in emission-reduction projects in non-Annex I countries by earning certified emission reduction credits toward their targets.

CHAPTER 3   THE INTERNATIONAL LEGAL AND POLICY FOUNDATIONS OF ESG

The Protocol also acknowledged the capacity of carbon sinks, such as forests, in absorbing carbon dioxide from the atmosphere. Forests sequestered approximately twice as much carbon as they emitted between 2001 and 2019.[32] Consequently, activities like reforestation could be used as offsets against emissions targets.

Impact of the Kyoto Protocol on Climate Change

The Kyoto Protocol sparked significant strides toward emissions reduction. It catalyzed a shift from a voluntary approach to a legally binding one, providing a tangible framework for climate action.

By creating a market for carbon credits, it also spurred innovation and investment in low-carbon technologies and practices. Moreover, the inclusion of carbon sinks in the Protocol led to increased attention and efforts toward forest conservation and sustainable land use.

Criticisms of the Kyoto Protocol

Despite its achievements, the Kyoto Protocol faced substantial criticism. The most significant contention was the lack of binding commitments for developing countries. The Protocol, while placing the burden of emissions reduction on developed countries, exempted developing nations from such obligations, leading to debates about equity and effectiveness.

Another criticism was the Protocol's limited environmental impact. Even if all Annex I Parties fully complied with their commitments, global emissions would continue to rise due to increasing emissions from non-Annex I parties, underscoring the need for a more inclusive and comprehensive agreement.

---

[32] Nancy L. Harris, David A. Gibbs, Alessandro Baccini, et al. *Global maps of twenty-first century forest carbon fluxes*, Nature Climate Change 11, 234–240 (2021) (https://doi.org/10.1038/s41558-020-00976-6).

Extension and Replacement

The Kyoto Protocol was extended in 2012 with the Doha Amendment at COP18, which introduced new emissions reduction targets for the second commitment period from 2012 to 2020. However, this extension was short-lived. In 2015, the Paris Agreement was adopted, effectively replacing certain aspects of the Kyoto Protocol. This new agreement incorporated all major GHG-emitting countries, promising to strengthen the global response to climate change.

US Position and Withdrawal

Notably, the United States, although once a signatory to the Kyoto Protocol, never ratified it. The United States argued that the Protocol was unfair as it only required developed nations to reduce emissions, potentially hurting the US economy. In 2001, the United States officially withdrew from the Protocol.

The Kyoto Protocol, despite its shortcomings, marked a juncture in international climate governance. It set the stage for global cooperation and initiated market-based mechanisms for emissions reduction. As the world continues to grapple with climate change, the Kyoto Protocol offers valuable lessons for future climate agreements—underscoring the need for inclusive, equitable, and robust action to safeguard our planet.

Paris Agreement (2015)

The Paris Agreement, a significant international pact aimed at tackling the escalating issue of climate change, has been the linchpin of global climate policy since 2015. The Paris Agreement was the fruit of decades of international cooperation and negotiation, aiming to mitigate the adverse effects of climate change. It was formally adopted by nearly 200 parties at the Paris Climate Conference, also known as COP21, in December 2015, and came into effect on November 4, 2016.

CHAPTER 3   THE INTERNATIONAL LEGAL AND POLICY FOUNDATIONS OF ESG

The Paris Agreement is unique as it is not only universal but also legally binding. It is the first-ever global climate change agreement that mandates all participating countries to contribute toward mitigating climate change. The text of the agreement between nations sets out a comprehensive framework to limit global warming to well below 2°C, while simultaneously pursuing efforts to cap it at 1.5°C. It also aims to enhance countries' resilience to the impacts of climate change and support their mitigation efforts.[33]

Key Components of the Paris Agreement

The Paris Agreement includes provisions on mitigation, transparency, adaptation, and financial support. It encourages countries to adopt robust measures to reduce emissions and to regularly report on their climate action progress.

The international fight against climate change was bolstered by the establishment of scientifically defined global warming maximum temperatures under the Paris Agreement. The threshold to hold the planetary temperature to a maximum increase above pre-industrial levels was a key outcome. Beyond that temperature cap, scientists have predicted dire and irreversible consequences for global systems.

Key provisions include

1. **Nationally Determined Contributions (NDCs)**

    One of the unique aspects of the Paris Agreement is its bottom-up approach, which allowed countries to set their own emission reduction targets, known as nationally determined contributions (NDCs).

---

[33] United Nations, *Paris Agreement*, Paris, December 12, 2015 (https://treaties.un.org/Pages/ViewDetails.aspx?src=TREATY&mtdsg_no=XXVII-7-d&chapter=27&clang=_en).

These NDCs represent each country's commitment to the objectives of the agreement and are expected to be regularly updated and strengthened.

The Paris Agreement acknowledged the role of non-signatories to the agreement, including the private sector, in addressing climate change. The agreement encourages aligned efforts to reduce emissions and mitigate climate change.

2. **Financial Support and Cooperation**

The Paris Agreement recognizes the need for financial aid and support to developing countries. Developed countries committed to mobilize USD$100 billion annually in aid to developing nations, with a higher goal set for the post-2025 period.

Impact on Climate Change

The Paris Agreement, despite its ambitious goals, has been criticized for being substantively inadequate to prevent a global temperature rise above the accord threshold. Current national commitments under the agreement are deemed insufficient to reach the agreed temperature objectives, prompting calls for more aggressive action. Critics argue that the Paris Agreement lacks an enforcement mechanism to ensure countries meet their commitments.

The Paris Agreement, as it stands, is not sufficient to avert the severe consequences of climate change. Therefore, it is necessary for countries to increase their emission reduction commitments and accelerate their efforts toward achieving a sustainable and climate-resilient future. It was a major step in the global fight against climate change, and while it has its shortcomings, it provides a robust framework for countries to enhance their efforts and work together toward a common goal. Its success or

CHAPTER 3   THE INTERNATIONAL LEGAL AND POLICY FOUNDATIONS OF ESG

failure will largely depend on the collective will of nations to embrace more ambitious climate actions and uphold their commitments to create a sustainable future for all.

The Paris Agreement was a catalyst for the corporate sector to set carbon emissions goals, which became known as "net-zero" goals and plans to get to a zero emissions level by a specified year. Recognizing that industry as a whole is the most significant contributor to GHG emissions, companies followed suit by setting their own net-zero goals. At first lauded for their net-zero commitments, scrutiny would follow given the relative lack of transition planning and true means to achieve meaningful reductions on set timetables.

COP28 in Dubai (2023)

The 28th Conference of the Parties to the United Nations Framework Convention on Climate Change (UNFCCC), commonly referred to as COP28, was held in Dubai and concluded in December 2023. Delegations from nearly 200 member nations to the UNFCCC convened in the ongoing effort to collaborate and combat the escalating threat of climate change. This was set against the concurrent confirmation by scientists that the year would be the hottest ever recorded.

The lead-up to the opening of COP28 was marked by criticism of the conference host and the conference presidency selection of the UAE energy minister. It should be noted that the UAE ranks seventh in global petroleum production at a 4% share of world total, behind the United States (21%), Saudi Arabia (13%), Russia (10%), Canada (6%), Iraq (5%), and China (5%), according to the US Energy Information Administration (EIA).[34]

---

[34] *Independent Statistic and Analysis*, U.S. Energy Information Administration (2022 statistics) (www.eia.gov/tools/faqs/faq.php?id=709&t=6).

CHAPTER 3   THE INTERNATIONAL LEGAL AND POLICY FOUNDATIONS OF ESG

The summit was a global call to action with mounting concern and pressure for government action to slow the rate of climate change and cap global temperature rise, per the COP21 Paris Agreement goals.[35] The COP28 Dubai Agreement[36] has been referred to as the "beginning of the end" of the fossil fuel era, with the inclusion of provisions calling for a just transition toward less carbon-intensive and more renewable energy sources. The conference was attended by some 85,000 people, including the largest cohort of businesses in the nearly three decades since the UNFCCC was established.

Key Outcomes of COP28

1. **Global Stocktake and Transition Away from Fossil Fuels**

    The central outcome of COP28 is the global stocktake, which is a comprehensive assessment of national climate action plans. The stocktake reiterated that global GHG emissions need to be cut by 43% by 2030, compared to 2019 levels, to limit global warming to 1.5°C.

    The stocktake was the most anticipated and scrutinized aspect of the delegation meetings. The final text, which was reached as COP28 came to a close, delivered a call to "transition away" from fossil fuels with the intention of "accelerating

---

[35] United Nations, *Paris Agreement*, Paris, December 12, 2015 (https://treaties.un.org/Pages/ViewDetails.aspx?src=TREATY&mtdsg_no=XXVII-7-d&chapter=27&clang=_en).

[36] United Nations, *Dubai Agreement text*, Dubai (December 12, 2023) (https://unfccc.int/sites/default/files/resource/cma2023_L17_adv.pdf).

and substantially reducing non-CO_2 emissions globally, including in particular methane emissions by 2030."[37] This was a significant move and a first in terms of agreed-upon text language by the member delegations. This decision emphasizes the urgent need for nations to translate their pledges into tangible and actionable GHG emissions reduction plans.

2. **Accelerating Renewable Energy and Energy Efficiency**

COP28 called for tripling renewable energy capacity and doubling energy efficiency improvements globally by 2030. This target aims to expedite the transition from fossil fuels and promote sustainable alternatives.

3. **Climate Finance**

Climate finance was one of the central themes of the conference. The Green Climate Fund (GCF)[38] received significant supplemental funds, with pledges coming in at USD$12.8 billion from 31 countries.

---

[37] Id.
[38] Green Climate Fund, UNEP (www.unep.org/about-un-environment/funding-and-partnerships/green-climate-fund).

4. **Loss and Damage Fund**

A major agreement was reached on the implementation and operationalization of the Loss and Damage Fund (LDF). This fund aims to provide technical assistance to developing countries that are particularly vulnerable to the adverse effects of climate change. Initial pledges from wealthy nations to support the fund totaled more than USD$650 million.

5. **Health and Inclusivity**

COP28 emphasized the importance of enabling all stakeholders to engage in climate action. While the origins of international and national environmental law were protection of human health and the environment from pollution, health impacts have often been left out of climate discussions. A significant outcome was the Declaration on Climate and Health,[39] which was endorsed by 142 member nations.

World leaders were joined by civil society, businesses, indigenous persons, youth, and philanthropic organizations in a shared determination to close the gaps to the climate goals by 2030, with a number of side announcements from the private and philanthropic sectors.

---

[39] United Nations, *COP28 UAE Declaration on climate and health*, Dubai, December 3, 2023 (www.who.int/publications/m/item/cop28-uae-declaration-on-climate-and-health).

6. **Strengthening Resilience to Climate Change**

   COP28 made progress in strengthening the resilience capabilities of vulnerable nations to the impending effects of climate change. Parties agreed on targets for the Global Goal on Adaptation (GGA) and its framework, reflecting a global consensus on adaptation targets.

Challenges and Steps Forward

Despite the significant outcomes, COP28 faced several challenges. The conference fell short of delivering decisive action on climate change, primarily due to the lack of adequate financial pledges and the slow pace of progress in many areas. Nonetheless, COP28 achieved notable progress in several areas. It marked a critical step forward in agreement on global climate policy, with the world's leaders formally recognizing the need for a planned transition away from fossil fuels.

The outcomes of COP28 have far-reaching implications, and they highlight the urgent need for nations to ramp up their efforts to combat climate change. The next two to three years will be determinative and will demonstrate whether we are positioned to meet our global goals by 2030. We expect to see ongoing engagement by the private sector in specific areas, including the continued maturity of voluntary carbon markets. We're in the most critical years for climate change, and it is incumbent upon the corporate sector to implement cogent climate transition plans toward a more secure, equitable, and livable future.

CHAPTER 3   THE INTERNATIONAL LEGAL AND POLICY FOUNDATIONS OF ESG

3.5 United Nations Sustainable Development Goals (SDGs)

Introduction to the Sustainable Development Goals

In the realm of global development, the Sustainable Development Goals (SDGs) have emerged as an indispensable framework for action. Formulated by the United Nations in 2015, the SDGs present a holistic plan for achieving sustainable global prosperity. This ambitious plan encompasses 17 goals, addressing a myriad of dimensions including economic growth, social inclusion, and sustainability. The aspiration is to realize these objectives by the year 2030, propelling humanity toward a future of peace, dignity, and equity on a healthy planet.

The SDGs evolved from decades of developmental work undertaken by member states and the United Nations. They can be tracked back to the Brundtland Commission and the Earth Summit in Rio, when more than 178 countries adopted Agenda 21 as the first comprehensive plan for global sustainable development. The next significant step was the Millennium Declaration at the Millennium Summit in 2000, which led to the formulation of eight Millennium Development Goals (MDGs) aimed at reducing extreme poverty by 2015.[40] The Johannesburg Declaration on Sustainable Development, prepared in 2002, reaffirmed the global commitments to poverty eradication and environmental protection.[41]

---

[40] *Millenium Summit of the United Nations*, New York, September 6–8, 2000 (www.un.org/en/development/devagenda/millennium.shtml).

[41] *World Summit on Sustainable Development*, Johannesburg, August 26–September 4, 2002 (www.un.org/en/conferences/environment/johannesburg2002#:~:text=The%202002%20World%20Summit%20on,account%20respect%20for%20the%20environment).

CHAPTER 3   THE INTERNATIONAL LEGAL AND POLICY FOUNDATIONS OF ESG

The United Nations Conference on Sustainable Development (Rio+20) held in 2012 led to the creation of a set of SDGs to build upon the MDGs.[42] The culmination of these efforts was the adoption of the 2030 Agenda for Sustainable Development, with 17 SDGs at its core, at the UN Sustainable Development Summit in September 2015 (Figure 3-1).[43]

Figure 3-1.   *United Nations Sustainable Development Goals*[44]

---

[42] *United Nations Conference on Sustainable Development, Rio+20*, June 20–22, 2012 (https://sustainabledevelopment.un.org/rio20).

[43] United Nations. *The Sustainable Development Agenda* (www.un.org/sustainabledevelopment/development-agenda/).

[44] United Nations. *Sustainable Development Goals* (www.un.org/sustainabledevelopment/). Permission to use the SDGs as depicted for this publication was granted by the United Nations. The content of this publication has not been approved by the United Nations and does not reflect the views of the United Nations or its officials or Member States.

CHAPTER 3   THE INTERNATIONAL LEGAL AND POLICY FOUNDATIONS OF ESG

Key Terms and Concepts

A few key terms and concepts are important to understand the SDGs. These include

- **Sustainable Development:** This reaffirms the Brundtland Report definition and importance of global sustainability to meet current generation needs without compromising future generations.
- **Global Partnership:** This highlights the need for cooperation and shared responsibility.
- **Agenda 2030:** This is the plan agreed upon by the member states, encompassing the 17 SDGs and the related 169 targets.

The Seventeen SDGs

The 17 SDGs are wide-ranging and interconnected. Here is a brief overview of each goal:

1. **No Poverty:** Aims to eradicate poverty
2. **Zero Hunger:** Objective to end hunger, achieve food security, improve nutrition, and promote sustainable agriculture
3. **Good Health and Well-being:** Seeks to ensure healthy lives and promote well-being
4. **Quality Education:** Aims to provide inclusive and equitable quality education and learning opportunities
5. **Gender Equality:** Strives to achieve gender equality and to empower women and girls

6. **Clean Water and Sanitation:** Objective to ensure availability and sustainable management of water and sanitation for all

7. **Affordable and Clean Energy:** Seeks to provide access to affordable, reliable, modern, and clean energy for all

8. **Decent Work and Economic Growth:** Strives to promote sustainable and inclusive growth, full and productive employment, and decent work for all

9. **Industry, Innovation, and Infrastructure:** Geared toward building resilient infrastructure and fostering innovation

10. **Reduced Inequalities:** Focused on reducing inequality within and between nations

11. **Sustainable Cities and Communities:** Seeks to make cities and human settlements safe, resilient, inclusive, and sustainable

12. **Responsible Consumption and Production:** Aimed at ensuring sustainable consumption and production patterns

13. **Climate Action:** Focused on taking urgent action to combat climate change and its impacts

14. **Life Below Water:** Seeks to conserve and sustainably use the oceans, seas, and marine resources for sustainable development

15. **Life on Land:** Aimed at protecting, restoring, and promoting sustainable use of terrestrial ecosystems, and halting land degradation and biodiversity loss

16. **Peace, Justice, and Strong Institutions:** Strives to achieve peaceful and inclusive societies, provide access to justice for all people, and to build accountable institutions

17. **Partnership for the Goals:** Strengthening the means of implementation and revitalizing the global partnership for sustainable development[45]

Progress Toward Targets

Monitoring and tracking progress toward the goals is paramount. Every year, the UN Secretary-General presents a progress report, based on the global indicator framework and data produced by both national and regional statistical and information systems. A comprehensive Global Sustainable Development Report is prepared every four years and informs the quadrennial SDG review deliberations by the UN General Assembly.

The United Nations released a mid-point report on the 2030 Agenda in 2023. The *United Nations Sustainable Development Goals Report 2023: Special Edition* documented that of the approximate 140 targets that can be measured, 50% were off track and 30% showed stalled progress or regression from the 2015 baseline.[46] The report warned of advancing global poverty, discrimination against women, and global temperature rise, and leaders issued a call to action to the public and private sectors to take action to meet the goals on the 2030 timeline.[47]

---

[45] United Nations. *The Sustainable Development Agenda* (www.un.org/sustainabledevelopment/development-agenda/).

[46] *United Nations Sustainable Development Goals Report 2023: Special Edition* (https://unstats.un.org/sdgs/report/2023/).

[47] United Nations Press Release, *World risks big misses across the Sustainable Development Goals unless measures to accelerate implementation are taken, U.N. warns* (July 10, 2023) (https://unstats.un.org/sdgs/files/report/2023/SDG_Report_2023_Press_Release_EN.pdf).

CHAPTER 3   THE INTERNATIONAL LEGAL AND POLICY FOUNDATIONS OF ESG

3.6 ESG Integration with the International Principles and SDGs

ESG factors are progressively being integrated with the SDGs. Businesses and organizations across the world are aligning their ESG strategies and reporting with the SDGs, recognizing that achieving these global goals can lead to better risk management, new opportunities, competitive advantage, and increased profitability. These efforts are primarily principles-based and use the SDG targets as a roadmap for action.

Participation of the private sector is pivotal in achieving the SDGs. Unlike the MDGs, which relied mainly on government funding and nonprofit organizations, the SDGs also call upon businesses to contribute toward changing unsustainable consumption and production patterns. Businesses stepping up to this challenge are demonstrating how private sector innovation and initiative can be a powerful force for sustainable development. Consumers, employees, and investors can and do seek to influence businesses, governments, and communities to prioritize and work toward the SDGs.

Investor Criteria

According to the Capital Group's ESG Study 2022, which "surveyed 1,130 institutional and wholesale investors, including pension funds, family offices and insurance companies, as well as funds of funds, retail/private banks and financial advisors, located in 19 markets around the world ... half (50%) of the investors surveyed say a fund's ability to meet the United Nations' Sustainable Development Goals (SDGs) is an important consideration when making fund selections."[48]

---

[48] Capital Group, *Study finds nearly two-thirds of investors globally prefer using active funds to integrate ESG* (May 18, 2022) (www.capitalgroup.com/about-us/news-room/esg-global-study-2022.html).

CHAPTER 3   THE INTERNATIONAL LEGAL AND POLICY FOUNDATIONS OF ESG

Signatories to the United Nations Principles for Responsible Investment (PRI) have a combined USD$121.3 trillion AUM.[49] PRI was established in 2006 and is a United Nations-coordinated international network of financial institutions collaborating in their commitment to implement "The Six Principles" agreed upon by all members.

Table 3-2. *The Six Principles of the United Nations Principles for Responsible Investment*[50]

1. We will incorporate ESG issues into investment analysis and decision-making processes.
2. We will be active owners and incorporate ESG issues into our ownership policies and practices.
3. We will seek appropriate disclosure on ESG issues by the entities in which we invest.
4. We will promote acceptance and implementation of the Principles within the investment industry.
5. We will work together to enhance our effectiveness in implementing the Principles.
6. We will each report on our activities and progress towards implementing the Principles.

---

[49] *Annual Report 2022*, United Nations Principles for Responsible Investment (www.unpri.org/annual-report-2022).

[50] *Investing with SDG Outcomes*, Principles for Responsible Investment (PRI) (2020) (www.unpri.org/download?ac=10795).

CHAPTER 3   THE INTERNATIONAL LEGAL AND POLICY FOUNDATIONS OF ESG

PRI provides a five-part framework for investors to apply to understand investment outcomes and align with the SDGs.[51] The framework is organized around: (1) identifying outcomes; (2) setting policies and targets; (3) reviewing how investors shape outcomes; (4) the contribution of the financial system in shaping collective outcomes; and (5) how global stakeholders collaborate to achieve outcomes in line with the SDGs.[52] Investors follow PRI's guidance for "Investing with SDG Outcomes" with the recognition that "[t]here is a continuous feedback cycle between ESG risks and opportunities and SDG-aligned outcomes."[53]

Ratings agencies like Moody's have begun incorporating the SDGs into their ESG screening tools[54] and into their corporate sustainability strategy.[55] The Global Head of Moody's ESG Measures, Sabine Lochmann, said at the time of the SDG Alignment Screening announcement: "With less than a decade left to achieve the 2030 Agenda for Sustainable Development, investors have a key role to play in providing the necessary finance to help meet the SDGs, which requires them to be presented as simple metrics."[56] Figure 3-2 depicts the individual icons for the Sustainable Development Goals.

---

[51] *Investing with SDG outcomes: a five-part framework*, United Nations Principles for Responsible Investment (June 14, 2020) (www.unpri.org/sustainable-development-goals/investing-with-sdg-outcomes-a-five-part-framework/5895.article).

[52] Id.

[53] Id. at 8.

[54] Mark Segal, *Moody's Launches SDG Alignment Screening Tool for Investors*, ESG Today (www.esgtoday.com/moodys-launches-sdg-alignment-screening-tool-for-investors/).

[55] *Moody's Sustainability* (https://sustainability.moodys.io/strategy#:~:text=Contributing%20to%20the%20UN%20SDGs&text=We%20are%20working%20to%20align,can%20be%20the%20most%20impactful.&text=In%202022%2C%20we%20joined%20the,enhanced%20Communication%20on%20Progress%20questionnaire).

[56] Id.

CHAPTER 3   THE INTERNATIONAL LEGAL AND POLICY FOUNDATIONS OF ESG

Figure 3-2. United Nations Sustainable Development Goals (Icons)[57]

Key Insights for ESG Integration

Transitioning to a business model that embeds SDG targets requires a deep understanding of ESG issues and a strategic approach to implementing changes. The start of that process involves reviewing the material ESG factors for the business and mapping those to relevant SDGs for each of the environmental, social, and governance pillars (Table 3-3). Companies can then begin to link their respective ESG goals to the Sustainable Development Goals and communicate how their actions support the global goals. We'll review SDG mapping in detail with respect to the environmental, social, and governance factors in Chapters 4–6.

---

[57] United Nations. *Sustainable Development Goals* (www.un.org/sustainabledevelopment/). Permission to use the SDGs as depicted for this publication was granted by the United Nations. The content of this publication has not been approved by the United Nations and does not reflect the views of the United Nations or its officials or Member States.

CHAPTER 3   THE INTERNATIONAL LEGAL AND POLICY FOUNDATIONS OF ESG

Table 3-3. *ESG Factors Mapped to SDGs*

Environmental	Social	Governance
SDGs 6, 7, 9, 11, 12, 13, 14, 15	SDGs 1, 2, 3, 4, 5, 6, 8, 9, 10, 12, 13, 16	SDGs 5, 8, 9, 10, 12, 13, 16, 17
Climate Change	Employee Welfare, Health & Safety	Accountable Policies, Procedures, & Practices
Air Emissions	Talent Development & Retention	Ethics & Anti-corruption
Water Use & Stewardship	Diversity, Equity, & Inclusion	Privacy & Data Security
Biodiversity & Natural Resources	Consumer Protection and Product Safety & Quality	Compliance, Risk, & Audit
Waste Management	Human Rights & Labor Practices	Management Responsibilities & Transparency
Energy Use	Philanthropy, Responsibility, & Volunteering	Board Oversight & Diversity
Responsible Resource Use	Community Relations	Corporate Culture
Green Innovation	Supply Chain Integrity	Stakeholder Engagement

Each environmental, social, and governance factor can be broadly mapped to the respective SDGs, providing a through line in support of the global goals. Here are some key insights for companies aiming to integrate ESG with international principles and SDGs:

- **Understanding ESG Issues**

    Understanding ESG issues is the first step toward integrating them into business operations. Companies need to identify which issues are most material to their business and where they can have the most significant impact.

- **Strategic ESG Integration**

    ESG integration with SDGs should be strategic. It should be charted to the company's overall business strategy and objectives. This will ensure that SDG-supporting activities contribute to the company's strategic goals and create value.

- **Stakeholder Engagement**

    Engaging stakeholders is necessary for successful integration. Companies should communicate their ESG initiatives to their stakeholders, including investors, employees, customers, and the community. They also need to consider their stakeholders' interests and expectations in their goal-setting process and strategy development.

- **Measurement and Reporting**

    Measuring and reporting on ESG performance and SDG alignment is essential for transparency and accountability. It allows companies to track their progress, demonstrate their commitment to ESG, and communicate their performance to stakeholders.

CHAPTER 3   THE INTERNATIONAL LEGAL AND POLICY FOUNDATIONS OF ESG

Challenges in Achieving the SDGs

Despite progress in some areas, significant challenges remain. Inequality, climate change, conflict, and the global pandemic recovery are some of the major hurdles. Gender inequality persists, and progress on some measures of poverty has been too slow. Global hunger is on the rise, and many people still lack access to basic services like clean water and sanitation. Achieving the SDGs by 2030 remains possible and will require accelerated efforts and renewed commitment from all stakeholders.

CHAPTER TAKEAWAYS

- International law is both foundational to ESG and an evolving field

- Familiarity with core international policies and conventions is fundamental

- International principles guide good governance and operating decisions

- ESG strategy and business practices should be aligned with the SDGs

- Navigating ESG factors, such as human rights, climate change, and natural resource protection, requires understanding the history, key treaties, and provisions of international law

CHAPTER 4

A Deep Dive into the "E" in ESG

Environmental issues represent some of the most pressing and comprehensive in the ESG framework. The environmental pillar is composed of topics such as climate change, resource depletion, pollution, and conservation. The environmental pillar factors have a direct impact on the planet's health and well-being, as well as the long-term viability of businesses. Those individual factors (Figure 4-1) represent how a business interacts with and impacts the environment.

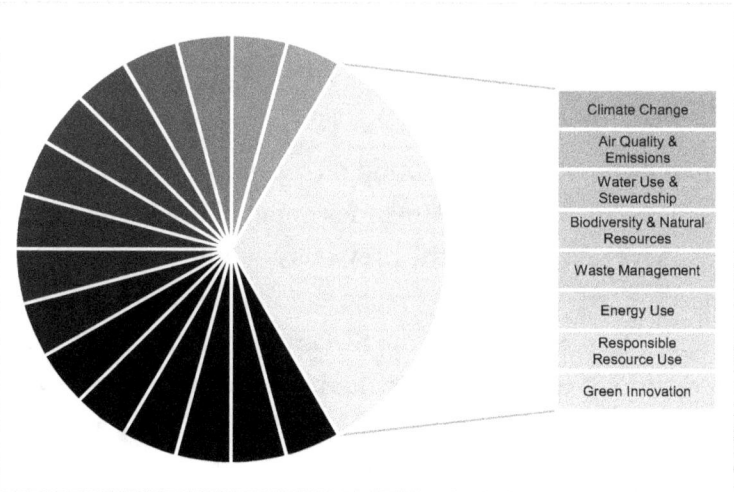

Figure 4-1. "E" Factors of ESG

© Kristyn Noeth 2024
K. Noeth, *The ESG and Sustainability Deskbook for Business*,
https://doi.org/10.1007/979-8-8688-0261-4_4

CHAPTER 4 A DEEP DIVE INTO THE "E" IN ESG

By addressing environmental issues, businesses can reduce their carbon footprint, adopt sustainable practices, promote energy efficiency, and preserve natural resources. Environmental reporting is rooted in satisfying environmental regulatory requirements. In focusing on developing a strong environmental track record, companies can comply with those regulatory requirements, mitigate risk, enhance corporate reputation, and foster innovation.

Favorable performance on environmental metrics can also enhance financial performance and attract capital investment opportunities. The commonly referred to environmental factors of ESG include

1. Climate change
2. Air quality and emissions
3. Water use and stewardship
4. Biodiversity and natural resources
5. Waste management
6. Energy use
7. Responsible resource use
8. Green innovation

This chapter provides background on the global environmental challenges that inform ESG and includes detailed briefings of each of the "E" factors. As we discussed in the previous chapter, the UN issued a call to the public and private sectors to take action to meet the UN Sustainable Development Goals (SDGs), and the SDGs are increasingly being incorporated into ESG ratings assessments. It is a meaningful exercise to review how a company's environmental, social, and governance practices can support the global goals.

CHAPTER 4   A DEEP DIVE INTO THE "E" IN ESG

The "Key Insights for ESG Integration" in Section 3.6 provides guidance on mapping and implementing the SDGs to a company's materials ESG factors. Figure 4-2 illustrates how companies might link their key environmental activities to the relevant global goals.

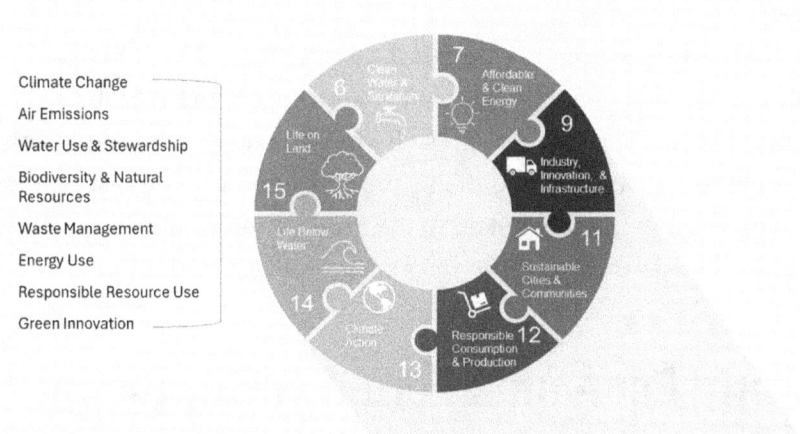

Figure 4-2. Mapping ESG "E" Factors to SDGs

4.1 Global Tipping Point

Any further delay in concerted global action will miss a brief and rapidly closing window to secure a livable future.[1]

—Hans-Otto Pörtner,
former Chair of the Intergovernmental Panel
on Climate Change (IPCC) Working Group

---

[1] *IPCC Sixth Assessment Report Press Release*, IPCC (February 28, 2022) (www.ipcc.ch/report/ar6/wg2/resources/press/press-release/).

CHAPTER 4   A DEEP DIVE INTO THE "E" IN ESG

The primary global climate goal, as discussed in Chapters 2 and 3, is to limit the Earth's warming to no more than 2°C and to pursue actions to limit the temperature to 1.5°C above pre-industrial levels.[2] Meeting that goal will require a *43% reduction from 2019 baseline greenhouse gas (GHG) emissions* by 2030. The past decade was the warmest on record in history, and as the United Nations has reported, we are now at a global pivot point.[3]

Human activity has already warmed the planet to an extent that has caused unprecedented change—some of which is permanent and cannot be remediated by adaptation measures.[4] Extreme weather conditions persist around the globe, causing the loss of life, property, and natural resources. Climate change is inextricably linked to the functioning of environmental systems. Climate risks translate into risks across air, water, and land.

4.2 World Economic Forum (WEF) Global Risks Report

The future belongs to those who understand that doing more of the same is not a viable strategy.

—WEF Global Risks Report 2023[5]

---

[2] United Nations. *Paris Agreement,* Paris, December 12, 2015 (https://treaties.un.org/Pages/ViewDetails.aspx?src=TREATY&mtdsg_no=XXVII-7-d&chapter=27&clang=_en).

[3] United Nations High Level Champions. *The Pivot Point* (2022) (https://climatechampions.unfccc.int/wp-content/uploads/2022/09/R2Z-Pivot-Point-Report.pdf).

[4] *World of Change: Global Temperatures,* National Aeronautics and Space Administration (https://earthobservatory.nasa.gov/world-of-change/global-temperatures).

[5] Id.

CHAPTER 4   A DEEP DIVE INTO THE "E" IN ESG

The World Economic Forum (WEF) publishes an annual **Global Risks Report**, which is developed in collaboration with Marsh McLennan and Zurich Insurance Group.[6] It is a source of intelligence and information on current and emerging risks faced around the world. The report is usually released prior to the WEF annual meeting in Davos and sets the stage for topical discussions during that convening.

The 2024 report is the nineteenth edition of the *Global Risks Report*. Saadia Zahidi, Managing Director, WEF, notes in her Preface that the report's "insights are underpinned by nearly two-decades of original data on global risk perception."[7] The reporting process involves consultation with more than 1,500 global experts, and assesses imminent risks over one-, two-, and ten-year horizons.[8]

Global Risks Report 2024

The World Economic Forum (WEF) recently released its *Global Risks Report* for the year 2024.[9] The report identifies a series of accelerating changes that could reshape the world in the coming decade. It then highlights the global risks that are likely to present material crises. The 2024 report builds upon the *Global Riks Report 2023*, which emphasized the interconnectedness of the global risks and which we also cover in this section.

---

[6] *Global Risks Report 2024*, MarshMcLennan (www.marshmclennan.com/insights/publications/2024/january/global-risks-report-2024.html#:~:text=It%20is%20developed%20by%20the,of%20the%20global%20risk%20landscape).
[7] *The Global Risks Report 2024, 19th Edition*, World Economic Forum (January 2024) (www3.weforum.org/docs/WEF_The_Global_Risks_Report_2024.pdf).
[8] Id.
[9] Id.

101

CHAPTER 4   A DEEP DIVE INTO THE "E" IN ESG

The *Global Risks Report 2024* identifies the scope of challenges that the world faces as a result of four structural forces: (1) climate change; (2) demographic bifurcation; (3) technological acceleration; and (4) geostrategic shifts.[10] Relative instability and a moderate risk of global catastrophes are anticipated in the short term. The long-term outlook is more turbulent over the next decade. The risks that pose the most significant threat in the long term include climate change, demographic changes, and technological and geopolitical shifts.

Top Global Risks 2024

The key findings section in the report leads with the heading, "[a] deteriorating global outlook," and follows that up in the second heading by saying that "[e]nvironmental risks could hit the point of no return."[11] In fact, environmental risks dominate across all three risk time frames (Figure 4-3). Across the two-year horizon, extreme weather events and pollution rank among the top ten risks.[12] However, the ten-year outlook ranks extreme weather events, critical change to Earth systems, biodiversity loss and ecosystem collapse, natural resource shortages, and pollution among the top ten global risks.[13] The pathway is clear: without a course correction, we will experience fundamental shortages and irreversible damage to the natural systems upon which we rely.

---

[10] Id.
[11] Id.
[12] Id.
[13] Id.

CHAPTER 4   A DEEP DIVE INTO THE "E" IN ESG

Figure 4-3. *World Economic Forum Global Risks Report 2024*[14]

Climate change is identified as a major systemic element of the global landscape that will shape the materialization and management of global risks over the next decade.[15] Extreme weather events like record-breaking heat conditions, drought, wildfires, and flooding have emerged as the top risks most likely to create a global crisis. The report warns of related threats to climate-vulnerable populations. Biodiversity loss and ecosystem collapse are other significant environmental risks highlighted by the report. The report suggests that younger respondents tend to rank these risks far more highly over the two-year period compared to older age groups.

---

[14] *Global Risks Report 2024,* MarshMcLennan (www.marshmclennan.com/insights/publications/2024/january/global-risks-report-2024.html#:~:text=It%20is%20developed%20by%20the,of%20the%20global%20risk%20landscape).

The Global Risks Report 2024, 19th *Edition,* World Economic Forum (January 2024) (https://www3.weforum.org/docs/WEF_The_Global_Risks_Report_2024.pdf). Permission to use material for this publication was granted by the WEF and is subject to the Creative Commons license (https://creativecommons.org/licenses/by-nc-nd/4.0/).

[15] Id.

CHAPTER 4   A DEEP DIVE INTO THE "E" IN ESG

Social and economic issues round out the top ten risks.[16] Misinformation and disinformation have also emerged as significant global risks. The report warns that these could further widen societal and political divides, undermine the legitimacy of newly elected governments, and incite violent protests and civil confrontation. Societal polarization is another top risk. The report warns that divisive factors such as political polarization and economic hardship could erode social cohesion and leave ample room for new and evolving risks to propagate.

Economic uncertainties, such as inflation and economic downturn, are notable entrants on the top risk rankings.[17] The report also highlights the risks related to conflict and security. It warns of the risk of the escalation or outbreak of interstate armed conflicts and the rise in conflict due to simmering geopolitical tensions combined with technological advances.

Global Risks Report 2023

The World Economic Forum's (WEF) *Global Risks Report 2023* identifies the unfolding risks that we are expected to encounter globally over the next decade.[18] The *Global Risks 2023* report draws on insights from over 1,200 experts and policymakers worldwide, providing a comprehensive analysis of the geopolitical, economic, and societal fissures that are set to trigger and exacerbate future crises.[19] By examining both short-term dangers and escalating risks over the next ten years, the report sheds light on the challenges that lie ahead.

---

[16] Id.

[17] Id.

[18] World Economic Forum, *Global Risks Report 2023* (www3.weforum.org/docs/WEF_Global_Risks_Report_2023.pdf).

[19] Id.

CHAPTER 4   A DEEP DIVE INTO THE "E" IN ESG

Measured across three time frames and five categories (economic, environmental, geopolitical, societal, and technological), the report ranks the key risks. Today, energy is the predominant crisis, along with food scarcity and the cost of living. The ten-year forecast tells us that by 2033, the impact and severity of climate and environmental issues will overtake the list and account for six of the top ten risks.[20]

The *Global Risks Report 2023* serves as a vital resource for understanding the risks that shape our world. By examining short-term dangers, long-term risks, and mid-term futures, the report offers a comprehensive analysis of the challenges we face.

1. **Short-Term Dangers:** The initial short-term time frame, spanning two years, focuses on the immediate impact of ongoing crises. As we navigate through the aftermath of the COVID-19 pandemic, we find ourselves grappling with a resurgence of "older" risks. These risks include inflation, trade wars, capital outflows from emerging markets, and geopolitical confrontations. Unsustainable levels of debt, low growth, and low global investment are cited among the factors that compound the challenges we face. The mounting pressure of climate change further contributes to an uncertain decade ahead.

2. **Long-Term Risks (Ten Years):** Certain risks are expected to intensify significantly over the next ten years. This outlook delves into the economic, environmental, societal, geopolitical, and technological risks that could compound future crises. Among these risks, competition for resources

---

[20] Id.

stands out. The report examines how the scarcity of food and minerals may trigger conflicts and exacerbate global challenges. It focuses on the need to address these issues proactively to mitigate potential crises in the long term.

3. **Mid-Term Futures (2030):** The report ventures into mid-term futures by exploring the interconnections between emerging risks. It presents a scenario of a "polycrisis" centered around natural resource shortages by the year 2030. This conceptual framework highlights the potential consequences of inadequate preparedness and the need for resilience-building efforts.[21]

Top Global Risks 2023

The report identifies several top global risks based on the findings of the underlying assessment across a wide range of areas, including economic and environmental risk. It identifies climate change and environmental risk as significant threats to the world's future.[22] This report delves into these risks and explores the far-reaching implications of rising global temperatures and environmental degradation.

The report emphasizes the need for immediate action to mitigate the effects of global climate change, such as rising temperatures, extreme weather events, and the loss of biodiversity. It highlights the dependent relationships between environmental risks and other global challenges, such as economic stability and societal well-being. To navigate the

---

[21] Id.
[22] Id.

complex landscape of global risks, it is essential to distill key risks from the report, as they provide actionable insights for policymakers, corporations, and a range of organizations within the risk rubric.

The findings also highlight the increasing competition for basic resources, including food and minerals. As the global population continues to grow, the strain on these resources intensifies. This competition can exacerbate geopolitical tensions and lead to conflicts if not adequately addressed. Developing sustainable resource management strategies and fostering international cooperation are requisites for ensuring resource availability for future generations.

Perhaps the most striking aspect of the risk ranking in the report is the interconnectedness of those risks. The report provides a critical analysis of the *polycrisis* we are facing—and discusses the crises in multiple global systems where the risks are more interdependent and damaging than ever to humanity.[23] Many of those are "old" and recurring risks such as energy, hunger, and inequality, which we have sought to resolve with the UN Sustainable Development Goals. They are now coupled with new geopolitical, societal, and economic risks and compounded by the climate crisis.

Takeaways from the Past Two Annual Global Risks Reports

WEF, echoing what has been stated by the United Nations, has indicated that we are at a strategic inflection point. The findings serve as a stark reminder of the intertwined environmental, social, and economic challenges that lie ahead. The past two years of reports emphasize the importance of proactive risk management and resilience-building efforts.

---

[23] Id.

CHAPTER 4   A DEEP DIVE INTO THE "E" IN ESG

Risk mitigation necessitates a new vanguard with cooperation among elected officials, community leaders, and corporate chiefs with a sharper lens on accountability. It emphasizes the need for collaboration, innovation, and sustainable practices to address global challenges effectively. The challenge is to navigate and mitigate risks to create a more resilient world. By embracing these principles and taking decisive action, we can shape a more resilient and sustainable future.

4.3 Climate Policy and the Intergovernmental Panel on Climate Change (IPCC)

The Intergovernmental Panel on Climate Change (IPCC) is a United Nations body and a primary entity and leading source of scientific information in the global conversation on climate change.[24] It is a beacon of scientific insights that enable governments to craft informed climate policies. The IPCC's rigorous assessments of climate change science, impacts, and mitigation options serve as a compass guiding international climate negotiations.[25] This section reviews the inception, structure, operations, and contributions of the IPCC in the realm of global sustainability and climate change policy.

The world's leading climate scientists have been warning us for more than 35 years of rising global temperatures. The IPCC was established by the United Nations Environment Programme (UNEP) and the World Meteorological Organization (WMO) in 1988. It was endorsed by the UN General Assembly and tasked with preparing a comprehensive review of climate change to specifically include recommendations on: (1) the state

---

[24] Intergovernmental Panel on Climate Change (www.ipcc.ch/).
[25] Id.

of knowledge on the science; (2) the social and economic impacts; and (3) potential response strategies for inclusion in a future, to-be-established international convention.[26]

IPCC issued its first report in 1990, and prepared a supplement for the Earth Summit in 1992, which was foundational and served as the basis for the United Nations Framework Convention on Climate Change (UNFCCC), as discussed in Chapter 3.[27] The IPCC continues to furnish governments with scientifically backed information on climate change. The organization currently has a roster of 195 member countries, demonstrating its global reach and influence.[28]

The IPCC's work has been instrumental in informing and shaping global climate change policies. Its reports have been used as key inputs in international climate negotiations, guiding the development of impactful climate policies. The IPCC's work also aids in identifying areas needing further research, thereby driving the direction of climate change science. The contribution of the IPCC was recognized when it received the Nobel Peace Prize jointly with former US Vice President Al Gore in 2007.[29]

Structure and Organization of the IPCC

The IPCC is composed of three working groups and a task force. Each division focuses on distinct aspects of climate change: (1) Working Group I examines the physical science; (2) Working Group II studies climate change

---

[26] United Nations General Assembly, *Protection of global climate for present and future generations of mankind: resolution adopted by the General Assembly* (1988) (https://digitallibrary.un.org/record/54234?ln=en).

[27] *Climate Change: The IPCC 1990 and 1992 Assessments*, IPCC (www.ipcc.ch/report/climate-change-the-ipcc-1990-and-1992-assessments/).

[28] *About the IPCC*, Intergovernmental Panel on Climate Change (www.ipcc.ch/about/#:~:text=The%20IPCC%20is%20divided%20into,with%20Mitigation%20of%20Climate%20Change).

[29] Nobel Peace Prize (2007) (www.nobelprize.org/prizes/peace/2007/summary/).

CHAPTER 4   A DEEP DIVE INTO THE "E" IN ESG

impacts, adaptation, and vulnerability; (3) Working Group III focuses on mitigation; and (4) the Task Force on National GHG Inventories develops methodology for calculating and reporting national GHG emissions.[30]

These working groups and the task force are fortified by a vast network of experts from around the globe who voluntarily contribute their time and expertise to the work and scholarship of the IPCC. The organization's commitment to transparency and objectivity is reflected in its open and rigorous review process for the public reporting of its assessments.

Reporting Cycles

The IPCC operates in assessment cycles, usually spanning six to seven years. Each cycle culminates in the publication of a comprehensive "Assessment Report," which provides an updated overview of the state of knowledge on climate change. These reports are drafted and reviewed in multiple stages and are instrumental in informing international negotiations on climate change.[31]

In addition to the assessment reports, the IPCC also publishes "Special Reports" and "Methodology Reports."[32] The Special Reports focus on specific topics related to climate change, providing nuanced insights into areas such as The Ocean and Cryosphere in a Changing Climate, Global Warming of 1.5°C, and Climate Change and Land.[33] The Methodology Reports offer guidance on estimating national greenhouse gas emissions and removals.[34]

---

[30] *About the IPCC*, Intergovernmental Panel on Climate Change (www.ipcc.ch/about/#:~:text=The%20IPCC%20is%20divided%20into,with%20Mitigation%20of%20Climate%20Change).
[31] Id.
[32] *Reports*, Intergovernmental Panel on Climate Change (www.ipcc.ch/reports/).
[33] Id.
[34] Id.

The IPCC conducts an exhaustive review of the scientific literature on climate change but does not engage in original research activities. The organization sets a deadline for the publication of scientific papers to be included in the reports, ensuring that the most recent and relevant knowledge is incorporated. The selection of authors for the IPCC reports is a meticulous process. Experts are chosen based on their knowledge and expertise in relevant fields, ensuring a comprehensive and authoritative assessment of the subject matter.[35]

Influence on Climate Policy

The IPCC reports are science-based and do not prescribe policy, making them an instrumental resource for policymakers worldwide. The reports not only outline the state of agreement in scientific circles but also identify areas where further research is required. They are a key input for international climate change negotiations, helping to shape informed climate policies.

As discussed in Chapter 2, the IPCC recently issued its Sixth Assessment Report and a related Synthesis Report.[36] The report reviewed the impacts of climate change on ecosystems and communities and examined global mitigation efforts. It also provides a comprehensive review of the current state of climate science, including the latest data trends on temperature rise and the mitigation and adaptation strategies needed to avoid impacts.

The report noted that "[c]limate change is already affecting every region on [e]arth, in multiple ways ... and [t]he changes we experience will increase with additional warming."[37] The Sixth Assessment provides

---

[35] Id.

[36] *Synthesis Report of the Sixth Assessment Report*, Intergovernmental Panel on Climate Change (www.ipcc.ch/ar6-syr/).

[37] Id.

CHAPTER 4   A DEEP DIVE INTO THE "E" IN ESG

an in-depth regional assessment of climate change, along with scientific information that can inform decision-making on climate change.[38] It calls attention to the significant disruption in nature affecting human life and stresses that vulnerable people and ecosystems least able to adapt are being hardest hit by the changing climate. It also highlights the importance of ambitious, accelerated action to adapt to climate change, while contemporaneously reducing greenhouse gas emissions.

The report provides new insights into nature's potential to not only reduce climate risks but also to improve people's lives. Actionable steps for sustained reductions in GHGs were presented in the report. Maarten Kappelle, Head of Scientific Thematic Investments at UNEP, said, "[a]s the report lays out, there are multiple options for policymakers to tackle the crisis."[39] As the world grapples with the ever-increasing challenges posed by climate change, the work of the IPCC remains as vital as ever.

4.4 Greenhouse Gases (GHGs)

The natural world is a complex system, interwoven with countless elements that maintain the balance required for diverse life forms to thrive. However, certain gases in our atmosphere, known as greenhouse gases (GHGs), are causing a significant disturbance in this balance. These gases, while pivotal for the sustenance of life on Earth, are now being excessively generated by human activities, leading to an alarming escalation in global temperatures—the phenomenon known as global warming, which is driving consequential climate change.[40]

---

[38] Id.

[39] *Time running out to defuse climate "time bomb,"* UNEP (March 21, 2023) (www.unep.org/news-and-stories/story/time-running-out-defuse-climate-time-bomb).

[40] *Energy and the environment explained, Greenhouse gases*, U.S. Energy Information Administration (www.eia.gov/energyexplained/energy-and-the-environment/greenhouse-gases.php).

Greenhouse gases are atmospheric components that trap heat or thermal radiation emitted by the Earth, thereby creating a "greenhouse effect" that warms the planet.[41] In essence, these gases allow sunlight to reach the Earth's surface but absorb the infrared radiation (heat) that the surface emits back, preventing it from escaping into space.[42]

List of Greenhouse Gases

GHGs are gases in the Earth's atmosphere that trap heat, contributing to the greenhouse effect and temperature rise. These gases are emitted by various natural and human-induced activities, with the latter being the primary driver of recent increases in GHG concentrations.

There are several types of GHGs, each with varying levels of impact on global warming, influenced by their abundance in the atmosphere, their capability to absorb heat, and their lifespan in the atmosphere.[43] Some occur naturally, while others are exclusively anthropogenic, or human-made. The following is a list of the key greenhouse gases:

- **Carbon dioxide (CO_2):** Carbon dioxide is the most prevalent GHG and the primary driver of global warming. It originates both naturally and from human activities such as the burning of fossil fuels (coal, oil, and natural gas) and deforestation.

- **Methane (CH_4):** Methane, a potent GHG, is released during the decomposition of organic matter and the production and transport of coal, oil, and natural gas. It is also a byproduct of livestock farming and rice cultivation. Methane is the second most significant

---

[41] Id.

[42] Id.

[43] Id.

contributor to global warming and accounts for approximately 30% of the warming that has occurred since pre-industrial times.[44]

- **Nitrous oxide (N_2O):** Nitrous oxide is generated predominantly from agricultural activities, particularly from the usage of synthetic and organic fertilizers and from the combustion of fossil fuels and biomass.

- **Water vapor (H_2O):** Water vapor is the most abundant and potent GHG. It increases as the Earth's atmosphere warms, creating a feedback loop that further amplifies global warming.

- **Industrial gases:** Industrial gases, such as hydrofluorocarbons (HFCs), perfluorocarbons (PFCs), sulphur hexafluoride (SF_6), and nitrogen trifluoride (NF_3), are exclusively human-made. These gases are used in various industrial applications, and, despite their low concentrations, they contribute significantly to global warming due to their high heat-trapping capabilities.[45]

---

[44] *Methane emissions are driving climate change. Here's how to reduce them*, United Nations Environment Programme (August 20, 2021) (www.unep.org/news-and-stories/story/methane-emissions-are-driving-climate-change-heres-how-reduce-them).

[45] *Global Greenhouse Gas Emission Data*, U.S. Environmental Protection Agency (www.epa.gov/ghgemissions/global-greenhouse-gas-emissions-data).

CHAPTER 4   A DEEP DIVE INTO THE "E" IN ESG

Primary Sources and Impacts of Greenhouse Gases

Human activities are the primary contributors to the rise in GHG concentrations in the atmosphere (Figure 4-4).[46] The combustion of fossil fuels, such as for use in energy generation and transportation, results in large-scale CO_2 emissions.[47] Methane emissions primarily originate from the energy sector and also from agricultural practices, including livestock farming and rice cultivation, as well as from natural gas systems and landfills.[48] Nitrous oxide emissions are largely due to agricultural practices, particularly the use of fertilizers, and from combustion of fossil fuels and biomass.[49]

---

[46] *The Causes of Climate Change*, National Aeronautics and Space Administration (https://climate.nasa.gov/causes/#:~:text=Human%20Activity%20Is%20the%20Cause,carbon%20dioxide%20(CO2)).

[47] Id.

[48] Id.

[49] Id.

CHAPTER 4   A DEEP DIVE INTO THE "E" IN ESG

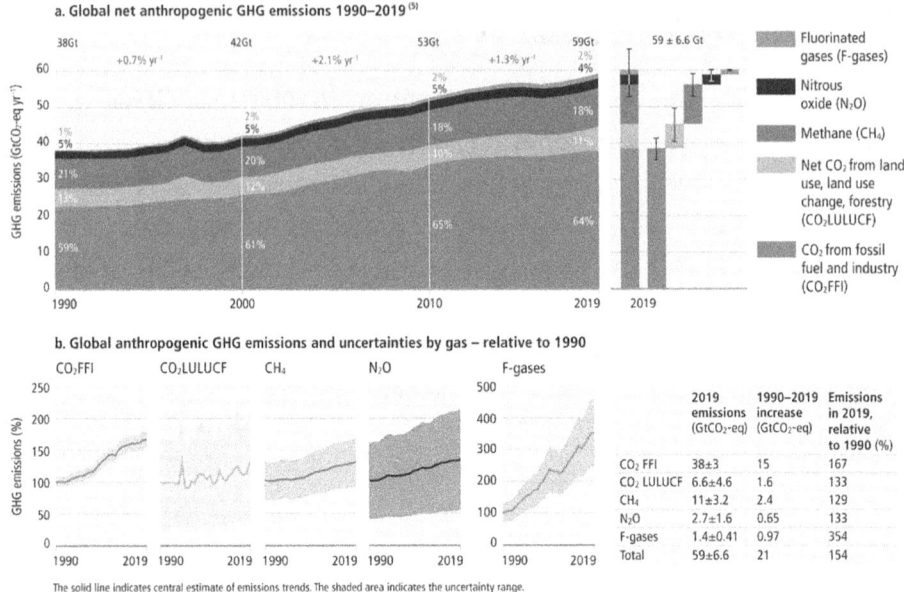

Figure 4-4. *IPCC Global Net Anthropogenic Emissions (1990-2019)*[50]

The excessive accumulation of GHGs in our atmosphere is causing a rapid increase in the planet's average temperature. This global warming is leading to a cascade of changes in our climate system, including rising sea levels and the increased frequency and intensity of extreme weather events. These changes pose significant risks to ecosystems and biodiversity, as well as to human societies and economies.

To mitigate the impact of GHGs and climate change, there is an urgency for a global shift toward sustainable practices. The most effective

---

[50] IPCC, 2022: Summary for Policymakers. In: Climate Change 2022: *Mitigation of Climate Change. Contribution of Working Group III to the Sixth Assessment Report of the Intergovernmental Panel on Climate Change* [P.R. Shukla, J. Skea, R. Slade, A. Al Khourdajie, R. van Diemen, D. McCollum, M. Pathak, S. Some, P. Vyas, R. Fradera, M. Belkacemi, A. Hasija, G. Lisboa, S. Luz, J. Malley (eds.)]. Cambridge University Press, Cambridge, UK and New York, NY, USA. doi: 10.1017/9781009157926.001.

way to reduce GHG emissions is to transition from fossil fuels to less carbon-intensive fuels and renewable energy sources such as solar and wind power.[51] Additionally, improvements in energy efficiency, changes in land use practices, and the development of technologies to capture and store carbon are all pivotal in the fight against climate change.[52]

Regulation of Greenhouse Gases

The escalation of GHG emissions has led to international agreements aimed at reducing these emissions, as highlighted in Chapter 3. The Kyoto Protocol, signed in 1997, legally bound signatory countries to reduce their GHG emissions by agreed-upon amounts. Nearly two decades later, the Paris Agreement requested signatories to declare a Nationally Determined Contribution (NDC), detailing their targets for mitigating climate change and their plans to achieve these targets. National and regional governments have set and adjusted emissions reduction targets in the intervening years. As we discuss in Chapters 2 and 7, many nations have undertaken climate disclosure and reporting regulation, as well as established rules for specific GHGs such as methane.

4.5 Top Greenhouse Gas (GHG)-Emitting Sectors and Countries

This section explores the industries and sectors that contribute most to the production of greenhouse gases and the countries with the highest levels of GHG emissions. We will also touch upon key concepts such as carbon intensity, offering a comprehensive understanding of the topic.

---

[51] *Greenhouse Gas Emissions*, US Environmental Protection Agency (www.epa.gov/ghgemissions/sources-greenhouse-gas-emissions).
[52] Id.

CHAPTER 4   A DEEP DIVE INTO THE "E" IN ESG

Carbon dioxide (CO_2) is the chief contributor to GHG emissions and is primarily released through the burning of fossil fuels such as coal, oil, and natural gas. This combustion occurs in various sectors, including energy production, transportation, and industry. In 2021, CO_2 emissions accounted for approximately 79% of total US anthropogenic GHG emissions.[53]

As mentioned in Section 4.4, other significant GHGs include methane, nitrous oxide, and fluorinated gases. Methane is released from landfills, coal mines, agriculture, and oil and gas operations.[54] Nitrous oxide is emitted through the application of nitrogen fertilizers and the burning of fossil fuels.[55] Fluorinated gases are synthetic, potent GHGs emitted from a variety of industrial applications.[56]

Major Contributing Sectors

A review of the percentage of global emissions data by sector (Figure 4-5) provides insight into the relative contribution to the world's total greenhouse gas emissions. The (1) industrial, (2) electricity, (3) agriculture, land use, and waste, (4) transport, and (5) buildings sectors typically generate the most GHG emissions. Figure 4-5 breaks each of those sectors down into sub-sectors as well. Understanding the relative emissions by sector and sub-sector guides the prioritization for targeted reductions.

---

[53] *Energy and the environment explained. Where greenhouse gases come from*, U.S. Energy Information and Administration (www.eia.gov/energyexplained/ energy-and-the-environment/where-greenhouse-gases-come-from. php#:~:text=Carbon%20dioxide,%2Dyear%20global%20warming%20potential)).
[54] Id.
[55] Id.
[56] Id.

CHAPTER 4   A DEEP DIVE INTO THE "E" IN ESG

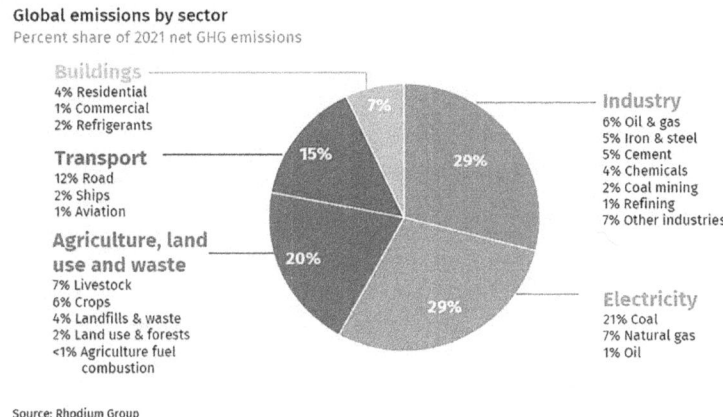

Figure 4-5. Global GHG Emissions by Sector (2021 data)[57]

Together, the industrial and electric power generating sectors make up more than half of total global GHG emissions, per data from the most recent total collection year of 2021.[58] General industry—including iron and steel, oil and gas production, cement, chemical manufacturing, coal mining, refining, and other industries—accounts for 29% of total GHG emissions.[59] Electric power generation accounts for 29% of global GHG emissions (coal combustion makes up the majority of that total).[60] Agriculture, land use, and landfills generate 20% of the global total GHG emissions.[61] Transportation follows at 15% and the building sector accounts for the remaining 7% of total GHG emissions.[62]

---

[57] Alfredo Rivera, Shweta Movalia, Emma Rutkowski, Yuren Rangel, Hannah Pitt, and Kate Larsen. *Global Greenhouse Gas Emissions: 1990-2021 and Preliminary 2022 Estimates*, Rhodium Group (September 19, 2023) (https://rhg.com/research/global-greenhouse-gas-emissions-2022/).
[58] Id.
[59] Id.
[60] Id.
[61] Id.
[62] Id.

CHAPTER 4   A DEEP DIVE INTO THE "E" IN ESG

Top Emitting Countries

The data showing GHG emissions by country (Figure 4-6) is also informative in illustrating the respective national shares of the global total. Historically, the top emitting countries include China, the United States, the European Union (supranational union of individual member counties), India, Russia, Brazil, and Japan.

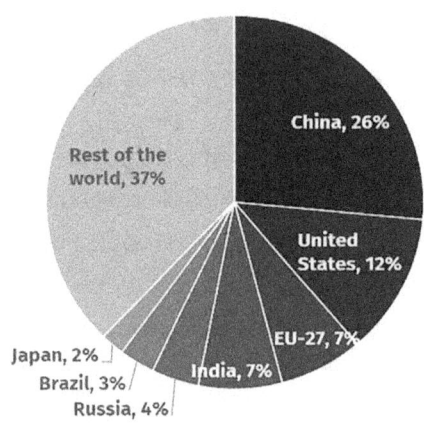

Figure 4-6. Net GHG Emissions by Country (2022 data)[63]

China is the world's largest emitter of GHGs, with the most significant percent share of the global total (26%), followed by the United States (12%), the EU, and India (each 7%).[64] Russia follows at 4%, Brazil at 3%, and Japan at 2%, with the remainder of the world accounting for a total of 36% of emissions.[65]

---

[63] Id.
[64] Id.
[65] Id.

CHAPTER 4   A DEEP DIVE INTO THE "E" IN ESG

Emissions Data Analysis Informs GHG Reduction Approaches

Emissions Per Capita

Scientists also look at additional data slices when assessing global emissions. For instance, a nation's per capita emissions data is a useful measurement. Carbon intensity is expressed as the amount of carbon emitted per unit of energy consumed or per unit of economic output.[66] It serves as an indicator of how "clean" or "dirty" an energy source or process is, or how sustainable or unsustainable an economic activity is.

Per capita emissions data correlate a nation's share of global emissions to their global population share. The United States has the highest per capita emissions ratio by far, according to a recent article in *Nature*, which utilized source data from the Global Carbon Budget for emissions calculations and the United Nations Population Division for population data.[67]

In fact, the United States produces almost 14% of global emissions yet has just 4% of the world population.[68] India has four times the population of the United States at nearly 18% of the world's 8 billion people yet less than 8% of total emissions.[69] The European Union ranks third with a 5.6% population share and 7.4% of emissions.[70] The list is rounded out by Russia

---

[66] *Low Carbon Fuel Standard*, California Air Resources Board (ww2.arb.ca.gov/sites/default/files/2020-09/basics-notes.pdf).

[67] Smitri Mallapaty, Jeff Tollefson, Carissa Wong, Sarah Wild, & Nisha Gaind, *How five crucial elections in 2024 could shape climate action for decades*, Nature (March 5, 2024) (www.nature.com/articles/d41586-024-00642-3).

[68] Id.

[69] Id.

[70] Id.

CHAPTER 4   A DEEP DIVE INTO THE "E" IN ESG

in fourth position at 1.8% of world population and 4.5% of emissions and Indonesia as the fifth largest per capita emitter with a 3.5% share of the Earth's population and 2% of global emissions.[71]

Figure 4-7 charts the respective carbon dioxide emissions per capita in the United States, Japan, China, the European Union, and India for the years 2000–2023, per the International Energy Agency (IEA). The emissions are measured as total carbon dioxide component (tons) per capita.

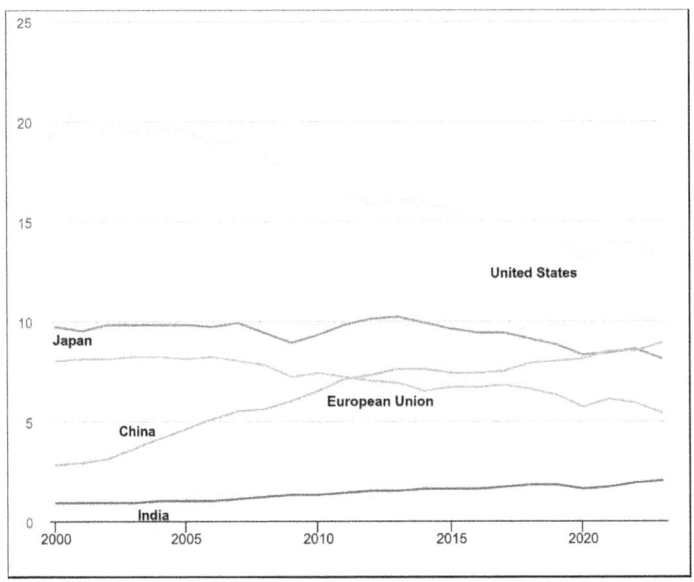

Figure 4-7. *IEA CO2 Total Emissions Per Capita by Region, 2000–2023*[72]

To combat climate change, it is vital to reduce GHG emissions. This requires a multifaceted approach, encompassing increased energy efficiency, a shift toward renewable energy, changes in agricultural practices, and enhanced carbon sequestration. Understanding the sources and drivers of GHG emissions is necessary to devise effective strategies to

---

[71] Id.

[72] *CO2 total emissions per capita by region, 2000-2023*, International Energy Agency (February 27, 2024) (www.iea.org/data-and-statistics/charts/co2-total-emissions-per-capita-by-region-2000-2023). License: CC BY 4.0.

CHAPTER 4 A DEEP DIVE INTO THE "E" IN ESG

combat climate change. While the challenge is significant, with concerted global action and the adoption of sustainable practices across all sectors, it is possible to transition to a low-carbon, sustainable future.

4.6 Carbon Markets
Understanding Carbon Markets: Compliance Market and Voluntary Market Models

Carbon markets, an innovative approach to combating climate change, have grown significantly over the past decade. They are built around the concept of "cap and trade," which is a market-based mechanism designed to reduce greenhouse gas emissions. The primary function of these markets is to set a cap on emissions and then allow entities to trade the rights to emit, instigating an economic incentive to reduce emissions. We'll review the historical background, development, expansion, challenges, and prospects of carbon markets to provide a comprehensive understanding of their utility in global sustainability efforts.

The foundation of carbon markets lies in the principle of cap and trade. In this system, a cap is a limit set on the total amount of specific air pollutants such as greenhouse gases (GHG) that may be emitted by factories, power plants, and other GHG-producing sources. Governments essentially established and ran the original carbon markets, which are now referred to as "compliance markets." The United States Congress established the world's first cap-and-trade program under the Clean Air Act Amendments of 1990, aimed at reducing sulfur dioxide (SO2) emissions from power plants to address acid rain.[73] The success of the Acid Rain Program's allowance trading market led to further development and expansion of additional pollutants by other jurisdictions.

---

[73] *Acid Rain Program*, US Environmental Protection Agency (www.epa.gov/acidrain/acid-rain-program).

Under the cap-and-trade scenario, companies are given a specific number of emission allowances, which represent a specific volume threshold for greenhouse gas emissions. The primary GHG covered is carbon dioxide (CO_2). Companies that reduce their emissions below their cap can sell their excess allowances to other companies, promoting an economic incentive to reduce emissions. The total number of allowances is reduced over time, gradually lowering the cap and ostensibly driving companies to innovate and adopt cleaner technologies.

The previous and historical model of pollution control was termed "command and control" whereby private sector companies were directed by regulatory agencies to adopt one-size-fits-all technologies such as air scrubbers that did not allow for flexibility or market activity. Cap-and-trade systems are market mechanisms that have realized direct environmental benefits in the reduction of criteria air pollutants. The frameworks were built to incentivize planned emissions reduction and enable innovation in lowering emissions and energy efficiency.

Development of Global and Regional Carbon Markets

The Kyoto Protocol, adopted in 1997, marked the birth of global carbon markets. The international agreement placed a heavier burden on developed nations, recognizing their primary responsibility for high levels of GHG emissions due to over a century of industrial activities. The Kyoto Protocol introduced the concept of carbon credits, or "Certified Emission Reductions (CERs)," which could be traded between countries, turning carbon reduction into a marketable commodity.[74]

Since the inception of the Kyoto Protocol, carbon markets have evolved significantly. Various regions and countries have established their own

---

[74] *Kyoto Protocol to the United Nations Framework Convention on Climate Change*, United Nations, December 11, 1997, Kyoto (https://treaties.un.org/Pages/ViewDetails.aspx?src=TREATY&mtdsg_no=XXVII-7-a&chapter=27&clang=_en).

CHAPTER 4   A DEEP DIVE INTO THE "E" IN ESG

carbon markets with different rules, caps, and trading mechanisms. In 2005, the world's largest cap-and-trade program at the time—the European Union (EU) Emissions Trading System (ETS)—was established, and it now covers around 40% of all GHG emissions generated in the European Union.[75] We'll discuss the EU's recently adopted Carbon Border Adjustment Mechanism (CBAM), which is a carbon tariff and a supplementary measure to the EU ETS, in Chapter 7. The United Kingdom developed its own ETS to replace its participation in the EU ETS and is exploring its expansion to extend its carbon pricing system to new sectors, to include GHG removals (GGRs), and to integrate nature-based removals.[76]

The Paris Agreement had a significant impact on the future direction of carbon markets. Unlike the Kyoto Protocol, which placed the burden of emissions reductions primarily on developed countries, the Paris Agreement requires all countries, both developed and developing, to contribute to the effort to combat climate change. Under the Paris Agreement, each country determines its Nationally Defined Contribution (NDC) and outlines the actions it will take to reduce its greenhouse gas emissions and the adaptation measures it will implement. These NDCs are fundamental to the functioning of carbon markets, as they set the national caps on emissions and drive demand and supply in the carbon markets.

China initiated its national emissions-trading system (ETS) to first focus on power sector emissions in 2021.[77] China's ETS has already become the world's largest carbon market, clocking in at three times the size of the EU's

---

[75] *What is the EU ETS?* European Commission (https://climate.ec.europa.eu/eu-action/eu-emissions-trading-system-eu-ets/what-eu-ets_en).

[76] Government of the United Kingdom. Proposals to expand the UK Emissions Trading Scheme, May 23, 2024 (https://www.gov.uk/government/news/proposals-to-expand-the-uk-emissions-trading-scheme).

[77] Jane Nakano and Scott Kennedy, *China's New National Carbon Trading Market: Between Promise and Pessimism*, Center for Strategic & International Studies (July 23, 2021) (www.csis.org/analysis/chinas-new-national-carbon-trading-market-between-promise-and-pessimism).

CHAPTER 4   A DEEP DIVE INTO THE "E" IN ESG

ETS. It is set to expand even further—by an estimated 70% in volume—as it adds heavy industry and manufacturing under its umbrella.[78]

In the United States, California also launched its own cap-and-trade program,[79] which offers a prime example of a successful regional carbon market. California has also instituted disclosure requirements for voluntary carbon offsets, which we'll discuss in Chapter 7.[80] Under a regional program, the regulating state sets a cap on total GHG emissions from regulated entities, and companies are allowed to trade emission permits. The program has been successful in reducing regional emissions, driving innovation, and promoting clean energy. More recently, the White House established "Principles for Responsible Participation" in high-integrity VCMs.[81] The accompanying Voluntary Carbon Markets Joint Policy Statement recognized the market potential for accelerating decarbonization in "unlocking capital and demand for real, additional, lasting, and independently verified emissions reductions and removals."[82]

---

[78] Chris Busch, *China's Emissions Trading System Will Be The World's Biggest Climate Policy. Here's What Comes Next*, Forbes (April 18, 2022) (www.forbes.com/sites/energyinnovation/2022/04/18/chinas-emissions-trading-system-will-be-the-worlds-biggest-climate-policy-heres-what-comes-next/?sh=1f000a0a2d59).

[79] *Cap-and-Trade Program*, California Air Resources Board (ww2.arb.ca.gov/our-work/programs/cap-and-trade-program).

[80] *AB-1305 Voluntary carbon market disclosures*, California Legislative Information (October 9, 2023) (https://leginfo.legislature.ca.gov/faces/billTextClient.xhtml?bill_id=202320240AB1305).

[81] The White House. Fact Sheet: Biden-Harris Administration Announces New Principles for High-Integrity Voluntary Carbon Markets, May 28, 2024 (https://www.whitehouse.gov/briefing-room/statements-releases/2024/05/28/fact-sheet-biden-harris-administration-announces-new-principles-for-high-integrity-voluntary-carbon-markets/).

[82] The White House and the U.S. Departments of Treasury, Agriculture, and Energy, the Senior Advisor to the President for International Climate Policy, the National Economic Adviser, and the National Climate Adviser. Voluntary Carbon Markets Joint Policy Statement and Principles, May 2024 (https://www.whitehouse.gov/wp-content/uploads/2024/05/VCM-Joint-Policy-Statement-and-Principles.pdf).

CHAPTER 4   A DEEP DIVE INTO THE "E" IN ESG

Challenges Facing Expansion of Carbon Markets

In recent years, the urgency of climate change has compelled businesses to scrutinize their carbon footprints. Among the solutions that have emerged is participation in Voluntary Carbon Markets (VCMs). Voluntary Carbon Markets operate on a voluntary basis, unlike Compliance Markets, which are created and regulated by governmental policies.

The structure of Voluntary Carbon Markets revolves around the issuance, purchase, and sale of carbon credits, with the supply primarily coming from private entities developing carbon projects or governments initiating programs certified by carbon standards. Companies use VCM platforms to buy and sell carbon credits. Entities that cannot fully eliminate their GHG emissions compensate by acquiring carbon credits.

One carbon credit represents the reduction, avoidance, or removal of one metric ton of carbon dioxide or its GHG equivalent.[83] When a company purchases carbon credits, they essentially compensate to some degree for their direct emissions. Corporate participation in carbon markets is increasing. Companies are also engaging in *carbon offsetting,* which involves financing climate mitigation projects to theoretically balance out the respective company's carbon footprint. More and more companies are purchasing carbon credits and utilizing carbon offsets to neutralize their emissions, driven by the need to meet their climate pledges and net-zero commitments.

It is estimated that the market for carbon credits has the potential to hit USD$50 billion in 2030.[84] Despite their potential, carbon markets

---

[83] *What are carbon markets and why are they important?* United Nations Development Programme (https://climatepromise.undp.org/news-and-stories/what-are-carbon-markets-and-why-are-they-important#:~:text=One%20tradable%20carbon%20credit%20equals,and%20is%20no%20longer%20tradable).

[84] *A blueprint for scaling voluntary carbon markets to meet the climate challenge,* McKinsey Sustainability (January 29, 2021) (www.mckinsey.com/capabilities/sustainability/our-insights/a-blueprint-for-scaling-voluntary-carbon-markets-to-meet-the-climate-challenge).

CHAPTER 4   A DEEP DIVE INTO THE "E" IN ESG

face numerous challenges at the present time. These include the oversupply and variability in the quality of carbon credits, the lack of a global regulatory framework, and the need for greater transparency and accountability. There is also ongoing debate about the balance between carbon offsetting and other decarbonization strategies and the use of carbon markets to achieve net-zero emissions targets. Adding to this is the lack of familiarity with carbon markets, which can create a barrier to entry as well as public scrutiny.

The Oxford Principles for Net Zero Aligned Carbon Offsetting was originally published in 2020 and revised in 2024.[85] The Oxford Principles provide a guidepost for carbon offsetting schemes in alignment with net-zero goals. The four primary elements elucidated in the updated principles, as paraphrased, are

1. Cutting emissions, ensuring environmental integrity of credits, and regularly revising the company offset strategy to keep pace with best practices

2. Transitioning to carbon removal offsetting for residual emissions to meet the net-zero target date of 2050

3. Shifting to carbon removals with durable storage to compensate for any residual emissions by the target date

4. Supporting the development of innovative and integrated approaches to achieve net zero[86]

---

[85] *The Oxford Offsetting Principles for Net Zero Aligned Carbon Offsetting (revised 2024)*, University of Oxford (February 2024) (www.smithschool.ox.ac.uk/research/oxford-offsetting-principles#:~:text=The%20Oxford%20Principles%20for%20Net,achieve%20net%20zero%20carbon%20emissions).
[86] Id.

CHAPTER 4   A DEEP DIVE INTO THE "E" IN ESG

While the challenges must be surmounted, carbon markets have the potential to play a part in achieving global emissions reduction targets. As regulatory frameworks evolve and the carbon markets continue to mature, it is likely that the scope and impact of these markets will continue to grow. Market-based mechanisms can incentivize emission reductions and the use of market tools can promote the adoption of cleaner, more sustainable technologies.

4.7 Pollution and Waste

Our modern, convenience-driven society has led to a surge in the production and use of various packaging materials. While these materials have become a ubiquitous part of our daily lives, they also pose significant environmental challenges. We'll look at the complexities of packaging and waste challenges and explore some of the contemporary solutions that are redefining the supply chain.

Packaging has various functions in the supply chain, including safeguarding goods during transit, reducing food waste, and offering convenience to consumers. The volume of packaging produced has significant environmental repercussions. Compounding this is the fact that mainstream packaging materials and waste, such as plastic, metal, and glass, are often not recycled properly. This leads to ecosystem damage and the pollution of natural environments, including our oceans.

According to the Organisation for Economic Co-operation and Development's (OECD) first *Global Plastics Outlook*, the world is producing twice as much plastic waste as two decades ago, with most of it ending up in landfills, incinerated, or leaking into the environment.[87]

---

[87] *Plastic pollution is growing relentlessly as waste management and recycling fall short, says OECD*, Organisation for Economic Co-operation and Development (www.oecd.org/newsroom/plastic-pollution-is-growing-relentlessly-as-waste-management-and-recycling-fall-short.htm).

Globally, only 9% of plastic waste is successfully recycled.[88] The OECD accentuates the need for international cooperation on reducing plastic pollution, including supporting lower-income countries in developing better waste management infrastructure.[89]

Sustainability Trends in Packaging

Faced with growing regulatory and public concerns around plastic and packaging waste, the packaging industry is poised for transformation. Increasingly, packaging producers and their value chains are having to innovate to meet new sustainability trends and demands. This change offers growth and partnership opportunities to existing businesses that are able to innovate and that proactively embrace sustainability, and it opens up opportunities across the industry.

Consumer awareness of packaging waste, particularly plastic waste, has significantly increased over the years. Today's consumers are more educated about the environmental impacts of their purchasing choices. Brands that prioritize more eco-friendly practices and more sustainable packaging are becoming more favorable consumer choices. Government bodies have also taken action, including regulations to discontinue single-use products and to reduce packaging waste.

With the intensifying demand for sustainable packaging, many fast-moving consumer goods companies are beginning to make bold commitments to address packaging waste. These companies are heavily focused on innovation and closely following consumer demand. Many are focused on redesigning traditional packaging which utilizes non-biodegradable material and exploring how to make packaging more sustainable to align with consumer demand and market forces.

---

[88] Id.
[89] Id.

Brands realize that sustainability can be a consumer differentiator and may lead to cost reductions in the long term through reduced material usage and streamlined operational efficiency. Transformation in packaging offers an opportunity for businesses to drive positive change and align with consumer values, reduce environmental impact, and harness a competitive advantage. We'll discuss circular models for packaging and product design later in this chapter.

4.8 Biodiversity and Natural Capital Depletion

Biodiversity and natural capital are the lifeblood of our planet—they are essential for the survival and prosperity of all species, including humans. However, our world is witnessing an alarming rate of biodiversity loss and natural capital depletion, posing a grave threat to the planet's health and economy.[90] The urgency to address this crisis has never been more evident.

Defining Biodiversity

Biodiversity is the "extraordinary variety of life on [e]arth—from genes and species to ecosystems and the valuable functions they perform."[91] Biodiversity is fundamental in maintaining the overall health and resilience of ecosystems. This biological wealth is vital for maintaining the

---

[90] *Living Planet Report 2022*, World Wildlife Fund (https://livingplanet.panda.org/en-US/#:~:text=The%20Living%20Planet%20Report%202022,in%20species%20populations%20since%201970).

[91] *What is Biodiversity?* National Museum of Natural History (https://naturalhistory.si.edu/education/teaching-resources/life-science/what-biodiversity#:~:text=Biodiversity%20is%20the%20extraordinary%20variety,the%20very%20stuff%20of%20life.%E2%80%9D).

health of our planet. However, biodiversity is under severe threat, with an estimated one million species at risk of extinction.[92] This can have far-reaching impacts on human well-being, including the potential collapse of food and health systems. Biodiversity also holds significant intrinsic, cultural, aesthetic, and recreational value.

Defining Natural Capital

Natural capital is the "stock of renewable and non-renewable resources (e.g., plants, animals, air, water, soils, minerals) that combine to yield a flow of benefits to people."[93] These resources provide what are often referred to as ecosystem services. In addition to providing food and water sources, ecosystem services include plant pollination and carbon sequestration.

Current State of Biodiversity and Natural Capital

Our planet is experiencing a global biodiversity crisis. The primary threats to biodiversity are: "habitat destruction, overexploitation, biological invasions, climate change, and pollution."[94] Terrestrial biodiversity loss is primarily driven by the production of three consumer staples—palm oil, beef, and

---

[92] *U.N. Report: Nature's Dangerous Decline 'Unprecedented'; Species Extinction Rates 'Accelerating,'* United Nations (May 6, 2019) (www.un.org/sustainable development/blog/2019/05/nature-decline-unprecedented-report/).

[93] *Investing in Nature: Private finance for nature-based resilience*, The Nature Conservancy (www.nature.org/content/dam/tnc/nature/en/documents/TNC-INVESTING-IN-NATURE_Report_01.pdf).

[94] Celine Bellard, Clara Marino, and Franck Courchamp, *Ranking threats to biodiversity and why it doesn't matter*, Nature Communications (May 16, 2022) (www.nature.com/articles/s41467-022-30339-y).

soy.[95] Marine biodiversity is also in peril, with more than one-third of marine mammals and nearly the same fraction of sharks, shark relatives, and reef-forming corals threatened with extinction.[96] The loss of marine life disrupts the balance of oceanic ecosystems, with far-reaching consequences.

Natural capital is also declining at an alarming rate. The World Economic Forum estimates that over half the world's GDP is moderately or heavily dependent on nature and its services, with "approximately USD$44 trillion of economic value generation moderately or highly depended on nature."[97] The World Bank estimates that ecosystem depletion could account for a 2.3% loss—about USD$2.7 trillion—to global GDP by 2030, with disparate impacts on low-income countries.[98]

The depletion of natural capital causes volatility in raw material prices, disrupts supply chains, and creates business risk. The systematic destruction of natural capital poses a significant systemic risk to the world economy. The reliance of the business world on the natural world cannot be overstated. Perhaps the term *interdependence* best describes our relationship with nature. The conversation shifted long ago from the distinction between preservation and conservation to balancing our reliance on the natural resources that make our world go around.

---

[95] *Natural Capital and Biodiversity: Reinforcing Nature as an Asset,* S&P Global (April 22, 2021) (www.spglobal.com/en/research-insights/featured/special-editorial/natural-capital-and-biodiversity-reinforcing-nature-as-an-asset#:~:text=Terrestrial%20biodiversity%20loss%20is%20tied,to%20deforestation%20globally%20each%20year).

[96] *Global Assessment Report on Biodiversity and Ecosystem Services,* Intergovernmental Science-Policy Platform on Biodiversity and Ecosystem Services (May 2019) (www.ipbes.net/node/35274).

[97] *Half of World's GDP Moderately or Highly Dependent on Nature, Says New Report,* World Economic Forum (January 19, 2020) (www.weforum.org/press/2020/01/half-of-world-s-gdp-moderately-or-highly-dependent-on-nature-says-new-report/).

[98] *Protecting Nature Could Avert Global Economic Losses of $2.7 Trillion Per Year,* The World Bank (July 1, 2021) (www.worldbank.org/en/news/press-release/2021/07/01/protecting-nature-could-avert-global-economic-losses-of-usd2-7-trillion-per-year).

CHAPTER 4   A DEEP DIVE INTO THE "E" IN ESG

A decline in biodiversity and natural resources has impacts on the business sector (Figure 4-8). Loss of resources and a lack of available raw materials cause supply chain disruptions. Those disruptions create corresponding business risk as well as broader societal impacts. The business sector has a direct interest in the protection of natural capital and biodiversity.

Figure 4-8. *Resource Loss and Business Sector Impacts*

The recent Montreal treaty on biological diversity, discussed in Chapter 3, marked a turning point in our approach to biodiversity protection. It introduced the Kunming-Montreal Global Biodiversity Framework,[99] with short- and long-term goals and targets "focus[ed] on ecosystem and species health including to halt human-induced species extinction, the sustainable use of biodiversity, equitable sharing of benefits, and on implementation and finance to include closing the biodiversity finance gap of $700 billion per year."[100]

---

[99] *The Biodiversity Plan for Life on Earth*, Convention on Biological Diversity (www.cbd.int/gbf/).

[100] *Kunming-Montreal Global Biodiversity Framework*, United Nations Environment Programme (December 19, 2022) (www.unep.org/resources/kunming-montreal-global-biodiversity-framework#:~:text=Four%20overarching%20goals%20to%20be,of%20%24700%20billion%20per%20year).

Climate change and biodiversity loss are deeply intertwined. Climate change exacerbates the loss of biodiversity, with scientists predicting that a business-as-usual warming scenario could drive one in six species to extinction.[101] In turn, the loss of biodiversity can worsen the impacts of climate change, as healthy ecosystems are essential to sequestering carbon and regulating the climate.

The ongoing biodiversity loss and depletion of natural capital have profound implications for the global economy and human welfare. The destruction of natural habitats increases the risk of disease and exacerbates the impacts of climate change. With robust supply chain monitoring and a reassessment of our connection to the natural world, we can hope to halt or even reverse biodiversity loss, safeguarding our planet's health for future generations.

Expansion of Natural Capital and Biodiversity Funds

Natural capital and biodiversity are fundamental to our existence and prosperity. They constitute the world stock of natural assets. However, these natural resources are under increasing threat due to rampant environmental degradation and climate change. The urgency to protect and restore our natural ecosystems has led to the emergence of natural capital and biodiversity funds.

Natural capital and biodiversity funds channel public and private investments to finance projects that promote the preservation of biodiversity and the enhancement of natural capital, contributing to

---

[101] Mark C. Urban, *Accelerating extinction risk from climate change*, Science (May 1, 2015) (www.science.org/doi/10.1126/science.aaa4984#:~:text=If%20 climate%20changes%20proceed%20as,climate%2Drelated%20loss%20of%20 biodiversity).

the sustainability of our planet and economies. Investments in natural capital and biodiversity can yield multiple benefits as they can help protect biodiversity, restore the environment, combat climate change, enhance livelihoods, and stimulate economic growth, particularly in marginalized communities. The funds typically invest in a diverse range of projects, including sustainable agriculture, forestry, clean energy, water management, and waste management, among others.

The past few years have shown a significant uptick in the establishment and growth of these funds. Several factors have contributed to their rising prominence. There is growing awareness and concern among investors and the wider public about the escalating environmental challenges. This has led to a shift in investment preferences, with more investors seeking opportunities that generate both environmental and financial returns.

Advancements in technology and data analytics are enabling better measurement and monitoring of the environmental impact of investments. This has helped to boost investor confidence in these funds. The ESG standards and reporting arena must continue to play catch-up on biodiversity. GRI is updating its biodiversity standards, and ISSB recently announced steps to include biodiversity standards in its reporting frameworks, and the newest arrival, the Taskforce on Nature-related Financial Disclosures (TNFD), recently released its inaugural framework. We'll discuss each of those organizations and their standards in Chapter 8.

The public and private sectors must continue to advance and incentivize better natural resource decision-making. Governments and regulatory bodies worldwide are enacting policies and regulations that encourage sustainable practices and green investments. These regulatory changes are creating favorable conditions for the growth of natural capital and biodiversity funds. Industry must elevate biodiversity and natural resource use, along with climate change impacts, in setting ESG strategies, targets, and policies. The investment sector must continue to progress impact mechanisms and prioritize nature-related risks and opportunities.

CHAPTER 4 A DEEP DIVE INTO THE "E" IN ESG

Several natural capital and biodiversity funds have made headlines in recent years due to their innovative approaches. One such fund is the Climate Fund for Nature launched by luxury fashion group Kering and cosmetics company L'Occitane Group.[102] It aims to mobilize resources to protect and restore nature, with an acceleration investment schedule equivalent to a total cumulative investment of USD$10 trillion by 2050. Another example that has received widespread media attention is the natural capital investment fund backed by HSBC Asset Management and climate change advisory firm Pollination, which has raised commitments of more than USD$650 million.[103]

There is a growing consensus that substantial funding and investment are needed to meet the global targets for biodiversity conservation and climate change mitigation. This is likely to drive the expansion of natural capital and biodiversity funds in the coming years. Moreover, as investors become more aware of the potential risks associated with the degradation of natural capital and biodiversity, they are likely to increasingly integrate these considerations into their investment decisions. This trend could further boost the growth of these funds.

Investors are pressing public companies for enhanced board oversight and new disclosures on these issues. Several coalitions of institutional investors have been formed, such as Nature Action 100,[104] to drive investor engagement and monitoring of corporate impacts on biodiversity and natural capital. This will support natural capital accounting and the expansion of nature credit markets.

---

[102] *Kering and L'Occitane Group join forces to finance nature protection at scale with the Climate Fund for Nature*, Kering (December 12, 2022) (www.kering.com/en/news/kering-and-l-occitane-group-join-forces-to-finance-nature-protection-at-scale-with-the-climate-fund-for-nature/).

[103] *Climate Asset Management closes over $650 million for Natural Capital projects*, Climate Asset Management (December 13, 2022) (https://climateassetmanagement.com/insight/climate-asset-management-closes-over-650-million-for-natural-capital-projects/).

[104] *Driving greater corporate ambition and action on tackling nature loss and biodiversity decline*, Nature Action 100 (www.natureaction100.org/).

Corporations can proactively address the financial risks stemming from biodiversity loss. The disclosure of nature-based metrics in corporate reporting is essential for investors to understand a company's exposure to biodiversity risk. It also allows for prudent and effective allocation of capital and resource planning. The Taskforce on Nature-related Financial Disclosures disclosure framework is a promising start to guide companies in reporting their nature-related financial risks and opportunities.

CASE STUDY: SCHNEIDER ELECTRIC

Schneider Electric is a French multinational company specializing in energy management and automation.[105] The company operates in more than 100 countries and has helped more than 1,000 suppliers deliver on the climate goal to reduce carbon dioxide emissions by 50% by 2025 through its decarbonization program.[106]

Schneider Electric has also made specific biodiversity commitments with respect to its facilities and operations. The company conducted an assessment of its biodiversity footprint in 2020 and has committed to achieve no net biodiversity loss by 2030, with those results to be independently validated. Interim goals include partnering with NGOs and local stakeholders so that 100% of sites will deploy biodiversity conservation and restoration programs and 100% of sites located in water-stressed areas will implement a water conservation plan, by 2025.

---

[105] *Innovate with Schneider Electric,* Schneider Electric (www.se.com/us/en/).
[106] *Accelerating sustainability for all,* Schneider Electric (www.se.com/us/en/about-us/sustainability/).

4.9 Responsible Sourcing

In this era of conscious consumerism and increased environmental awareness, the term *responsible sourcing* has emerged in the discourse of sustainable business practices. This section aims to explore the importance of responsible sourcing for the environment, human rights, and biodiversity, and to discuss methods for responsible sourcing for businesses.

Responsible sourcing refers to "a voluntary commitment by companies to take into account social and environmental factors when managing their relationships with suppliers."[107] It involves the integration of ethical considerations, sustainable practices, and social consciousness into procurement and overall supply chain management. Responsible sourcing necessitates that business transactions between a buyer and its suppliers are conducted in a manner that mitigates negative impacts on society and the environment.

Establishing a Responsible Sourcing Strategy

Businesses have a responsibility to understand their impact and strive to limit any negative repercussions. This includes conducting due diligence on and being updated on their suppliers' operating standards and business practices. Key benefits of adopting responsible sourcing practices include

- **Impact on the environment:** Responsible sourcing contributes to minimizing the environmental footprint of businesses. It helps in identifying and mitigating responsible sourcing risks such as environmental pollution, excessive resource consumption, and greenhouse gas emissions.

---

[107] *ICC Guide to Responsible Sourcing*, International Chamber of Commerce (https://iccwbo.org/news-publications/policies-reports/icc-guide-to-responsible-sourcing/).

- **Impact on human rights:** By adopting responsible sourcing practices, companies can require that their suppliers adhere to human rights principles. This includes zero tolerance for child labor, forced labor, unsafe working conditions, and any form of abuse or discrimination. We'll further discuss human rights practices in Chapter 5.

- **Impact on biodiversity:** Responsible sourcing also helps in promoting biodiversity by encouraging suppliers to adopt sustainable harvesting practices. This ensures that the extraction of raw materials does not adversely affect the ecological balance in the area of operation.

There are key elements that will help structure and organize the initiative. Many organizations already have responsible sourcing policies in place, but their effectiveness varies greatly. Below are the core elements of a cogent, responsible sourcing program:

- **Transparent and measurable compliance:** A responsible sourcing strategy should be able to demonstrate compliance with measurable criteria aligned with global best practices. These criteria should cover key areas such as human and workplace rights, environmental protection, and business integrity.

- **Traceability:** Traceability (tracking resources from originations and knowing who the suppliers are at every step) is significant for connecting metrics or Key Performance Indicators (KPIs). Opaque supply chains, where some suppliers are unknown, prevent claims of responsible sourcing.

- **Assurance mechanism:** A credible demonstration of responsibility should include verifiable means of assurance. This mechanism should balance the need for credible data and due diligence with the need to keep costs realistic.

Companies may choose to manage responsible sourcing in-house. This approach requires building a data foundation, generating insights by identifying risks and data gaps, and reporting and steering to create transparency on responsible sourcing for cross-functional stakeholders. Alternatively, companies can partner with an accredited assurance company or a qualified NGO that specializes in responsible sourcing. This option can provide substantial capabilities, including tight auditing and certification processes, brand recognition, and connections to global supplier networks. We'll discuss the function of NGOs as certifying and verifying organizations in Chapter 14.

The growing focus on sustainability and ethical practices has put responsible sourcing at the forefront of business strategies. With increasing ESG regulations, responsible sourcing is likely to continue to rise on the corporate agenda. Responsible sourcing is not just an ethical choice—it is a strategic business decision that can lead to substantial benefits.

4.10 Water Use and Stewardship

Water is an essential resource for all life forms and for our global economy. It's not only the cornerstone of healthy ecosystems but is also required for the basic functioning of various industries, from agriculture and manufacturing to energy production. However, the growing demand and competition for use, climate change impacts, and water quality decline are exerting immense pressure on the world's finite water resources, posing significant risks for businesses and society, and thus creating both risks and opportunities for investors.[108]

There are growing concerns about water availability, responsible water use, and the impact of businesses drawing upon local water sources. It's important to understand water use and stewardship in the ESG context and why it is of prime importance to corporations and investors.

Water Stewardship as a Business Imperative

The concept of water stewardship goes beyond merely managing water resources. It involves the socially and culturally equitable, environmentally sustainable, and economically beneficial use of water, achieved through a stakeholder-inclusive process.[109] Essentially, it is about comprehensive and sustainable water management, considering both local watershed factors and wider stakeholder interests.

Water stewardship and assessing a business's water risk are important aspects of corporate water strategy. Risks are primarily physical, reputational, and regulatory.[110] Water is a shared resource, and its management requires the involvement of multiple stakeholders.

---

[108] *Accelerating Corporate Water Stewardship and Resilience: 2022 Impact Report*, CEO Water Mandate, UN Global Compact (https://ceowatermandate.org/wp-content/uploads/2023/07/CEO-Water-Mandate-2022-Impact-Report.pdf).

[109] *Progress by innovation*, United Nations Industrial Development Organization (www.unido.org/our-focus/safeguarding-environment/resource-efficient-and-low-carbon-industrial-production/industry-and-adaptation/water-stewardship#:~:text=Stewardship%20means%20the%20responsible%20planning,site%20and%20catchment%20based%20actions).

[110] *Assessing Your Business' Water Risks*, CEO Water Mandate, UN Global Compact (https://university.ceowatermandate.org/university/101-the-basics/lessons/assessing-your-business-water-risks/).

Assessing risk when entering new markets is necessary as water access and availability are local issues, and business operations in a water-abundant region will have different concerns than a facility in a water-scarce region.

Water stewardship commitments have gained significant importance in ESG reporting. A company's water management practices, adherence to regulations, and stakeholder engagement form pivotal aspects of ESG reporting. Efficient water use enhances operational efficiency and resilience, while innovative water conservation technologies foster adaptability. Prioritizing water stewardship strengthens community relations and reputation.[111]

A collective water risk management approach understands that water risks are multifaceted and can pose substantial challenges to businesses and their supply chains. For instance, the agricultural sector, being the world's largest consumer of water, is highly exposed to water risks due to climate impacts and competition for water use. Companies with supply chains dependent on agriculture, such as food, beverage, and textiles, and extractive industries that use large water volumes, are directly exposed to these risks.

Given the intensifying water-related challenges and concerns about water scarcity and quality, more companies are recognizing the importance of sustainable water management and are making efforts to address water-related challenges in their ESG initiatives. Responsible water use and stewardship are essential to corporate social responsibility, sustainability, and long-term profitability. By integrating water strategies into their ESG strategy and reporting, companies can showcase their commitment to a sustainable future, attract responsible investment, and contribute to global sustainability goals.

---

[111] *The Business Case for Water Stewardship,* CEO Water Mandate, UN Global Compact (https://university.ceowatermandate.org/university/ 101-the-basics/lessons/the-business-case-for-water-stewardship/).

CHAPTER 4   A DEEP DIVE INTO THE "E" IN ESG

4.11 Circular Economy

The circular economy is a system where materials never become waste and nature is regenerated. In a circular economy, products and materials are kept in circulation through processes like maintenance, reuse, refurbishment, remanufacture, recycling, and composting.[112]

—Ellen MacArthur Foundation

The circular economy necessitates shifting from the traditional linear model to a circular model. That means moving from the take-make-dispose approach to a model that focuses on reducing waste, reusing materials, and recycling. The circular economy is an innovative and transformative system where waste is not an end product but a resource for new production. The central concepts are reusing, recycling, and regenerating resources. This results in a more sustainable and regenerative economic model that can benefit not only the economy but also people and the planet. The model created by the European Parliament Research Service illustrates the circular system (Figure 4-9).

---

[112] *Circular economy introduction*, Ellen MacArthur Foundation (www.ellenmacarthurfoundation.org/topics/circular-economy-introduction/overview#:~:text=The%20circular%20economy%20is%20a,remanufacture%2C%20recycling%2C%20and%20composting).

CHAPTER 4   A DEEP DIVE INTO THE "E" IN ESG

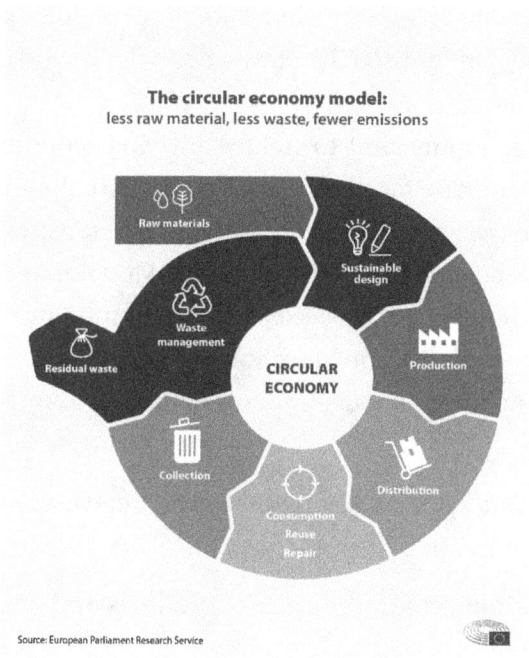

Figure 4-9. *The Circular Economy Model*[113]

The concept of a circular economy was first modeled by David W. Pearce and R. Kerry Turner in 1989. They described a system where waste at the extraction, production, and consumption stages is turned into inputs, marking a significant shift from the traditional linear or open-ended economic system to the circular economic system.[114] The concept has gained immense popularity in recent years. The Ellen MacArthur Foundation, European Parliament, and many other organizations are

---

[113] *Circular economy: definition, importance and benefits*, European Parliament (May 24, 2023) (www.europarl.europa.eu/topics/en/article/20151201ST005603/circular-economy-definition-importance-and-benefits).
[114] David W. Pearce and R. Kerry Turner, *Economics of Natural Resources and the Environment* (December 1, 1989) (https://doi.org/10.56021/9780801839863).

CHAPTER 4   A DEEP DIVE INTO THE "E" IN ESG

promoting the circular economy model and its principles given the potential to transform industrial and waste processes and open new market opportunities.

The circular economy aims to create a thriving economy through a regenerative system. The model is a stark contrast to the traditional linear economy model, where resources are extracted, made into products, and ultimately discarded as waste. This model "tackles global challenges like climate change, biodiversity loss, waste, and pollution by decoupling economic activity from the consumption of finite resources."[115]

The circular economy model is focused on three fundamental principles:

1. Designing systems where there is no net waste or pollution

2. Maintaining continuous use of products and materials

3. Regenerating natural systems[116]

A circular economy of scale would support our efforts to slow down climate change, halt biodiversity loss, and protect water resources. Applying circular principles could reduce the environmental impact of material extraction and processing. A circular economy could save approximately USD$100 billion in annual waste management costs, according to the United Nations Environment Programme.[117] The United Nations

---

[115] Id.

[116] *Towards a circular economy: Key Drivers*, The Circular Economy in Cities and Regions: Synthesis Report, OECD (www.oecd-ilibrary.org/sites/7bf512c1-en/index.html?itemId=/content/component/7bf512c1-en).

[117] *Global Waste Management Outlook 2024*, United Nations Environment Programme (February 28, 2024) (www.unep.org/resources/global-waste-management-outlook-2024).

International Resource Panel concluded that the extraction and processing of materials, fuels, and food "contribute half of total global greenhouse gas emissions and over 90 percent of biodiversity loss and water stress."[118]

Environmental Justice

> *The circular economy, when designed in a thoughtful and inclusive manner, has the potential to protect the environment, improve economics, and elevate social justice. Sustainability from its foundation requires social equity. How we extract, use, and dispose of our resources can affect already vulnerable communities disproportionately.[119]*
>
> —US EPA

According to the US EPA, environmental justice is defined as "the fair treatment and meaningful involvement of all people regardless of race, color, national origin, or income, with respect to the development, implementation, and enforcement of environmental laws, regulations, and policies. This goal will be achieved when everyone enjoys: [t]he same degree of protection from environmental and health hazards, and [e]qual access to the decision-making process to have a healthy environment in which to live, learn, and work."[120]

---

[118] *U.N. calls for urgent rethink as resource use skyrockets*, United Nations Environment Programme (March 12, 2019) (www.unep.org/news-and-stories/press-release/un-calls-urgent-rethink-resource-use-skyrockets).

[119] *What is a Circular Economy?* US Environmental Protection Agency (www.epa.gov/circulareconomy/what-circular-economy#:~:text=The%20circular%20economy%2C%20when%20designed,its%20foundation%20requires%20social%20equity).

[120] *Environmental Justice*, US Environmental Protection Agency (www.epa.gov/environmentaljustice).

The environmental benefits of the circular economy are regularly touted, but the social implications are often overlooked. To address systemic inequalities, environmental justice should be central to the circular economy. This involves ensuring that burdens and benefits are distributed equally. Circular practices can address historic patterns of resource exploitation in marginalized communities and support inclusive communities, local economic opportunity, responsible use, and waste reduction models.

Future of the Circular Economy

A just transition from a linear economy to a circular economy requires fairness and inclusivity, mitigation of impacts on the labor force and on communities, providing opportunities for fair work, and reducing inequality. [121] The International Labour Organization has issued *Guidelines for a just transition towards environmentally sustainable economies and societies for all*, as a policy framework to guide governments and employers on managing the just transition to achieve both climate goals and the Sustainable Development Goals.[122] Reviewing and implementing those guidelines is an imperative for businesses seeking to adopt circular practices that incorporate environmental and social justice considerations.

The circular economy provides a promising solution to the interconnected challenges of climate change and the loss of biodiversity and natural capital. It offers a way for us to reshape our economy and our society to be more sustainable and resilient. The circular economy

---

[121] *Guidelines for a just transition towards environmentally sustainable economies and societies for all*, International Labour Organizations (www.ilo.org/wcmsp5/ groups/public/@ed_emp/@emp_ent/documents/publication/wcms_432859.pdf).
[122] Id.

has profound implications for urban development. It encourages concentration and mixed-use development. Unlike mass production, reuse, repair, refurbishment, repurposing, and remanufacture are small-scale, labor-intensive, specialist activities conducted near to the market and connected by small scale or micro logistics.

The implementation of the circular economy approach will generate many jobs in small-scale enterprises. The vision for the future of the circular economy is to create a world where waste is a thing of the past and resources are used efficiently and sustainably. It's about shifting from a wasteful system to a regenerative system. The circular economy model can realize significant business benefits in optimizing resource utilization and waste reduction.

There has been considerable activity on the regulatory front to support the circular economy. China was an early actor in the circular economy and enacted the Circular Economy Promotion Law in 2008.[123] The European Union launched a new Circular Economy Action Plan in 2020, which was an update on its 2014 plan, as "one of the main building blocks of the European Green Deal, Europe's new agenda for sustainable growth."[124] We'll discuss more of the related regulatory developments in Chapter 7.

---

[123] *China Circular Economy Promotion Law*, The World Bank Group (ppp.worldbank.org/public-private-partnership/library/china-circular-economy-promotion-law).

[124] *Circular economy action plan*, European Commission (https://environment.ec.europa.eu/strategy/circular-economy-action-plan_en#:~:text=The%20EU's%20new%20circular%20action,new%20agenda%20for%20sustainable%20growth).

CHAPTER 4   A DEEP DIVE INTO THE "E" IN ESG

CHAPTER TAKEAWAYS

- Set specific GHG emissions targets and actualize transition plans
- Promote the use of renewable energy
- Advocate for strong climate policy
- Invest in and support climate solutions
- Conduct a risk assessment of environmental systems and resource use
- Implement industry-specific decarbonization pathways
- Review environmental justice practices
- Identify opportunities for green innovation and environmental stewardship

CHAPTER 5

A Deep Dive into the "S" in ESG

We will now review the social factors which constitute one of the three pillars of ESG. The social pillar encompasses a wide range of concerns, such as fair practices; diversity, equity, and inclusion; human rights; community relations; and social responsibility. A shorthand for the social pillar is it involves business activities that impact people including employees, customers, community stakeholders, and members of the larger society.

5.1 Social Factors and Their Impacts

Social issues are important as they directly impact a company's reputation, brand value, and long-term sustainability. By addressing social concerns, companies can build trust, enhance stakeholder relationships, attract top talent, mitigate risks, and drive innovation. Figure 5-1 shows the "S" issues as part of the larger and interconnected whole of ESG. Furthermore, ESG-oriented investors increasingly prioritize companies that demonstrate a commitment to being socially responsible, making social issues a key factor in attracting capital and securing financial success.

CHAPTER 5   A DEEP DIVE INTO THE "S" IN ESG

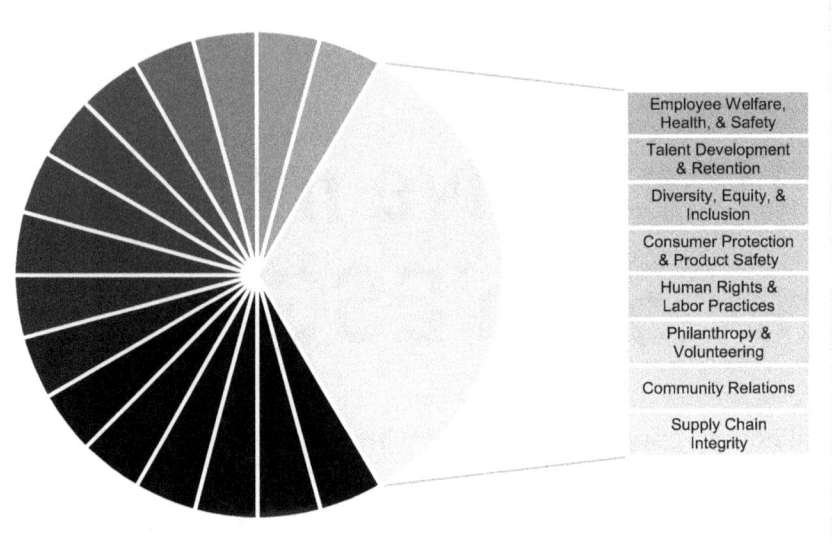

Figure 5-1. "S" Factors of ESG

The social part of the ESG pie continues to come into sharper focus with businesses, investors, and regulators. The primary social factors of ESG include

1. Employee Welfare, Health, and Safety
2. Talent Development and Retention
3. Diversity, Equity, and Inclusion
4. Consumer Protection and Product Safety and Quality
5. Human Rights and Labor Practices
6. Philanthropy, Responsibility, and Volunteering
7. Community Relations
8. Supply Chain Integrity

CHAPTER 5   A DEEP DIVE INTO THE "S" IN ESG

We'll review the core social issues of ESG and provide issue briefings on each "S" factor in this chapter and how business social practices impact society. Assessing how the company's social practices can support the relevant UN Sustainable Development Goals (SDGs) yields societal benefits. The "Key Insights for ESG Integration" in Section 3.6 provides guidance on mapping and implementing the SDGs to a company's materials ESG factors. Figure 5-2 is an example of how a business might map its material social issues to the relevant global goals.

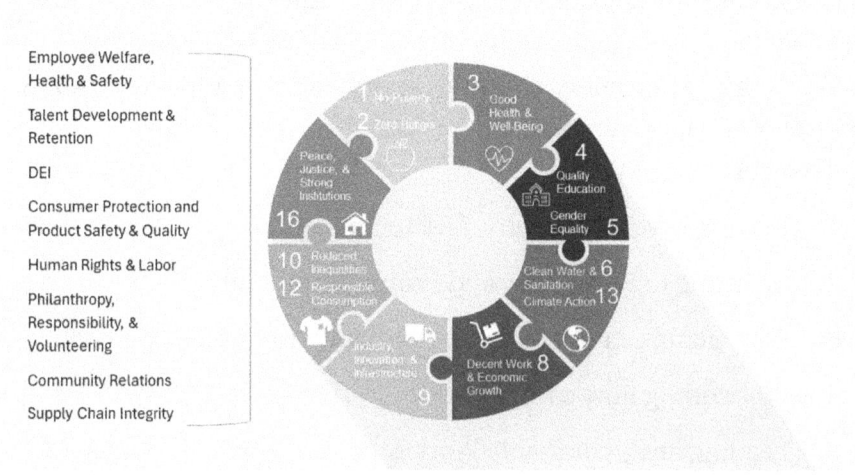

Figure 5-2. Mapping ESG "S" Factors to SDGs

5.2 Human Capital and Employee Welfare

Evolution of Human Capital Management in the Era of ESG

In recent years, the business landscape has witnessed a dynamic shift toward a more comprehensive approach to corporate success. In this context, measuring and reporting on Human Capital Management (HCM) has become more important than ever before. While perhaps not the

CHAPTER 5   A DEEP DIVE INTO THE "S" IN ESG

best title, HCM encapsulates the economic value that employees bring to a company, contributing to its overall productivity and growth. The US Securities and Exchange Commission Investor Advisory Committee has stated that "[h]uman capital can be considered the collective knowledge, skills, and experiences of the workforce that powers economic growth."[1]

Human Capital Management has emerged as a key driver of transformation success, especially in the context of ESG. The focus on people, their skills, knowledge, and well-being has a vital place in driving and enabling the success of ESG initiatives. One of the key dimensions of HCM is fostering diversity and inclusion within the organization. Studies have shown that organizations with high levels of racial diversity generate substantially higher revenues and sales compared to companies with lower levels of racial diversity.[2]

Key aspects of HCM include

1. Driving Diversity, Equity, and Inclusion
2. Investing in, Retaining, and Promoting Talent
3. Embracing Change Management and Progress
4. Embedding Purpose in Work
5. Leading an Accountable Workforce
6. Providing a Healthy, Safe, and Equitable Workplace

Effective HCM includes redesigning the objectives, outcomes, services, and activities performed by people, technology, and processes to drive ESG and sustainability outcomes. Organizations should identify necessary

---

[1] *Recommendation of the SEC Investor Advisory Committee's Investor-as-Owner Subcommittee regarding Human Capital Management Disclosure*, US Securities and Exchange Commission Investor Advisory Committee (September 14, 2023) (www.sec.gov/files/20230914-draft-recommendation-regarding-hcm.pdf).

[2] Cedric Herring, *Does Diversity Pay? Race, Gender, and the Business Case for Diversity*, American Sociological Review (2009) (https://psycnet.apa.org/record/2009-06909-003).

shifts for future roles, skills, and talent development required to deliver on their ESG strategies and expected outcomes. This involves integrating employee welfare considerations into the organization's culture, ways of working, governance, and operations to drive trust and transparency within internal and external communities. This opens up opportunities and enables workforce accountability.

Leveraging data and technology to elevate defined HCM metrics with the ESG program is another important component of success and meaningful reporting. Organizations should develop solid business strategies to make progress toward defined human capital goals. These strategies cannot be successful without an intentional focus on work, the workforce, and the workplace. We'll discuss government regulation of Human Capital Management in Chapter 7.

5.3 Diversity, Equity, and Inclusion

Many organisations are waking up to the fact that embracing accessibility leads to multiple benefits—reducing legal risks, strengthening brand presence, improving customer experience and colleague productivity.[3]

—Paul Smyth, Head of Digital Accessibility, Barclays

Diversity, Equity, and Inclusion (DEI) is a core concept of HCM, and its importance warrants a separate section of discussion. Companies that prioritize DEI often score higher on social metrics, as they are viewed as having a competitive advantage and demonstrate a commitment to ethical

---

[3] *Accessibility guide for Suppliers*, Barclays (May 2020) (https://home.barclays/content/dam/home-barclays/documents/who-we-are/our-suppliers/Accessibility-guide-for-Suppliers-May2020.pdf).

CHAPTER 5   A DEEP DIVE INTO THE "S" IN ESG

business practices and societal improvement. Moreover, those companies also have superior financial returns[4] and higher revenue correlated with innovation.[5]

There are varying definitions of diversity, equity, and inclusion. The following are some generally accepted explanations for the terms:

- **Diversity** is the representation of various backgrounds, experiences, and perspectives within an organization. It acknowledges the differences among individuals, including but not limited to race, ethnicity, age, gender, sexual orientation, and socioeconomic status.

- **Equity** refers to the fair treatment, access, and advancement of individuals within an organization. It ensures that each person has the opportunity to thrive, irrespective of their background or identity.

- **Inclusion** involves creating an environment where all individuals feel valued and included and can fully participate. It's about fostering a sense of belonging and enabling individuals to bring their authentic selves to the workplace. It is important to always include **accessibility**, which refers to the design and development of facilities, technology, programs, and services so that all people can independently use them, as it is fundamental to full inclusion.

---

[4] *Diversity wins: How inclusion matters*, McKinsey & Co. (May 19, 2020) (www.mckinsey.com/featured-insights/diversity-and-inclusion/diversity-wins-how-inclusion-matters).

[5] *The business case for diversity in the workplace is now overwhelming*, World Economic Forum (April 29, 2019) (www.weforum.org/agenda/2019/04/business-case-for-diversity-in-the-workplace/); *How Diverse Leadership Teams Boost Innovation*, Boston Consulting Group (January 23, 2018) (www.bcg.com/publications/2018/how-diverse-leadership-teams-boost-innovation).

CASE STUDY: TESCO

Tesco was founded in 1919 and is now a publicly traded company and one of the largest groceries in the UK.[6] Tesco first introduced emissions reductions targets and started measuring and reporting on emissions in 2006. Eleven years later, the company announced science-based targets for its operations and supply chain and switched to 100% renewable energy in the UK. Tesco has also made strides in packaging and food waste, and healthier food products. The company reports gender pay gap data and was the first company to set science-based targets in line with the Paris Agreement.

Tesco was an early digital adopter and one of the first grocers to enter the online grocery market. In 2001, the company worked with the Royal National Institute of Blind People (RNIB), a charity supporting people across the United Kingdom with vision loss, to create and beta test an online grocery interface that was more accessible to visually impaired persons.[7] Tesco won RNIB's first award for the accessible site, and following on the success of the pilot, the company redesigned all of its online services for accessibility and found that as a result, its annual revenue from site sales markedly increased.[8]

Corporate culture reflects the identity and values of an organization. It influences everything, from employee retention to daily interactions. Successful integration of DEI into the corporate culture fosters an environment where employees feel valued and respected. This sense of

---

[6] *About Us,* Tesco (www.tescoplc.com/).

[7] *Sustainability,* Tesco (www.tescoplc.com/sustainability/); *Case Study of Accessibility benefits: Tesco,* W3C Web Accessibility Initiative (www.w3.org/WAI/business-case/archive/tesco-case-study).

[8] Id.

CHAPTER 5   A DEEP DIVE INTO THE "S" IN ESG

belonging is decisive in engaging employees, enhancing team cohesion, enabling innovation, and establishing the desired company culture.[9]

By incorporating DEI into the culture of the organization, businesses can enhance their ability to attract and retain top talent, build effective teams, and drive business growth. It is also important for leaders to lead by example and to be held accountable for driving DEI at all levels within the company. According to MSCI (originally formed by Morgan Stanley, and later spun off, as Morgan Stanley Capital International), of the 2,811 companies listed in the MSCI ACWI Index (All Country World Index of large and mid-cap companies), only 5.8% had female CEOs and 16.9% had female CFOs, in 2022.[10]

DEI typically pertains to a company's workforce and falls within the social issues of ESG. It's important to note that there are principal aspects of corporate governance—or the "G" factors of ESG—that involve diversity, equity, and inclusion metrics. Metrics like board diversity, succession planning, and executive compensation fall under this umbrella. We'll discuss board diversity in the next chapter.

---

[9] *The power of inclusion: How DEI initiatives boost employee engagement*, British Council (November 20, 2023) 9 https://corporate.britishcouncil.org/insights/power-inclusion-how-dei-initiatives-boost-employee-engagement#:~:text=DEI%20initiatives%20foster%20a%20sense,increased%20commitment%20to%20their%20organisation).

[10] Tanya Matanda, Carrie Wang, and Olga Emelianova, *Women on Boards 2022 Progress Report*, MSCI (www.msci.com/documents/10199/36771346/Women_on_Boards_Progress_Report_2022.pdf).

CHAPTER 5 A DEEP DIVE INTO THE "S" IN ESG

5.4 Gender Equality and Equity

At the current rate, it will take an estimated 300 years to end child marriage, 286 years to close gaps in legal protection and remove discriminatory laws, 140 years for women to be represented equally in positions of power and leadership in the workplace, and 47 years to achieve equal representation in national parliaments.[11]

—United Nations

Gender equality, a fundamental human right under the Universal Declaration of Human Rights and the subject of Sustainable Development Goal 5, is the state of equal ease of access to resources, opportunities, and rewards regardless of gender.[12] It involves the elimination of all forms of discrimination against women and girls to create a society where they can participate fully in social, political, and economic life. Only 15.4% of the Goal 5 indicators, including eliminating discrimination and violence against women and ensuring women's full and effective participation and equal opportunities, were on track for Agenda 2030.[13]

According to the World Economic Forum's *2023 Global Gender Gap Report*, which examined gender parity in 146 countries, it could take another 131 years to close the gender gap.[14] At the current rate of progress,

---

[11] *Goal 5: Achieve gender equality and empower all women and girls*, United Nations (www.un.org/sustainabledevelopment/gender-equality/).

[12] *Frequently asked questions about gender equality*, United Nations Population Fund (www.unfpa.org/resources/frequently-asked-questions-about-gender-equality#:~:text=Gender%20equality%20requires%20equal%20enjoyment,to%20economic%20and%20social%20resources).

[13] Id.

[14] *Global Gender Gap Report 2023*, World Economic Forum (June 20, 2023) (www.weforum.org/publications/global-gender-gap-report-2023/digest/#:~:text=The%20global%20gender%20gap%20score,compared%20to%20last%20year%27s%20edition).

159

CHAPTER 5   A DEEP DIVE INTO THE "S" IN ESG

we won't achieve global gender parity until 2154. The report provided parity scores for individual nations. Iceland (91.2%) ranked highest, with Norway, Finland, Sweden, and New Zealand rounding out the top five with scores over 80%, and the United States dropping to 43rd place on the list from 27th the prior year with a parity score falling 2.1% to 74.8%.[15]

A diverse and inclusive society that values and respects all genders equally is essential for social development. It fosters innovation, enhances performance, and drives economic growth. In contrast, gender inequality can hinder productivity and economic development and result in social instability.[16]

Gender Pay Gap

The gender pay gap refers to the difference between the earnings of men and women.[17] Despite the pronounced strides in societal evolution, the gender pay gap remains an intractable issue in the 21st century, affecting economic growth and development globally. There is a stark difference in earnings between men and women. The gender pay gap is a measure of inequality.

Globally, women earn 77 cents for each dollar men earn for work of equal value.[18] The gap widens further for women of color, women with

---

[15] Id.

[16] *Frequently asked questions about gender equality*, United Nations Population Fund (www.unfpa.org/resources/frequently-asked-questions-about-gender-equality#:~:text=Gender%20equality%20requires%20equal%20enjoyment,to%20economic%20and%20social%20resources).

[17] Rakesh Kochhar, *The Enduring Grip of the Gender Pay Gap*, Pew Research Center (March 1, 2023) (www.pewresearch.org/social-trends/2023/03/01/the-enduring-grip-of-the-gender-pay-gap/#:~:text=The%20gender%20pay%20gap%20E2%80%93%20the,every%20dollar%20earned%20by%20men).

[18] *Equal pay for work of equal value*, UN Women (www.unwomen.org/en/news/in-focus/csw61/equal-pay).

disabilities, immigrants, and women with children. That creates a "lifetime of income inequality between men and women and more women are retiring in poverty."[19]

In the United States, the wage gap is 82 cents—a figure that has hardly changed in the past 20 years, despite a lot of talk about equalizing pay.[20] The Equal Pay Act was enacted in 1963,[21] but it is estimated that it will take until 2056 before women earn as much as men for the same work in the United States, according to the Center for American Progress.[22]

From a broader perspective, women make up almost half of the world's working-age population. According to World Bank data, only about 50% of these women participate in the labor force, compared to 80% of men worldwide.[23] While the world has made significant progress in education and workforce participation, substantial barriers to entry remain.

Women regularly earn less than their male counterparts, despite having similar education levels and performing the same work. Women also generally spend less time in paid work, leading to lower pensions and a higher risk of poverty in older age. In the workforce, few women

---

[19] Id.

[20] Rakesh Kochhar, *The Enduring Grip of the Gender Pay Gap*, Pew Research Center (March 1, 2023) (www.pewresearch.org/social-trends/2023/03/01/the-enduring-grip-of-the-gender-pay-gap/).

[21] *The Equal Pay Act of 1963*, US Equal Employment Opportunity Commission (www.eeoc.gov/statutes/equal-pay-act-1963).

[22] *What To Know About the Gender Wage Gap as the Equal Pay Act Turns 60*, Center for American Progress (June 8, 2023) (www.americanprogress.org/article/what-to-know-about-the-gender-wage-gap-as-the-equal-pay-act-turns-60/#:~:text=CAP%20analysis%20also%20shows%20that,worked%2C%20will%20not%20achieve%20parity).

[23] *Female labor force participation*, The World Bank Gender Data Portal (January 9, 2022) (https://genderdata.worldbank.org/data-stories/flfp-data-story/#:~:text=Women%20are%20less%20likely%20to,do%20work%2C%20they%20earn%20less).

rise to senior positions or start their own businesses compared to men. A well-known data point that speaks to inequality is that women-owned businesses receive only 1.9% of all venture capital funding.[24]

The gender pay gap presents not just a social justice issue but also an economic one. The disparity in wages limits the pool of talent available to employers, leading to lower productivity and economic growth. It can also impact corporate reputation, employee morale, and overall productivity. Closing the gender pay gap is imperative to achieving gender equality and promoting economic growth. When more women participate in the labor force, everyone, including men, benefits. Gender equality and equity in the workplace enhance performance, decision-making, and financial results.

The corporate sector has to make up ground to close the gender pay gap and promote gender equality. Corporations can take steps to eliminate the gender pay gap, such as conducting pay audits, increasing pay transparency, and implementing equitable pay policies. They can adopt inclusive hiring practices, implement DEI policies, promote women to leadership positions, conduct unconscious bias training at all levels, and screen for bias in the talent review process.

CASE STUDY: ELEMIS

Elemis is a British skincare brand and a certified B Corp.[25] The company aims to meet high standards of social and environmental performance, transparency, and accountability. Elemis works with farmers to improve ingredient sourcing and traceability, trials biodegradable materials, and continues to improve

---

[24] Lan Pham, *Women receive just 1.9% of VC funding. Here's why and what founders can do*, Fast Company (October 23, 2023) (www.fastcompany.com/90970411/women-receive-just-1-9-of-vc-funding-heres-why-and-what-founders-can-do#:~:text=The%20fact%20that%20a%20few,you're%20a%20female%20founder.&text=In%20the%20world%20of%20VC,receive%20just%201.9%25%20of%20funding).

[25] *Our Story*, Elemis (https://us.elemis.com/our-story).

packaging circularity. The company is also a certified Living Wage provider, carries out pay audits, provides pay equity training for managers involved in pay reviews, and issues a public Gender Pay Gap Report.[26]

Gender equality and the gender pay gap are consequential issues that fit into the broader ESG framework. Companies can enhance their ESG performance, attract investment, and drive growth by taking meaningful steps to champion gender equality and close the gender pay gap. A commitment to transparency and action can lead to a fairer and more inclusive workforce. We'll discuss regulatory efforts to address the gender pay gap in Chapter 7.

5.5 Human Rights and Labor Practices

An estimated 50 million people are in modern slavery, including 28 million in forced labour. Almost one in eight of all those in forced labour are children. Most cases of forced labour (86%) are found in the private sector.[27]

—United Nations

As we discussed in Chapter 3, companies worldwide must recognize their responsibility to uphold human rights and ensure fair labor practices. That means respecting and protecting human rights globally across operations and supply chains. This responsibility includes implementing policies safeguarding human rights; identifying, addressing, and

---

[26] *Our Story*, Elemis (https://us.elemis.com/our-story); Living Wage Foundation UK (www.livingwage.org.uk/).
[27] *International Day for the Abolition of Slavery, 2 December*, United Nations (www.un.org/en/observances/slavery-abolition-day).

remediating human rights risks and any violations; and engaging with stakeholders to take stock of and make improvements to company human rights practices.

Human rights are the basic rights and freedoms to which every individual is entitled, regardless of nationality, race, ethnicity, gender, religion, or any other characteristic. These rights encompass a vast range of aspects, including social justice, equality, and fair treatment. ***Labor practices*** such as providing fair wages and safe working conditions are closely linked with human rights and employee welfare. In the corporate sphere, respecting human rights and fair labor practices are fundamental both in compliance with the law and as a reflection of a company's responsible business practices. Workplace-related human rights and labor issues can significantly impact a company's reputation and performance.

Supply chains are a predominant source of risk to human rights and fair labor issues. Companies must conduct thorough due diligence to identify and address these risks and promote fair labor practices across their value chain. Failing to do so runs the risk of noncompliance with laws, loss of supplier and other contracts, and severe negative impacts on social and reputational standing with far-reaching consequences. Ethical labor practices can improve a company's brand value and customer loyalty, reduce legal and reputational risks, and enhance employee morale, productivity, and retention.

While many global companies have zero-tolerance policies for human rights violations, that is often not enough to ward off human rights abuses. The World Benchmarking Alliance conducted its first Corporate Human Rights Benchmark (CHRB) in 2017. The CHRB is an annual analysis of the human rights practices in a selection of over 100 major companies, typically in the apparel, agricultural, and extractive industries.[28] In the

---

[28] *Corporate Human Rights Benchmark*, World Benchmarking Alliance (www.worldbenchmarkingalliance.org/publication/chrb/about/).

2018 CHRB, the organization found that most of those companies were failing to meet the UN human rights principles.[29] The 2023 CHRB found that while some companies have made progress, more than 60% of apparel companies scored less than 20/100 on human rights indicators and 38% scored less than 20/100 on gender equality indicators.[30]

Strong due diligence, supplier reviews, grievance mechanisms, whistleblower protections, providing a living wage and workers' rights, allowing freedom of association and collective bargaining, and third-party monitoring are integral to best practices in fair labor and protecting human rights. It is important to remember that much of international law and policy is customary, which is why international organizations and business groups have adopted guidance documents for human rights practices. We review the regulations governing human rights and what is most often termed "modern slavery" under the law in Chapter 7. Several ESG reporting initiatives and frameworks have been established to guide businesses in disclosing their human rights and labor practices, which we'll cover in Chapter 8.

5.6 Supply Chain Transparency

Supply chain transparency involves shedding light on the entire process of a product's journey, from raw materials to the end customer. It provides a clear view of the operations, relationships, and transactions within

---

[29] Umberto Bacchi, *Most big companies failing U.N. human rights test, ranking shows*, Reuters (November 11, 2018) (www.reuters.com/article/us-global-rights-forced-labour/most-big-companies-failing-u-n-human-rights-test-ranking-shows-idUSKCN1NH02F/?edition-redirect=uk).

[30] *Fashion brands putting millions of workers at risk with gaps on human rights and gender equality, shows new research*, World Benchmarking Alliance (November 17, 2023) (www.worldbenchmarkingalliance.org/research/2023-corporate-human-rights-benchmark-insights-report/).

CHAPTER 5   A DEEP DIVE INTO THE "S" IN ESG

a supply chain and is a vital part of sustainable business practices. In addition to the environmental aspects of responsible sourcing that we discussed in Chapter 4, we'll also discuss the social aspects of transparency in the supply chain.

Supply chain transparency is a two-pronged concept involving visibility and disclosure.[31] ***Visibility*** is about being able to accurately track and trace the journey of a product or a component through the supply chain.[32] It provides insights into where a product comes from, the conditions under which it was produced, and how it reached the end customer. ***Disclosure*** is about communicating the information gathered through visibility efforts to stakeholders.[33] This can include regulatory bodies, customers, and the public.

The complexity of modern supply chains is one of the biggest obstacles to transparency. Many products are produced through global supply chains that involve numerous suppliers, manufacturers, and distributors spread across different countries. Each of these entities has its own practices and standards, making it difficult to gather consistent and accurate data. Supply chain transparency goes beyond a business knowing its immediate suppliers. It's about achieving end-to-end visibility, breaking down silos, and sharing data seamlessly among all partners. It involves understanding the social impacts of a company's supply chain, including materials sourcing, labor practices, human rights, health and safety, and community impacts.

There is a lack of standardization in the way supply chain data is collected and reported. Different companies, and even different departments within the same company, might use different metrics and

---

[31] Tam Harbert, *Supply chain transparency, explained*, MIT Sloan Management Review (February 20, 2020) (https://mitsloan.mit.edu/ideas-made-to-matter/supply-chain-transparency-explained).
[32] Id.
[33] Id.

methodologies. This makes it difficult to compare data and get a unified view of the supply chain. Many companies are still relying on outdated technology and manual processes to manage their supply chains. This can result in errors, inefficiencies, and a lack of real-time visibility. Without the right technology, it's virtually impossible to protect proprietary information and achieve a high level of supply chain transparency. The adaptation of artificial intelligence (AI) and blockchain technology is enabling better tracking, traceability, and verification in supply chains. The accuracy and reliability of the information disclosed present another risk that can result in loss of trust, negative publicity, and liability.

Maintaining a Transparent Supply Chain

Consumers are becoming more aware and concerned about the origin of materials and where and how their products are made. Consumers are shifting their spending toward products making positive ESG-related claims, as products making ESG assertions averaged 28% growth over a five-year period compared to products with no ESG association, according to a study conducted jointly by NielsenIQ and McKinsey of 600,000 products representing USD$400 billion in annual retail revenues from 2017 to 2022.[34]

A transparent supply chain aligned with an overall ESG strategy has many benefits to business value, including

1. Operating efficiency and cost reduction
2. Access to capital
3. Optimized supplier relationships

---

[34] *Consumers care about sustainability—and back it up with their wallets*, NielsenIQ and McKinsey & Co. (February 2023) (https://nielseniq.com/global/en/insights/report/2023/consumers-care-about-sustainability-and-back-it-up-with-their-wallets/).

4. Improved risk management
5. Compliance with laws
6. Creating ESG-aligned products
7. Minimizing business disruption
8. Maintaining ethical business practices
9. Protecting corporate reputation and brand value

5.7 Consumer Protection, Product Safety, and Quality

Consumer safety and protection is a top priority for governments worldwide. However, there is a disparity among countries in terms of their product safety frameworks. While more developed countries have implemented comprehensive safety frameworks that include laws, enforcement institutions, recall mechanisms, and communication campaigns, developing nations with weaker systems are less equipped to handle unsafe products.[35]

International cooperation is indispensable to enhance product safety for all consumers and guard against unsafe products crossing borders. In 2020, the United Nations Conference on Trade and Development (UNCTAD) adopted its first recommendation on product safety, which seeks to curtail the international trade of unsafe products by strengthening cooperation among consumer product safety authorities and raising awareness among businesses and consumers.[36]

---

[35] *Unsafe products exact a high price on consumers globally*, United Nations Conference on Trade and Development (July 19, 2022) (https://unctad.org/news/unsafe-products-exact-high-price-consumers-globally#:~:text=As%20consumers%20try%20to%20reduce,of%20over%20%243%2C000%20per%20capita).
[36] Id.

CHAPTER 5   A DEEP DIVE INTO THE "S" IN ESG

The current global scenario is marked by the most significant cost-of-living crisis in a generation, triggered by escalating food and energy prices amid tightening financial conditions.[37] This creates a demand for cheaper products, which, unfortunately, may be met by manufacturers who place unsafe products on the market. The United States reports 43,000 fatalities and 40 million injuries per year tied to consumer products, with annual costs exceeding $3,000 per capita.[38]

Product governance—"a company's management of the entire lifecycle of its products and services, to prevent adverse and unexpected consequences for its customers and end-users"[39]—is an important ESG consideration and indicator of investment risk. Product safety and quality include mitigating the use of harmful ingredients or chemicals, ensuring safe product design and components, and minimizing health effects for consumers.[40]

As ESG topics continue to shape investment decisions and ESG ratings take into account product safety, quality, and consumer protection, companies are reviewing their product development, marketing, and compliance strategies. MSCI identifies the key product safety and quality risks under their ESG Ratings Methodology as

---

[37] *Cost of living crisis hits poorest the hardest, warns UNCTAD*, United Nations (July 19, 2022) (https://news.un.org/en/story/2022/07/1122842).

[38] *Unsafe products exact a high price on consumers globally*, United Nations Conference on Trade and Development (July 19, 2022) (https://unctad.org/news/unsafe-products-exact-high-price-consumers-globally#:~:text=As%20consumers%20try%20to%20reduce,of%20over%20%243%2C000%20per%20capita).

[39] Kristoffer Inton and Ava Gams, *Why Product Safety Risk Matters for Stocks*, Morningstar Sustainable Investing (October 25, 2022) (www.morningstar.com/sustainable-investing/why-product-safety-risk-matters-stocks).

[40] *Designing safe and sustainable products requires a new approach for chemicals*, European Environment Agency (February 7, 2023) (www.eea.europa.eu/publications/designing-safe-and-sustainable-products-1).

CHAPTER 5   A DEEP DIVE INTO THE "S" IN ESG

- "Damage to brand value from loss of consumer trust or negative publicity.
- Increased costs to comply with additional regulatory requirements or litigation.
- Increased costs of implementing large-scale product recalls or to fulfill warranties."[41]

Those risks should inform business strategy in prioritizing consumer protection, product safety, and product quality. We'll discuss the regulatory system in place in Chapter 7.

5.8 Philanthropy, Responsibility, and Volunteering

> *The "S" in ESG includes activities such as volunteerism, grant-making, and community partnerships, and how to measure the effects of these programs is of vital concern to society, investors, and other stakeholders.*[42]
>
> —Chief Executives for Corporate Purpose (CECP)

Companies are increasingly measuring social impacts and the creation of social value and incorporating those factors into their ESG strategy and long-term business planning. According to Deloitte's *2019 Global Human*

---

[41] MSCI ESG Ratings Methodology: Product Safety & Quality Key Issue, MSCI ESG Research LLC (October 2023) (www.msci.com/documents/1296102/34424357/ MSCI+ESG+Ratings+Methodology+-+Product+Safety+%26+Quality+Key+Issue. pdf/0909d087-16fa-e928-0f1d-9b23b0f23ad0?t=1666182601603).

[42] *Giving in Numbers, 2023 Edition*, Chief Executives for Corporate Purpose (CECP) (https://cecp.co/wp-content/uploads/2023/10/GIN2023_FINAL.pdf).

Capital Trends, CEOs cited societal impact most often as the top factor used to evaluate corporate annual performance, and above customer satisfaction, employee satisfaction/retention, financial performance, and regulatory adherence.[43]

According to research by Chief Executives for Corporate Purpose (CECP), 83% of companies confirm that they consider the investor perspective when reporting social results in the company's sustainability reporting.[44] In the era of ESG, companies are measuring what CECP refers to as *Total Social Investment,* which is defined as the sum of "all monetary resources (operational expenses, staff time, and more) the company used for 'S' in ESG efforts" and excludes traditional community investment numbers.[45] In 2021, the median Total Social Investment for companies surveyed was reported to be USD$34.3 million.[46]

Corporate philanthropy typically takes the form of charitable grants through either the company's CSR program or the company's corporate foundation if the company has established such an entity, or both. In the United States, companies typically make grants to qualified charitable organizations which are nonprofit corporate entities that are tax-exempt under section 501(c)(3) of the IRS Code.[47] Giving is aligned with the company's CSR mission statement, and it is often part of a partnership with the charitable organization and may be coupled with in-kind giving and employee volunteering to the organization.

---

[43] *2019 Deloitte Global Human Capital Trends*, Deloitte Insights (www2.deloitte.com/us/en/insights/focus/human-capital-trends/2019.html).

[44] *Giving in Numbers, 2023 Edition*, Chief Executives for Corporate Purpose (CECP) (https://cecp.co/wp-content/uploads/2023/10/GIN2023_FINAL.pdf).

[45] Id.

[46] Id.

[47] 26 U.S. Code §501 – Exemption from tax on corporations, certain trusts, etc. Legal Information Institute, Cornell Law School (www.law.cornell.edu/uscode/text/26/501).

Corporate matching gift programs are formal programs established by companies to match employee donations, typically up to a maximum dollar amount, to qualifying charitable organizations. Approximately 65% of Fortune 500 companies offer matching gift programs, and an estimated USD$2-3 billion is donated through matching gift programs annually.[48] Total charitable giving by corporations in the United States was USD$21.08 billion in 2021.[49]

The Society for Human Resource Management defines employee-sponsored volunteerism as "organizational support, often in the form of paid leave or sponsorship, for employees pursuing volunteer opportunities or performing community services."[50] Company programs typically allow employees to take one or two days per year to volunteer with community and nonprofit organizations. Many corporate programs also include skills-based volunteering opportunities in which a business often partners with a local civic or charitable organization and matches employees with specific opportunities to contribute their skill sets. Many businesses encourage those activities and track those volunteer hours to include that metric in their annual CSR and ESG reports as well as report volunteer hours as part of their Human Capital Management Disclosures. We'll discuss the importance of institutional philanthropy in Chapter 15.

---

[48] *Matching Gift Statistics*, Double the Donation (https://doublethedonation.com/matching-gift-statistics/).

[49] *Annual Report on Philanthropy*, Giving USA (2022) (https://givingusa.org/wp-content/uploads/2022/06/GivingUSA2022_Infographic.pdf).

[50] *What is employee-sponsored volunteerism?* Society for Human Resource Management (www.shrm.org/resourcesandtools/tools-and-samples/hr-qa/pages/employersponsoredvolunteerism.aspx).

CHAPTER 5   A DEEP DIVE INTO THE "S" IN ESG

5.9 Community Relations

Community relations refers to a company's efforts to establish and maintain positive relationships with the local communities where they operate. It's typically part of a company's corporate social responsibility approach. Community engagement is also a strategic imperative that contributes significantly to a business's social license to operate and long-term success.

Businesses today are recognizing the potential benefits of engaging with and being part of their communities, from enhancing their reputation and operations to gaining access to local knowledge and resources. Effective community engagement can also facilitate local talent acquisition and workforce development. Engaging with community members opens the door to valuable insights into local conditions, needs, and preferences. This local knowledge can greatly enhance a company's decision-making processes and strategic planning.

Supporting local education, health care, and social welfare programs contributes to community well-being and social equity. It also builds trust and goodwill among community members. Proactive community engagement can help to identify and mitigate potential conflicts and risks. This reduces the possibility of operational disruptions and reputational damage. For example, many market-leading companies choose to forgo local tax breaks and other government incentives as a mark of goodwill when they site new facilities. They also do effective advance work with the states and localities to understand how they can be good corporate citizens in the new neighborhood.

Companies that do not do this effectively and are not seen as good partners who build trust at the outset often see their planned projects and facilities subject to public scrutiny and ultimately denied by community action. This was the case with e-retail giant Amazon's failed plan to site a new headquarters in the Borough of Queens in New York City in 2019. While the development plan called for 25,000 new area jobs, it also included USD$3 billion in government subsidies and meant the likely displacement of local businesses as well as other urban planning concerns. Amazon ultimately scrapped the plan owing to the extensive community outcry against it.[51]

CASE STUDY: NATURA & CO.

Natura & Co. is a São Paulo-based, multinational personal care cosmetics group company.[52] The company devised a carbon "insetting" project within its production campaign. Known as "Circular Carbon," the project renumerates communities for environmental conservation efforts that offset deforestation in the Amazon region.[53] The project serves the dual goals of enhancing community relations and seeking to mitigate impacts to one of the global regions most at risk for deforestation.

Community relations done well can also open new market opportunities. Engagement with communities in decision-making processes that impact them fosters transparency and accountability and provides feedback to the business on the community's needs and expectations. Effective practices for community engagement include the following:

- **Defining the purpose** of community engagement and setting measurable goals. This helps to track progress and evaluate the impact of the initiative.

---

[51] J. David Goodman, *Amazon Pulls Out of Planned New York City Headquarters*, The New York Times (February 14, 2019) (www.nytimes.com/2019/02/14/nyregion/amazon-hq2-queens.html#:~:text=Amazon%20on%20Thursday%20canceled%20its,%243%20billion%20in%20government%20incentives).

[52] *Governance*, Natura & Co. (https://ri.naturaeco.com/en/a-natura-co/the-group/g-governance/).

[53] *Commitment to Life, Sustainability Vision 2030*, Natura & Co. (www.naturaeco.com/sustainability-vision-2030/); *Natura's Carbon Neutral Progamme*, United Nations Climate Change (https://unfccc.int/climate-action/momentum-for-change/climate-neutral-now/natura).

CHAPTER 5 A DEEP DIVE INTO THE "S" IN ESG

- *Engaging with community members* as partners, not as beneficiaries. This dialogue fosters open communication and mutual respect.

- *Conducting a stakeholder analysis* and regularly updating the company's stakeholder map. Identify and understand the diverse needs, interests, and concerns of various stakeholder groups within the community.

- *Designing tailored engagement strategies* that align with the specific needs and priorities of each stakeholder group.

- *Keeping communities informed* about activities, decisions, and outcomes that impact the local area. This maintains transparency and accountability.

- *Regularly assessing the effectiveness* of community engagement efforts. Ask for feedback, identify areas for improvement, and adapt strategies accordingly.

CHAPTER TAKEAWAYS

- Assess Human Capital Management (HCM) metrics
- Champion Diversity, Equity, and Inclusion (DEI)
- Close the gender pay gap and erase inequality
- Build and maintain a transparent supply chain with safe and quality products
- Develop meaningful charitable partnerships, including volunteering and giving opportunities
- Engage positively with local communities

CHAPTER 6

A Deep Dive into the "G" in ESG

Governance elements compose the third pillar of ESG. The governance pillar incorporates topics such as board composition, executive compensation, transparency, ethics, and risk management. It is generally regarded as a reflection of how a company is managed.

6.1 Governance Factors and Their Impacts

Effective governance practices help establish a strong corporate culture, foster trust among stakeholders, mitigate risk, and inform decision-making. The "G" pillar reflects the internal management and corporate governance practices of a company. The "G" factors (Figure 6-1) are the primary foundations for good corporate governance, and performance on those factors reflects how well a company is operated and managed.

CHAPTER 6   A DEEP DIVE INTO THE "G" IN ESG

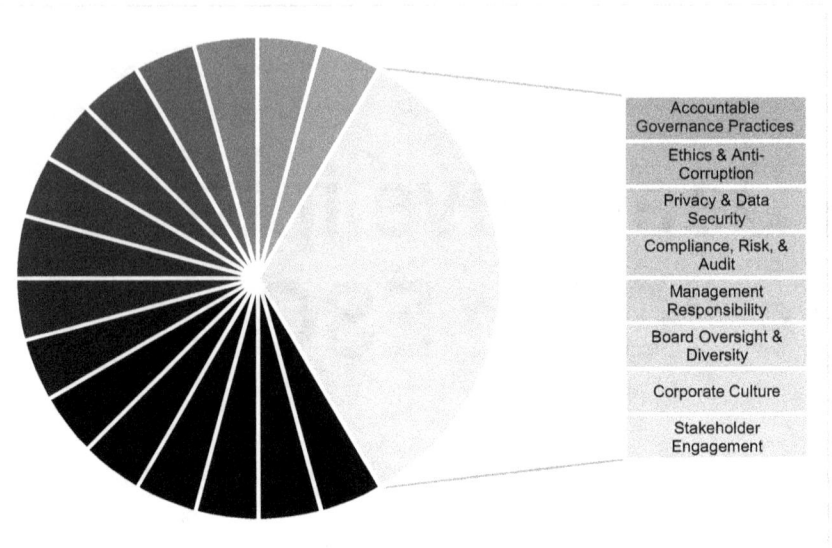

Figure 6-1. *"G" Factors of ESG*

By sufficiently addressing governance issues, companies can enhance their reputation, attract investors, comply with regulations, indicate a well-managed business, and promote long-term sustainability. The commonly accepted governance factors of ESG include

1. Accountable policies, procedures, and practices
2. Ethics and anti-corruption
3. Privacy and data security
4. Compliance, risk, and audit
5. Management responsibilities and transparency
6. Board oversight and diversity
7. Corporate culture
8. Stakeholder engagement

CHAPTER 6   A DEEP DIVE INTO THE "G" IN ESG

We'll walk through briefings on each of those governance factors, and as we do, it will be useful to consider the factors in the context of global business and how a company's practices can support the UN Sustainable Development Goals (SDGs). As discussed in previous chapters, business engagement with the SDGs has become more commonplace and is increasingly considered in ESG ratings assessments.

Linking the governance factors that are core to the business to the SDGs can inform how the business supports the relevant global goals. Figure 6-2 illustrates how a business might map its governance activities to the relevant global goals.

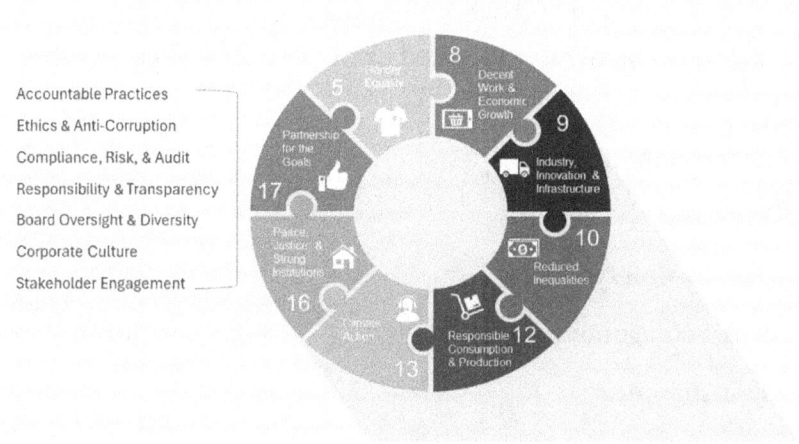

Figure 6-2. *Mapping ESG "G" Factors to SDGs*

CHAPTER 6 A DEEP DIVE INTO THE "G" IN ESG

6.2 Transparent and Accountable Governance Practices

> *Research shows that good citizenship and good governance qualities account for nearly 30% of corporate reputation, more than any other factors besides products and services.*[1]
>
> —Deloitte, *The Wall Street Journal*

Corporate governance is the framework of policies, guidelines, and processes derived from applicable laws, regulations, rules, and standards that (a) guide how a company is directed and controlled, and (b) inform a company's management and oversight practices, decision-making, and culture. The five generally accepted principles of good corporate governance are

1. Accountability
2. Fairness
3. Responsibility
4. Risk Management
5. Transparency

Ratings providers score ESG particulars and assess corporate governance. Companies seeking to achieve solid ESG ratings must adopt relevant corporate governance materials that mark their commitment to environmental, social, and governance best practices. Not only does good ESG governance guide corporate operations and conduct, but it also provides internal and external benefits, including

---

[1] Deloitte Content, *Measuring the Business Value of Social Impact Efforts*, The Wall Street Journal (September 21, 2020) (https://deloitte.wsj.com/cio/measuring-the-business-value-of-social-impact-efforts-01600714926).

1. **Managing Investor/Shareholder Expectations:** Investor expectations for companies (even pre-commercial), particularly those for ESG and socially responsible funds, are growing. Shareholder pressure and inquiries for ESG governance are commonplace and an increasing basis for shareholder resolutions.

2. **Providing Transparency to Stakeholders:** Good governance is the key to sound management and effective boards. It provides a baseline for reporting and compliance, particularly given the evolving global regulatory and standards framework. It enables transparent and positive engagement with stakeholders, including customers, community, employees, regulators, and investors/shareholders.

3. **Enabling Corporate Strategy Implementation:** ESG governance is an initial step needed to guide the implementation of the company's ESG strategy and its ESG priorities and initiatives. Corporations have to respond to various due diligence requests during their life cycle—from investors in a Series A to regulators in an IPO and beyond—and governance is always pivotal in evidencing the implementation of strategy.

4. **Showing the Company as a Responsible Business:** Companies must walk the talk and root climate, stewardship, diversity, inclusion, and other core elements into corporate culture. Good ESG governance underscores a company's commitment to being a responsible business.

CHAPTER 6   A DEEP DIVE INTO THE "G" IN ESG

Corporate governance materials are the building blocks of ESG. Corporations must have an ESG strategy and effective policies and procedures in place to support that strategy. That requires rooting corporate policies in specific legal requirements, adopting meaningful statements on key ESG subjects relevant to business and operations, requiring compliance with those written documents for employees, contractors, and suppliers, and socializing the governance materials with internal and external stakeholders. Businesses can review A Best Practice List of ESG Corporate Governance Materials (Table 6-1) to assess where they may have gaps in coverage depending on the nature of their operations and footprint.

Table 6-1. *A Best Practice List of ESG Corporate Governance Materials*

1. Environmental, Social, & Governance (ESG) Strategy
2. Environmental, Social, & Governance (ESG) Policy
3. Climate Change Statement
4. Biodiversity Statement
5. Water Stewardship Statement
6. Responsible Sourcing Statement
7. Diversity, Equity, & Inclusion (DEI) Policy
8. Human Rights, Modern Slavery, & Human Trafficking Policy
9. Political Activities and ABAC Policy
10. Community Engagement, Employee Volunteerism, & Charitable Giving Policy
11. Conflicts of Interest Policy
12. Whistleblower Policy

Embedding ESG into the company management responsibilities and operations requires updating relevant corporate policies and statements to be fit-for-purpose and customizing those to business needs. Good corporate governance is key to effective management and to board oversight of ESG. It also serves as baseline for reporting and compliance, as well as positive engagement with stakeholders, signals good management practices to investors and shareholders, and underscores a company's commitment to ESG.

CASE STUDY: L'ORÉAL

L'Oréal is the world's largest cosmetics company and is headquartered in France. L'Oréal has long implemented responsible sourcing and purchasing policies, and, in 2010, launched an accompanying Solidarity Sourcing program. Solidarity Sourcing is a fair trade and inclusive purchasing program in which L'Oréal sourcing teams establish partnerships with suppliers to employ people from vulnerable communities to enable access to fair work and income. To date, 85,544 people have gained access to work in 401 projects across 70 countries. L'Oréal's stated goal is to reach 160,000 beneficiaries by 2030.

6.3 Management Responsibilities and Transparency

Corporate management has a duty to lead on ESG and to do so in a transparent manner in compliance with requisite laws and regulations, which imparts integrity and trust. The elements of ESG represent a broad range of issues that can impact a company's performance and public perception. These can include everything from a company's carbon footprint and waste management practices to its labor standards and diversity policies. In recent years, there has been a significant shift in how companies view and integrate ESG features into their strategies.

CHAPTER 6   A DEEP DIVE INTO THE "G" IN ESG

Too often, the overlooked part of ESG—the "governance" element—can greatly influence many other aspects of a company's success, including its compliance, sustainability, executive performance, and brand reputation in the marketplace. Environmental issues like the climate crisis and social issues like human rights, labor relations, data and privacy, and DEI are dominating corporate board agendas and investor conversations. Good governance is what helps ensure that environmental and social issues are managed effectively from the top down. Leaving corporate governance out of the ESG calculus is a red flag, particularly from the risk, compliance, and ratings perspective.

It is management's responsibility to review and revise corporate governance as needed so that it is fit-for-purpose and to keep the board of directors regularly informed on ESG issues. Today, a company's ESG performance is not just a reflection of its corporate responsibility—it can also impact its bottom line. Studies have suggested that companies with strong ESG practices tend to have "a lower cost of capital, lower volatility, and fewer instances of bribery, corruption and fraud."[2] They also tend to be more efficiently operated and more resilient during market downturns.

Executive Compensation

Recognizing the importance of ESG, many companies are starting to link executive compensation to ESG performance. According to a recent survey conducted by the IBM Institute for Business Value of 3,000 CEOs from more than 30 countries across 24 industries in 2023, about half of CEOs reported that their compensation is now tied to sustainability goals, up from only 15% one year prior.[3] There are several reasons why companies

---

[2] *ESG and Performance*, MSCI (www.msci.com/esg-101-what-is-esg/esg-and-performance).
[3] *CEO decision-making in the age of AI: Act with intention*, IBM Institute for Business Value (June 27, 2023) (www.ibm.com/downloads/cas/1V2XKXYJ).

might choose to link executive compensation to ESG performance. For one, it can help align the interests of executives and shareholders. If executives stand to gain financially from improving the company's ESG performance, they may be more motivated to prioritize these issues.

Linking pay to ESG can signal to stakeholders that the company is committed to ESG. This can enhance the company's reputation, build trust with stakeholders, and potentially drive financial performance. It can also help companies attract and retain talent. Today's employees want to work for companies that share their values. By tying pay to ESG performance, companies can show that they are serious about these issues. Companies should be mindful to choose the right ESG metrics, set appropriate targets, and ensure that they can accurately measure and report performance.

The trend of linking executive compensation to ESG performance is likely to continue. According to a study conducted by The Conference Board, a large majority (73%) of S&P 500 companies are now tying executive compensation to some form of ESG performance.[4] The study found that the most significant increase on any ESG topic was in companies' use of diversity, equity, and inclusion (DEI) goals, which rose from 35% in 2020 to 51% in 2021.[5] As attention to climate change continues to grow, the percentage of companies "that tied carbon footprint and emission reduction goals to executive pay" also increased, from 10% in 2020 to 19% in 2021.[6]

Corporate leaders are prudent to regularly review their management responsibilities and transparency practices. Table 6-2 provides a checklist of best practice steps for effective management in the ESG arena.

---

[4] *Linking Executive Compensation to ESG Performance,* The Conference Board (www.conference-board.org/pdfdownload.cfm?masterProductID=41301).
[5] Id.
[6] Id.

Table 6-2. Best Practice Corporate Leadership ESG Checklist: Ten Key Steps for Effective ESG Management

1. **Strategic Integration:** Incorporating ESG factors into the company's core business strategy and operations to create long-term value and ensure sustainable growth.
2. **Governance Development:** Establishing robust policies and procedures that guide the organization's ESG efforts, including environmental stewardship, social responsibility, and ethical governance.
3. **Risk Management:** Identifying, assessing, and managing ESG-related risks that could impact the company's financial performance or reputation.
4. **Compliance and Oversight:** Monitoring compliance with relevant ESG-related laws, regulations, and voluntary standards and providing oversight to ensure that the company's operations align with its ESG commitments.
5. **Stakeholder Engagement:** Actively engaging with stakeholders, including investors, employees, customers, and communities, to understand their expectations and concerns.
6. **Transparency and Reporting:** Ensuring transparency in ESG practices and performance through regular reporting and using accepted frameworks.
7. **Performance Measurement:** Setting clear ESG goals and metrics and regularly measuring performance against these benchmarks to drive continuous improvement.
8. **Accountability to the Board:** Keeping the board of directors apprised of ESG strategy and metrics so it may carry out its oversight duties effectively.
9. **Fostering Corporate Culture:** Fostering a diverse, equitable, and inclusive corporate culture that values ESG principles and encouraging employees at all levels to contribute to sustainability and responsible business practices.
10. **Investment Decisions:** Considering the ESG impact of investment decisions, including capital expenditures, mergers, and acquisitions.

6.4 Board Duties and Oversight

As we progress into the 21st century, the role of corporate boards in overseeing environmental, social, and governance matters has evolved significantly. It's no longer a topic that can be shuffled to the side or treated as an afterthought. Instead, ESG has become a predominant consideration in corporate strategy and risk management and embedded into the fiduciary duties of directors.

In the past, boardrooms were primarily concerned with financial performance and regulatory compliance, as exemplified by the implementation of the Sarbanes-Oxley Act and the Dodd-Frank Act in response to corporate scandals and financial crises, respectively. However, recent years have seen a shift toward a more comprehensive and proactive approach. Board members are now expected to actively engage with a wide array of ESG issues, from environmental sustainability and social responsibility to corporate governance and stakeholder engagement. This chapter will delve into the evolving landscape of board oversight of ESG issues, offering insights into how boards can effectively navigate this complex terrain.

The maturity of board oversight of ESG issues has been largely driven by market forces rather than regulatory requirements. In fact, the rise of ESG can be seen as a shift from a shareholder-centric model of corporate governance to a more inclusive, stakeholder-centric model. The importance of ESG issues for boards cannot be overstated. ESG performance directly affects a company's long-term profitability, reputation, and sustainability. It also has far-reaching implications for the company's relationships with various stakeholders, from investors and employees to customers and communities.

As we've discussed, companies with strong ESG performance are often seen as having more attractive investment opportunities, as they are likely to be better prepared for future challenges and opportunities. They may also enjoy a more positive reputation, stronger customer loyalty, and higher employee engagement. Moreover, boards that fail to adequately address ESG issues may expose the company to significant risks, such as reputational damage, legal penalties, and loss of investor confidence.

The push for ESG integration has come from various quarters, including investors, employees, consumers, communities, and governments. These stakeholders have recognized the long-term value of ESG performance and are increasingly demanding companies act responsibly and sustainably. This has led to a significant change in the board's role and involvement in ESG matters. Instead of merely ensuring compliance with laws and regulations, boards are now expected to demonstrate leadership on ESG matters, set the company's strategic direction, and oversee its ESG performance.

Today, boards of directors are expected to

- Understand the company's ESG profile and the key risks and opportunities

- Engage with management to adopt a clear and coherent ESG strategy

- Approve ambitious but achievable ESG targets

- Be regularly informed of and work to mitigate ESG risks

- Oversee the company's ESG performance on a regular basis

- Ensure transparency and accountability in ESG reporting

It's worth noting that the integration of ESG factors into board oversight is not a separate exercise. Rather, it is an integral part of the board's broader responsibilities, which include setting the company's strategic direction, overseeing its performance, and ensuring its long-term sustainability. Table 6-3 contains a checklist for boards to establish a fit-for-purpose framework of ESG oversight.

Table 6-3. *Board of Directors ESG Oversight Checklist: Ten Key Steps to a Fit-for-Purpose Framework*

1. Define the **scope of ESG board oversight** and document key responsibilities.
2. Assess the **board committee structure** and **map it against key ESG responsibilities**.
3. **Amend the board governance documents** as necessary to incorporate the key ESG responsibilities.
4. **Assess board composition** to ensure representative climate, human capital, and other ESG competencies.
5. **Conduct annual board training** and include a substantive session on ESG.
6. Request management **refresh the ESG strategy** and **report on integration** with business strategy as needed.
7. **Review ESG commitments, implementation plans, and progress toward targets** on a regular basis.
8. Maintain a **line of communication** to the board from the **corporate officer responsible for ESG**.
9. Receive **regular updates on ESG regulations and disclosure requirements** as they continue to evolve.
10. Inform investors/shareholders and other stakeholders on the **framework for board oversight** of ESG.

6.5 Board Composition and Capabilities

The composition of the board and the capabilities of individual directors are indispensable to effective ESG oversight. Given the complexity and importance of ESG issues, it is necessary for boards to have the right mix of skills, experience, and perspectives. For instance, boards should have

CHAPTER 6 A DEEP DIVE INTO THE "G" IN ESG

members with subject-matter expertise in areas such as climate change, sustainability, human rights, and corporate governance. They also need directors with a deep understanding of the company's industry, markets, and stakeholders.

A recent study by WTW (formerly Willis Towers Watson) and the NASDAQ Center for Board Excellence surveyed 349 board members across 44 countries and found that 75% of the directors agreed that a cogent ESG strategy creates stronger financial performance and that alignment of ESG with business strategy is critical.[7] That same research revealed that almost half (48%) of board members reported that they lack the skills and expertise to address climate issues.[8]

Public companies have a host of additional legal requirements beyond what is required of private companies. These laws and governing documents spell out governance requirements, covering board responsibilities, director qualifications, director independence, board committee structure, and many other aspects of governance. For example, companies with securities listed on the New York Stock Exchange (NYSE) or the National Association of Securities Dealers Automated Quotations (NASDAQ) Stock Market, and not subject to specific exemptions, must comply with the Securities Exchange Act of 1934, as amended (the "Exchange Act");[9] the rules of the US Securities and Exchange Commission (the "SEC");[10] the Dodd-Frank Wall Street Reform and Consumer

---

[7] *Fostering Corporate Governance and Enhancing Board Effectiveness Survey Findings*, WTW and NASDAQ Center for Board Excellence (July 11, 2023) (www.wtwco.com/en-us/insights/2023/07/infographic-boards-see-improvement-opportunities-in-climate-and-governance-areas).
[8] Id.
[9] *Securities Exchange Act of 1934, As Amended*, US Government Printing Office (www.govinfo.gov/content/pkg/COMPS-1885/pdf/COMPS-1885.pdf).
[10] *Rules and Regulations for the Securities and Exchange Commission and Major Securities Laws*, US Securities and Exchange Commission (www.sec.gov/about/laws/secrulesregs).

Protection Act of 2010 ("Dodd-Frank");[11] the Sarbanes-Oxley Act of 2002, as amended ("SOX");[12] and the respective listing standards of the NYSE or NASDAQ, which primarily pertain to corporate and board governance.[13]

Effective board leadership and committee structure are key to successful ESG oversight. The board chair or lead independent director should be able to guide the board through the complexities of ESG issues, foster a culture of open discussion and critical thinking, and ensure that the board's decisions are informed, strategic, and forward-looking.

Board Diversity

The linkage between diverse boards and profitability has been substantiated by a number of research studies.[14] Companies with diverse boards receive higher scores on ESG performance metrics more often than those with non-diverse boards.[15] A diverse board is not only representative of societal diversity but is also better equipped to steer companies through challenging times, resulting in increased profitability, greater innovation, and lower risk.

---

[11] *Dodd-Frank Wall Street Reform and Consumer Protection Act*, US Government Printing Office (www.govinfo.gov/content/pkg/COMPS-9515/pdf/COMPS-9515.pdf).

[12] *Sarbanes-Oxley Act of 2002, As Amended*, US Government Printing Office (www.govinfo.gov/content/pkg/COMPS-1883/pdf/COMPS-1883.pdf).

[13] *Company Resources: Listing Standards, Manuals, and Fees*, NYSE (www.nyse.com/listings/resources); *Rulebook – The NASDAQ Stock Market*, NASDAQ (https://listingcenter.nasdaq.com/rulebook/nasdaq/rules).

[14] *Diversity wins: How inclusion matters*, McKinsey & Co. (May 19, 2020) (www.mckinsey.com/featured-insights/diversity-and-inclusion/diversity-wins-how-inclusion-matters).

[15] Cristina Banahan and Gabriel Hasson, *Across the Board Improvements: Gender Diversity and ESG Performance*, ISS Corporate Solutions, Harvard Law School Forum on Corporate Governance (September 6, 2018) (https://corpgov.law.harvard.edu/2018/09/06/across-the-board-improvements-gender-diversity-and-esg-performance/).

CHAPTER 6   A DEEP DIVE INTO THE "G" IN ESG

Research conducted by Institutional Shareholder Services (ISS) found that "[b]oards with at least two women outperform the average Russell 3000 returns over 3-, 4-, and 5-year periods, while male-dominated boards underperform the index over the same periods."[16] ISS, along with another major proxy voting advisory firm, Glass Lewis, has adopted proxy voting guidelines in support of board diversity.

Large asset managers like State Street and BlackRock have also called on corporations to diversify their boards. State Street Global Advisors' voting guidelines call for a vote against a nominating committee chair at Russell 3000 companies that lack at least 30% gender diversity and against chairs at S&P 500 companies who fail to provide data on the race and ethnicity of board members and do not have a director from an underrepresented community.[17] BlackRock's voting guidelines encourage corporate boards to have at least two women and one member from an underrepresented group, with the expectation that S&P 500 companies should aim for at least 30% diversity on their boards.[18]

However, progress to achieve diversity on boards has been quite slow at both public and private companies. A 2021 study of 500 private, venture-backed companies representing about $USD140 billion in funding and with more than 180,000 employees revealed that not only are women far underrepresented at 14% of total board members and that nearly 40% of

---

[16] *The Content of Their Character: How Diversity & Inclusion Continue to Drive Change*, ISS (October 27, 2021) (www.issgovernance.com/library/the-content-of-their-character/).

[17] *Guidance on Diversity Disclosures and Practices*, State Street Global Advisors (May 2023) (www.ssga.com/library-content/pdfs/asr-library/proxy-voting-guidance-diversity-disclosures.pdf).

[18] *Proxy voting guidelines for U.S. securities*, BlackRock Investment Stewardship (January 2023) (www.blackrock.com/corporate/literature/fact-sheet/blk-responsible-investment-guidelines-us.pdf).

companies do not have a single woman on their board, but only 3% of board members are women of color.[19]

As we discuss throughout the book, the developing regulatory paradigm is changing the way companies are governed. Global regulations and the new standards by the International Sustainability Standards Board (ISSB) include required disclosures on board oversight, board expertise, and board review and approval of corporate sustainability targets. We'll discuss the regulations and stock exchange rules governing board diversity in Chapter 7 and the related standards in Chapter 8.

Given the recent activity, there has been a corresponding uptick in the number of shareholder proposals on board diversity topics. Those include requesting specifics on a company's efforts to diversify their boards and seeking amendment to nominating and governance policies regarding candidate criteria for open board seats.

Boards should strive for diversity, as diverse boards are more likely to bring a wider range of perspectives to the table, leading to better risk management, decision-making, and problem-solving. Having the right people in the boardroom is not enough. Boards also need to invest in ongoing education and training to ensure that their members stay up-to-date with the latest developments in ESG issues.

Board Committee Structure

In terms of committee structure, there is no one-size-fits-all approach. Some boards maintain ESG oversight responsibilities at the board level, while others delegate them to specific committees. The right approach is informed by various factors, including the company's size, structure, industry, and ESG priorities, as well as the board's composition, culture, and workload.

---

[19] *2021 Study of Gender Diversity on Private Company Boards*, Crunchbase and Him For Her (March 29, 2022) (https://news.crunchbase.com/business/him-for-her-2021-diversity-study-private-company-boards/).

CHAPTER 6   A DEEP DIVE INTO THE "G" IN ESG

For larger, publicly traded companies, it often makes sense to establish an ESG steering committee and to incorporate specific oversight duties into the relevant board committee charter. The most common and foundational committee structure for publicly traded companies is to have (1) a Nominating and Governance Committee, (2) a Compensation Committee, and (3) an Audit Committee. The board may establish other committees either of a permanent or temporary nature, and each committee may have subcommittees, depending on the company profile. Many boards have opted to establish an ESG Steering Subcommittee of their existing Nominating and Governance Committee to coordinate with and disseminate information across the respective board committees, to serve a clearinghouse function of emerging ESG issues for the company, to act as a liaison to corporate leadership, and to assess board performance on ESG.

Figure 6-3 outlines the steps which boards can take to modernize their committee structure and respective committee scope of coverage to ensure appropriate oversight of ESG issues.

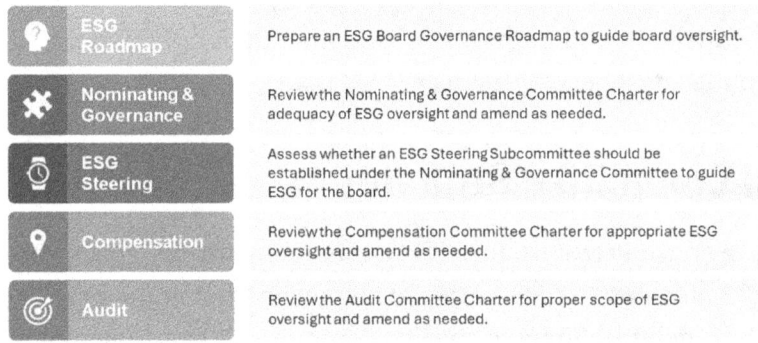

Figure 6-3. *Best Practice for Updating the Board Committee Structure*

Regardless of the approach, it is important for boards to clearly define the roles and responsibilities of each committee, ensure effective coordination and communication among committees, and regularly review and adjust the committee structure as needed. More companies have been identifying and precisely laying out their board ESG oversight responsibilities. That includes reporting those publicly in their annual reports, governance materials posted on their company website, and through the regulatory reporting cycles.

It is also useful for a board of directors to develop an ESG Board Governance Roadmap (Table 6-4). A roadmap outlines the approach to allocating board oversight responsibilities of ESG. Cogent governance structure and accountability are key aspects of good board governance, and companies must individually assess how best to align ESG topics within their governance framework, considering aspects such as the status of the corporation as a private or publicly traded entity; nature of the business; industry sector; existing policies and processes; significance and priority of ESG issues; and risk management profile.

Table 6-4. Best Practice Example of an ESG Board Governance Roadmap

Nominating & Governance Committee	Compensation Committee	Audit Committee
Oversight of substantive ESG board expertise and training	Oversight of ESG goals and metrics in executive compensation plans	Oversight of ESG reporting and disclosures (e.g., climate risk and human capital)
Oversight of ESG capacity in board skills matrix	Oversight of the DEI Policy	Oversight of ESG in risk management program and mitigation strategies

(continued)

Table 6-4. (continued)

Nominating & Governance Committee	Compensation Committee	Audit Committee
Oversight of board diversity	Oversight of DEI programs and initiatives (e.g., pay equity)	Oversight of controls and procedures
Oversight of company ESG strategy and alignment with business strategy	Oversight of talent and culture (including employee engagement, retention, talent development, recruitment, benefit and compensation programs, and succession planning)	Oversight of the Political Activities & ABAC Policy and the Community Engagement, Employee Volunteerism, & Charitable Giving Policy
Oversight of stakeholder engagement (including polititcal) & reputation risk		Oversight of the Whistleblower Policy, the Conflicts of Interest Policy, and the Cybersecurity Policy
Oversight of the Human Rights Policy		Oversight of compliance with ESG-related laws, regulations and rules
Oversight of ESG-related management committees and board subcommittees		Oversight of ESG-related litigation risk

Board Performance

Performance evaluation is a key tool for boards to assess the company's ESG performance, as well as their own performance, in overseeing ESG issues. This involves setting clear ESG targets, monitoring progress toward these targets, and holding management accountable for the results. Performance evaluations should also extend to the board itself.

Boards should conduct regular self-assessments to evaluate their effectiveness in overseeing ESG issues, identify areas for improvement, and implement necessary changes. The trajectory of board oversight of ESG issues is evolving rapidly, and directors must stay ahead of the curve. Directors should keep abreast of emerging ESG trends, risks, and opportunities, continuously review updates on ESG strategy and progress against targets, and be ready to respond to changing stakeholder expectations and regulatory requirements.

Perhaps most importantly, boards should embrace ESG not as a peripheral issue but as a strategic imperative. Directors should recognize that ESG is not just about doing the right thing for society and the environment, but also about creating long-term value for the company and its stakeholders. Board oversight of ESG issues is a complex and challenging task, but also a unique opportunity. By embracing ESG, boards can help their companies navigate current challenges, build sustainable and resilient businesses, and create lasting value for all stakeholders.

CHAPTER 6   A DEEP DIVE INTO THE "G" IN ESG

6.6 Business Ethics and Values

As the definition of good corporate governance expands to embrace ethical and stakeholder-centred issues, so, too, does the significance of corruption risks vis-à-vis the "G" in ESG. Corruption not only affects a company's exposure to legal risk and mismanagement, it also has an impact on the integrity of political partners and the welfare of surrounding communities. [20]

—World Economic Forum

Business ethics and **business values** are two elements that are increasingly recognized as fundamental governance factors. When it comes to ESG, business ethics and values are foundational in establishing and maintaining governance standards. These standards are increasingly important for corporations in navigating a complex global business landscape marked by heightened scrutiny from stakeholders and regulatory bodies.

Business ethics refers to the application of ethical principles in the context of business operations. It involves the consideration of what is right or wrong, fair or unfair, and just or unjust in a business setting. This touches on areas such as corporate governance, insider trading, anti-bribery and corruption, discrimination, corporate social responsibility, and fiduciary responsibilities.

Business values are the guiding principles that shape the behavior and decision-making processes within a corporation. These values are often explicitly articulated within an organization's mission statement, code of conduct, or corporate strategy, serving as a compass for employees

---

[20] *Investing in Integrity in an Increasingly Complex World: The Role of Anti-Corruption amid the ESG Revolution*, World Economic Forum (June 2022) (www3.weforum.org/docs/WEF_Investing_in_Integrity_GFC_2022.pdf).

and executives alike. They can encompass a wide range of concepts, such as integrity, accountability, excellence, respect for individuals, and commitment to customers.

Ethical decision-making is at the core of ESG governance, and it informs sound corporate policies and procedures. Corporate policies and programs focused on anti-bribery and corruption (ABAC), and on political activity compliance, are instrumental to ensuring compliance with the many national, state, and local laws and regulations on these topics and to good business ethics. They promote transparency, integrity, and fair competition and prevent corruption, conflicts of interest, and undue influence to ensure a level playing field and foster trust in public institutions.

ABAC laws and regulations are enacted by governments worldwide to combat unethical practices in the business and public sectors. They aim to prevent bribery, such as offering or accepting gifts or favors in exchange for influence over actions, and corruption, which refers to the abuse of public power for personal gain. Prohibited activities are typically enumerated in the laws, as are penalties for violations and guidelines for implementation.

Political activity laws and regulations are enacted by governments worldwide to ensure fairness and transparency in the political process. They typically apply to campaign contributions, lobbying activities, and political advertising. Political activity laws and regulations often require disclosure of financial contributions to political campaigns and place limits on the amounts that can be donated and the types of corporate entities that can donate. They also establish rules for lobbying and define what activities constitute "lobbying."

Transparency and accountability are other key ethical considerations in ESG. Companies should be transparent in how they govern, manage, and oversee ESG risk. Corporate culture is shaped by business values as well as accountability. The tone at the top of an organization significantly influences its corporate culture and ESG practices. Leaders who demonstrate their commitment to ethical behavior set the expectation for the organization.

6.7 Compliance, Risk, and Audit

Integrating ESG responsibilities into company compliance, risk management, and audit functions is essential for modern corporations. Compliance and risk management functions are often referred to as the second line of defense, and audit is the third line of defense for a company. Business units are the first line of defense. As ESG ratings platforms and regulatory frameworks advance, the functions of compliance, risk, and audit take on added importance to minimize risk and liability and to protect the company's public profile.

Conducting ESG assessments to identify risks and opportunities and incorporating ESG into risk appetite statements is an opportunity for corporations to make fundamental choices and changes as to how they approach their long-term ESG business strategy. This enables the perspective of ESG risk in terms of financial, operational, and reputational risk. The inclusion of ESG considerations in existing governance structures, as discussed in this chapter, is a foundational step from which to build the second-line risk, compliance, and audit infrastructure.

Monitoring and tracking ESG performance are essential aspects of ESG risk management. This involves defining relevant key risk indicators (KRIs) and metrics that allow the monitoring and measurement of ESG risk exposure and performance. The dynamic nature of ESG risks necessitates continuous updating of these risks in the risk management framework. Regular review and revision of the ESG risk management strategies are required to adapt to the evolving ESG risks and emerging best practices.

Challenges to effective ESG compliance and risk management result when companies do not have a comprehensive ESG strategy and program, do not revise and adopt robust ESG corporate governance, and do not assess their internal resources as to ESG capabilities for both talent and technology. In terms of talent, companies that have prioritized building up the ESG expertise of the business line may not have yet focused on supplementing the ESG expertise of compliance and legal teams.

According to a recent Deloitte survey, eight in ten senior executives reported that they are not completely confident that their current staffing levels on ESG issues are sufficient.[21] That is a gap that must be closed.

Technology is another gap, and tech solutions can significantly aid in the effective management of ESG risks and compliance with ESG laws and regulations. Technology solutions can facilitate the collection and analysis of relevant ESG data, enabling proactive risk mitigation efforts and the accuracy of reporting. This fosters more reliable communication of ESG performance to internal and external stakeholders.

Corporate internal audit operates under the oversight of the audit committee of the board of directors. The purpose of internal audit is to independently assess risk, monitor compliance, evaluate findings, and report on those topics to the board audit committee. The integration of ESG into audit functions includes testing ESG-related controls established by the business team and separately by the compliance team, primarily for design and effectiveness; providing assurance of ESG-related data and related reporting; and validating the accuracy, reliability, and completeness of ESG disclosures.

With the shift from voluntary to mandatory reporting models, companies will have to undertake assurance as required by new regulations. The International Auditing and Assurance Standards Board (IAASB) launched a new initiative in 2023, titled the International Standard on Sustainability Assurance (ISSA) 5000.[22] The ISSA 5000 is a proposed global standard

---

[21] *US Public Companies Prepare for Increasing Demand for High Quality ESG Disclosures*, Deloitte (March 14, 2022) (www2.deloitte.com/us/en/pages/about-deloitte/articles/press-releases/us-public-companies-prepare-for-increasing-demand-for-high-quality-esg-disclosures.html).

[22] *Proposed International Standard on Sustainability Assurance 5000, General Requirements for Sustainability Assurance Engagements,* International Auditing and Assurance Standards Board (August 2, 2023) (www.iaasb.org/publications/proposed-international-standard-sustainability-assurance-5000-general-requirements-sustainability).

tailored to meet the rising demand for reliable and transparent sustainability reporting. The standards highlight the significance of conducting risk management and diligence procedures with the aim of elevating the quality and trustworthiness of sustainability information, thereby developing greater accountability and transparency in reporting and disclosure.[23]

The IAASB employed a collaborative approach in the development of ISSA 5000. Consultations were held with leading global regulatory and standard-setting organizations, including the International Organization of Securities Commissions Oversight (IOSCO), the International Forum of Independent Audit Regulators (IFIAR), the Financial Stability Board (FSB), the International Ethics Standards Board for Accountants (IESBA), the International Sustainability Standards Board (ISSB), and the Global Reporting Initiative (GRI), as well as IAASB members and other key stakeholders.[24]

6.8 Stakeholder Engagement

Stakeholder engagement is another aspect of management responsibility for and board oversight of ESG issues. Stakeholder groups include employees, investors, shareholders, customers, communities, and regulators. By engaging with various stakeholders, companies can gain valuable insights into their concerns, expectations, and perspectives, which can inform their decision-making and strategy development.

It is prudent for companies to build capacity for stakeholder engagement. Stakeholder mapping and outreach planning can be pivotal to brand and project success, as we noted in Section 5.9, "Community Relations." Figure 6-4 displays the principles and steps for capacity building across a company's primary stakeholder groups.

---

[23] *Understanding International Standard on Sustainability Assurance 5000,* International Auditing and Assurance Standards Board (www.iaasb.org/focus-areas/understanding-international-standard-sustainability-assurance-5000).

[24] Id.

CHAPTER 6   A DEEP DIVE INTO THE "G" IN ESG

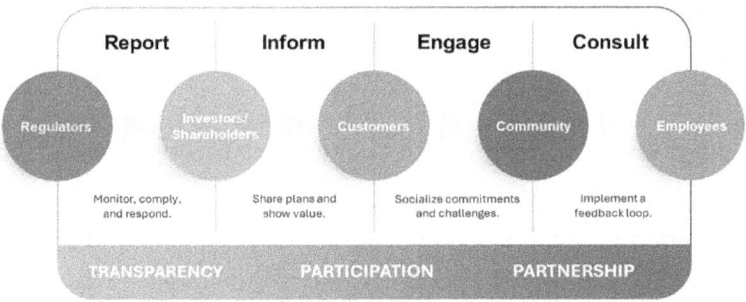

Figure 6-4. *Stakeholder Capacity Building*

While the primary responsibility for stakeholder engagement usually rests with management, boards also have a role to play in stakeholder engagement planning to review its efficacy. Boards should oversee the company's stakeholder engagement strategy, ensure that it aligns with the company's ESG priorities, and engage directly with key stakeholders when appropriate. That includes ensuring that the company's ESG disclosures are transparent, accurate, and meaningful and that they address the information needs of various stakeholders, which is a task that is often delegated to the audit committee.

Investors continue to seek clarity that the board is making well-informed decisions on ESG issues and that they are balancing the interests of stakeholders. They expect companies to articulate the ESG issues that are material and strategically important. Boards should also utilize a decision-making framework that takes stakeholder impact into account and that requires that the board receive regular updates and information on stakeholder engagement.

CHAPTER 6   A DEEP DIVE INTO THE "G" IN ESG

6.9 Shareholder Activism, Proxy Voting, and Asset Management

In the United States, proxy voting is the practice through which shareholders vote by proxy, meaning they cast their votes on proposals via a representative. The proxy season refers to the time frame in which publicly traded companies report on the current state of corporate affairs to their shareholders during an annual meeting and through related communications, including an annual report and proxy statements.

State law requires public companies to hold an annual shareholders meeting, and the Securities and Exchange Commission (SEC) mandates that registered companies file their proxy statements in advance of the annual meeting, as the proxy details the resolutions upon which shareholders will vote. Most companies schedule their annual meeting at an interval, allowing for financial statements to be prepared and audited after the close of their fiscal year. The corporate and investment communities anticipate the announcement of engagement priorities from BlackRock and other asset managers, as well as voting guidelines from the likes of Glass Lewis and ISS.

Shareholder activists have capitalized on the proxy system as a tool to advance ESG priorities. The shareholder proposal has always been an arrow in the quiver of changemakers and value creators to catalyze boards and management toward better corporate governance, but, recently, the proxy process has been co-opted by some groups seen as agitators and headline grabbers. For example, the volume—both in number and media noise—of anti-ESG resolutions, which often attack anti-discrimination policies, has risen in recent years. There continues to be a lack of traction on those resolutions as they fail from the mainstream investor perspective, yet they reap a disproportionate share of the headlines.

CHAPTER 6   A DEEP DIVE INTO THE "G" IN ESG

ESG-related shareholder proposals dominated the 2023 proxy season, as they have in past years.[25] The number of proposals each year continues to surpass the previous year's total. That is owing to the pressing nature of the current environmental and social issues, the augmented media coverage, and the fact that the SEC revised its rules governing the process to narrow the grounds by which companies may exclude shareholder proposals.[26]

The primary ESG topics that shaped the most recent proxy season are

1. **Climate Change**, with a particular focus on GHG emissions reduction plans

2. **Political Influence**, in the form of values alignment, lobbying and third-party spending, and political disclosure

3. **Social Policy**, including diversity and human rights

4. **Executive Compensation** and say-on-pay proposals

We've seen more prescriptive proposals seeking to direct board and management decision-making repeatedly fail. The most meritorious resolutions are those that push for transparency, disclosure, and enhanced board oversight of key risks. That aligns with the principles of ESG.

Investor engagement in the 2024 proxy season is on pace to match the same level as in prior years. As ESG has matured in the corporate and investment sectors, the shareholder proposals seem to be keeping pace. Shareholder resolutions are increasing focus on climate-related resilience and business transformation. Investors understand the physical and transition risks associated with climate change and the potential for business disruption without effective planning. The ongoing push for more

---

[25] *ESG proxy voting & how it's shaping the boardroom*, Diligent (October 17, 2023) (www.diligent.com/resources/blog/esg-proxy-voting).
[26] *Rule 14a-8*, US Securities and Exchange Commission (www.sec.gov/divisions/corpfin/rule-14a-8.pdf).

CHAPTER 6 A DEEP DIVE INTO THE "G" IN ESG

detail on human capital disclosures also ranks on the ESG agenda this season and it is expected that we'll continue to see anti-ESG proposals.

An early entry in the 2024 proxy season was the filing of shareholder proposals by the New York City Comptroller and trustees of the New York City Employees' Retirement System (NYCERS), Teachers' Retirement System (TRS), and Board of Education Retirement System (BERS), to require banks including JPMorgan Chase, Morgan Stanley, Bank of America, Citigroup, Goldman Sachs, and Royal Bank of Canada to report their clean energy to fossil fuel financing ratios.[27] JPMorgan Chase (JPM) agreed with the three pension funds having investments in the bank totaling USD$487 million to disclose its financing ratio and became the first bank to commit to do so. The NYC Comptroller subsequently announced agreement was also reached with Citigroup and Royal Bank of Canada.[28] In its coverage of the JPM agreement, *Fortune* declared that "[t]he dean of Wall Street CEOs is green" in reference to the well-known bank leader Jamie Dimon.[29]

The 2023 proxy season was successful for climate change proposals. According to Ceres, of the 256 climate-related shareholder proposals, investors withdrew nearly one-third after negotiating agreements with

---

[27] *NYC Comptroller Lander & NYC Pension Funds Launch Shareholder Drive to Hold Banks Accountable for Transition Away from Financing of Fossil Fuels,* New York City Comptroller Brad Lander (January 31, 2024) (https://comptroller.nyc.gov/newsroom/nyc-comptroller-lander-nyc-pension-funds-launch-shareholder-drive-to-hold-banks-accountable-for-transition-away-from-financing-of-fossil-fuels/).

[28] City Comptroller Brad Lander. NYC Comptroller Lander and NYC Public Pension Boards Reach Agreement on Climate Finance Disclosures with JPMorgan Chase, Citi, and Royal Bank of Canada, New York (April 4, 2024) (https://comptroller.nyc.gov/newsroom/nyc-comptroller-lander-and-nyc-public-pension-boards-reach-agreement-on-climate-finance-disclosures-with-jpmorgan-chase-citi-and-royal-bank-of-canada/).

[29] Amanda Gerut, *Jamie Dimon takes a stand by signing JPMorgan up as the first big bank to reveal a key clean energy metric to investors,* Fortune (March 4, 2024) (https://fortune.com/2024/03/04/jamie-dimon-takes-a-stand-by-signing-jpmorgan-up-as-the-first-big-bank-to-reveal-a-key-clean-energy-metric-to-investors/).

CHAPTER 6   A DEEP DIVE INTO THE "G" IN ESG

the target companies.[30] That indicates the growing number of companies willing to directly engage and make change to address climate risk and how their policies and operations line up with their climate commitments and transition planning.

An often-cited example of climate success is the voting record of the California State Teachers' Retirement System (CalSTRS). CalSTRS is the largest educator pension fund in the world, with more than USD$315 billion in assets.[31] In 2021, CalSTRS pledged to achieve net zero in its investments by 2050 as climate risk "broadly impacts CalSTRS's investment portfolio."[32] The organization has taken action to achieve that goal by holding its portfolio companies accountable for minimizing climate risk and implementing net-zero transition plans. The commitment led to CalSTRS voting against boards of directors at 2,035 companies because they did not meet "basic disclosure expectations."[33]

The CalSTRS voting record signals the firm expectation of companies and their boards to adequately disclose and mitigate climate risk. A primary shift over the past decade that coincides with increased ESG shareholder activism is that boards have become more informed on ESG issues and the key ESG risks to long-term business strategy, more aware of their fiduciary obligations with respect to overseeing ESG issues and risk, and more attuned to all stakeholders in overseeing ESG goals, metrics, and strategy. Those shifts, along with the push for diversity in the ranks of corporate directors, have in many instances created meaningful change in ESG practices and enhanced long-term value for publicly traded companies.

---

[30] *Engagement Tracker*, Ceres (https://engagements.ceres.org/).

[31] *CalSTRS escalates efforts to hold global companies accountable for not adequately disclosing climate change risks; votes against 2.035 boards of directors in proxy season 2023*, CalSTRS News release (August 10, 2023) (www.calstrs.com/calstrs-escalates-efforts-to-hold-global-companies-accountable-for-not-adequately-disclosing-climate-change-risks-votes-against-2-035-boards-of-directors-in-proxy-season-2023).

[32] Id.

[33] Id.

207

6.10 Greenwashing

In the contemporary corporate landscape, where the call for sustainable business practices is growing louder, **greenwashing** has emerged as a significant concern. The concept of greenwashing generally refers to the practice of businesses making exaggerated, misleading, or unfounded claims about their products, services, or operations being sustainable or environmentally friendly.

This deceptive marketing strategy is designed to appeal to increasingly conscious consumers and investors who are interested in supporting businesses that prioritize sustainability. The practice involves the use of vague, unsubstantiated, and sometimes outright false claims regarding a company's commitment to environmental stewardship. Greenwashing can occur unintentionally due to a lack of knowledge or understanding on the part of management, but it can also be carried out intentionally to mislead consumers and investors.

The term "greenwashing" was first coined by environmentalist Jay Westerveld in the 1980s, when he published an essay in response to a hotel's campaign urging guests to reuse towels to conserve energy.[34] Westerveld criticized the hotel's claim as a form of greenwashing, noting that the claim wasn't substantiated as the hotel hadn't measured the environmental impacts of the campaign.

Greenwashing has evolved over the years, becoming more sophisticated and pervasive across various industries. With the rise of corporate ESG initiatives, the concept has expanded to include social and governance factors. Consequently, greenwashing now encompasses not only misleading environmental claims but also deceptive statements about social responsibility and corporate governance.

---

[34] Jim Motavalli, *A History of Greenwashing: How Dirty Towels Impacted the Green Movement*, Daily Finance (February 12, 2011) (https://web.archive.org/web/20150923212726/www.dailyfinance.com/2011/02/12/the-history-of-greenwashing-how-dirty-towels-impacted-the-green/).

CHAPTER 6   A DEEP DIVE INTO THE "G" IN ESG

Companies polish their green resumes beyond fact for various reasons, primarily driven by the increasing demand for sustainability in the marketplace. GreenPrint's Business of Sustainability Index found that consumers are willing to pay more for eco-friendly and sustainable products and broke that percentage down by generation: 75% of Millennials; 64% of Gen X; 63% of Gen Z; and 57% of Boomers.[35]

Appearing "green" can be good for business; however, instead of actually being green and investing in genuine sustainable practices, some companies opt for the easier route of greenwashing to appeal to this market trend. Many of these claims run afoul of the laws and regulations enacted by governments to protect customers and to ensure marketers make claims that are true and substantiated, such as the US Federal Trade Commission's *Green Guides*.[36]

Greenwashing can manifest when a company uses vague language, obfuscates damaging environmental practices by pointing out only select non-damaging practices, cherry-picks data in a similar vein, or applies misleading labels that lack substation like *sustainable* or *eco-friendly*. Companies committed to avoiding the perception of greenwashing need to prioritize transparency and accountability in their sustainability efforts. This includes providing clear and specific information about impacts, using credible third-party verifications and certifications, and ensuring that all ESG claims are supported by accurate data.

While greenwashing remains a significant challenge, increased awareness among consumers and investors, coupled with stricter regulations, can help mitigate its impact. As the demand for genuine

---

[35] *GreenPrint Survey Finds Consumers Want to Buy Eco-Friendly Products, but Don't Know How to Identify Them*, GreenPrint news release (March 22, 2021) (www.businesswire.com/news/home/20210322005061/en/GreenPrint-Survey-Finds-Consumers-Want-to-Buy-Eco-Friendly-Products-but-Don%E2%80%99t-Know-How-to-Identify-Them).
[36] *Green Guides*, US Federal Trade Commission (www.ftc.gov/legal-library/browse/rules/green-guides).

sustainability continues to rise, companies that engage in greenwashing risk damaging their reputations and losing the trust of their stakeholders. We'll discuss the regulatory efforts to curb greenwashing and the liabilities companies face for greenwashing in the next chapter.

The World Economic Forum (WEF) recognized how foundational governance is to ESG and enumerated a set of factors, sub-factors, and key indicators (Table 6-5) for business integration.

CHAPTER TAKEAWAYS

Table 6-5. World Economic Forum List of "G" Factors, Sub-factors, and KPIs in ESG[37]

Factors	Example sub-factors, key indicators
Business ethics	Purpose, values, culture, integrity beyond compliance, ESG integration, pursuit of and reporting on KPIs
Board composition	Competencies, diversity, structure, committees, oversight capacity, independence
Corporate leadership	Tone, knowledge, experience, power allocation, compensation, decision-making processes, independence and empowerment of compliance function
Risk and crisis management	Preparedness, mitigation, past performance, regulatory compliance, segregation of duties, audit independence, shareholder rights, information governance, cybersecurity

(continued)

---

[37] *Defining the 'G' in ESG Governance Factors at the Heart of Sustainable Business*, World Economic Forum (June 2022) (www3.weforum.org/docs/WEF_Defining_the_G_in_ESG_2022.pdf). Permission to use material for this publication was granted by the WEF and is subject to the Creative Commons license (https://creativecommons.org/licenses/by-nc-nd/4.0/).

Table 6-5. (*continued*)

Factors	Example sub-factors, key indicators
Resource allocation	Capital allocation, personnel allocation, mergers and acquisitions
Incentive structures	Compensation, promotion, reporting structures, defined prohibited misconduct, disciplinary measures
Political responsibility	Lobbying, amicus briefs, campaign finance, political contributions
Transparency	Ownership, subsidiaries/holdings, open contracting, lobbying, charitable donations, countries of operation, verifiability of disclosures
Anti-corruption and integrity	Training and communications, whistle-blower protocols, due diligence, risk assessments, public procurement, government relations, gifts and entertainment, conflicts of interest, remuneration and payment procedures, record-keeping, financial controls, reporting and accounting, contractual obligations, public commitments, past incidents, internal investigation and remediation
Tax strategy	Tax compliance, anti-tax avoidance, tax disclosures
Fair competitive practices	Anti-collusion, anti-exclusion, anti-monopoly, anti-coercion, market-based pricing
Stakeholder engagement	Understanding corporate impact and stakeholder priorities, pursuing stakeholder-centred practices
Supply/value chain management	ESG integration, transparency, contractual obligations, countries of operation

CHAPTER 7

The Global ESG Regulatory Paradigm

We've covered the prevailing international law on primary ESG subject matter in prior chapters. In this chapter, we'll aim to provide the global context on national environmental, and social, governance law and regulation. To do so, we'll review a selection of the most significant laws that have been in the books since before the term ESG came into use and which form the foundational legal requirements of ESG. We'll also highlight some of the primary new laws and regulations in sustainability and ESG.

7.1 Development of the ESG Regulatory Environment

The global ESG regulatory environment has developed by leaps and bounds in recent years and continues to evolve. Initially, as ESG was primarily investor-focused and finance-related, regulation of ESG issues such as climate risk was limited to the field of financial regulation. However, with growing awareness of the impact of businesses on the environment and society, governments and regulatory bodies worldwide have taken steps to incorporate ESG considerations into a suite of requirements across the legislative, regulatory, and executive framework.

CHAPTER 7   THE GLOBAL ESG REGULATORY PARADIGM

We'll generally refer to those requirements collectively as *regulatory* in nature, as that is the common collective vernacular.

The development of the global ESG regulatory environment can be attributed to several key factors. First, there has been a growing recognition that businesses have a responsibility to address environmental and social challenges. This has been fueled by increasing concerns over climate change, resource depletion, and social inequality. Second, the investment community has long influenced ESG regulations. Institutional investors, such as pension funds and asset managers, have increasingly considered ESG components in their investment decisions. This has led to a demand for standardized ESG reporting and disclosure, prompting regulators to take action. Third, international frameworks and agreements significantly shape the global ESG regulatory landscape. For example, the United Nations Sustainable Development Goals (SDGs) and the Paris Agreement have provided a common language and goals for governments, businesses, and investors to align their efforts.

As a result of these factors, governments and regulators around the world are developing and implementing more comprehensive ESG regulations. The scope of these regulations has become wider and it encompasses areas such as climate change mitigation, human rights, labor standards, diversity and inclusion, supply chain management, and corporate governance. As we've discussed, the ESG regulatory environment is still evolving. One overarching trend is the increasing focus on mandatory reporting and disclosure requirements. Many jurisdictions have introduced or are considering mandatory ESG reporting frameworks to ensure transparency and comparability of ESG data. This enables investors and other stakeholders to make informed decisions and to hold businesses accountable for their ESG performance.

The growing push for standardization and harmonization of ESG reporting frameworks has also put pressure on regulators, and many have begun to respond by incorporating those standards into regulation and legislation. Various organizations and initiatives are working toward

CHAPTER 7 THE GLOBAL ESG REGULATORY PARADIGM

developing globally accepted standards for ESG reporting, which would enhance comparability and consistency across different jurisdictions. We'll cover the standards and frameworks in more detail in Chapter 8.

The ESG regulatory arena will continue to develop and expand as governments, businesses, and stakeholders recognize the importance of addressing environmental, social, and governance issues. Companies are wise to seek guidance and counsel on the applicability of these laws and regulations to their operations. Employees in corporate functions with ESG responsibility, corporate leadership, and corporate boards should have a working knowledge of the ESG legal framework under which their companies must operate and comply. It is prudent to conduct training sessions and to provide regular briefings on ESG requirements across the company.

Building Upon Existing Environmental, Social, and Governance Regulations

It is important to keep in mind that while there is tremendous activity in the burgeoning ESG legal and regulatory field, many of the core environmental, social, and governance requirements of ESG are already codified in long-standing international and national law. National, state, and local law encompass a broad range of legislation, regulations, executive actions, and common law within a specific country. It is a body of law to which a company must adhere in all relevant jurisdictions. These requirements can be mapped to each of the E, S, and G factors of ESG.

- *Environmental regulation* is designed to address various environmental challenges, including pollution control, natural resource management, endangered species protection, and waste management.

- *Social regulation* focuses on promoting social welfare, equality, justice, and human rights within society.

- ***Governance regulation*** aims at establishing requirements for corporate governance and protecting against bribery, corruption, and undue political influence.

It's important for businesses to take an inventory of all the laws and regulations they are subject to and to map those requirements for compliance. The requirements may impose administrative, civil, and/or criminal penalties and sanctions for non-compliance and provide mechanisms for monitoring and reporting violations. They also represent some of the foundational legal requirements for activities underlying ESG metrics. For example, companies typically must measure, monitor, and report air emissions and water use under various national, state, regional, and local laws and permitting regulations.

To illustrate this point, a short list of major federal laws just in the United States that incur administrative, civil, and criminal penalties for violations is provided. There are large sections of law libraries worldwide devoted to each of the three pillars of environmental, social, and governance law. Here is a list of just ten laws in the United States applicable to business operations that are foundational to ESG issues, and which predate ESG by decades.

1. **Clean Air Act (1970):** This law regulates air emissions from stationary and mobile sources, sets National Ambient Air Quality Standards (NAAQS), and grants authority to the Environmental Protection Agency (EPA) to establish emission standards.[1]

---

[1] *Clean Air Act, as amended,* Pub. L. 117-286 (December 27, 2022), US Government Printing Office (www.govinfo.gov/content/pkg/COMPS-8160/pdf/COMPS-8160.pdf).

2. **Clean Water Act (1972):** The act governs discharges of pollutants into US waters, sets water quality standards, and provides funding for wastewater treatment plants.[2]

3. **Endangered Species Act (1973):** This law protects endangered and threatened species and their habitats. It prohibits actions that may harm these species and requires federal agencies to ensure their actions do not put them in jeopardy.[3]

4. **Resource Conservation and Recovery Act (1976):** The act regulates the management of hazardous and non-hazardous solid waste. It regulates waste generation, transportation, treatment, storage, and disposal.[4]

5. **Toxic Substances Control Act (1976):** This law regulates the manufacture, import, processing, distribution, and disposal of chemicals in the United States.[5]

---

[2] *Federal Water Pollution Control Act,* Pub. L. 95-217 (December 27, 1977) US Government Printing Office (www.govinfo.gov/link/statute/91/1609).

[3] *Endangered Species Act of 1973, as amended,* Pub. L. 93-205 (December 27, 2022) US Government Printing Office (www.govinfo.gov/content/pkg/COMPS-3002/pdf/COMPS-3002.pdf).

[4] *Resource Conservation and Recovery Act of 1976,* Pub. L. 94-580 (October 21, 1976), US Government Printing Office (www.govinfo.gov/content/pkg/STATUTE-90/pdf/STATUTE-90-Pg2795.pdf).

[5] *Toxic Substances Control Act, as amended,* Pub. L. 94-469 (December 27, 2022), US Government Printing Office (www.govinfo.gov/content/pkg/COMPS-895/pdf/COMPS-895.pdf).

6. **Comprehensive Environmental Response, Compensation, and Liability Act (1980):** This law, commonly called Superfund, provides a federal program to remediate hazardous waste sites and hold responsible parties accountable for the cleanup costs.[6]

7. **National Environmental Policy Act (1970):** The act requires federal agencies to make an assessment of the potential environmental impacts of their proposed actions, which weighs into the decision of whether to move forward on a project. It also promotes transparency and public involvement in the decision-making process.[7]

8. **Safe Drinking Water Act (1974):** The act sets standards for drinking water quality, establishes requirements for public water systems, and provides funding for infrastructure improvements.[8]

9. **Energy Policy Act (2005):** This law promotes energy conservation, renewable energy, and the development of advanced technologies for energy production.[9]

---

[6] *Comprehensive Environmental Response, Compensation, and Liability Act of 1980, as amended,* Pub. L. 115-141 (March 23, 2018), US Government Printing Office (www.govinfo.gov/content/pkg/COMPS-886/pdf/COMPS-886.pdf).

[7] *National Environmental Policy Act of 1969, as amended,* Pub. L. 91-190 (June 3, 2023), US Government Printing Office (www.govinfo.gov/content/pkg/COMPS-10352/pdf/COMPS-10352.pdf).

[8] *Safe Drinking Water Act, as amended,* Pub. L. 117-286, US Government Printing Office (www.govinfo.gov/app/details/COMPS-892).

[9] *Energy Policy Act of 2005,* Pub. L. 109-58 (August 8, 2005), US Government Printing Office (www.govinfo.gov/content/pkg/PLAW-109publ58/pdf/PLAW-109publ58.pdf).

CHAPTER 7   THE GLOBAL ESG REGULATORY PARADIGM

10. **Foreign Corrupt Practices Act (1977):** This law prohibits US citizens and entities from bribing foreign government officials to benefit their interests.[10]

National governments are also responding to the science on greenhouse gas emissions and taking new approaches to regulating GHG emissions. The United States Environmental Protection Agency Administrator signed a final rule on methane emissions and other sources of air pollution from oil and gas operations on November 30, 2023.[11] In its announcement of the final rule, the agency referred to methane as a "climate 'super pollutant' that is more potent than carbon dioxide and is responsible for approximately one third of current warming resulting from human activities."[12] In conjunction with its release of the methane rule, EPA issued an updated social cost analysis for greenhouse gases.[13]

---

[10] *Foreign Corrupt Practices Act of 1977, as amended,* Pub. L. 95-213 (November 10, 1998), US Government Printing Office (www.govinfo.gov/content/pkg/COMPS-9569/pdf/COMPS-9569.pdf).

[11] *Standards of Performance for New, Reconstructed, and Modified Sources and Emissions Guidelines for Existing Sources: Oil and Natural Gas Sector Climate Review, Final Rule,* 40 CFR Part 60, United States Environmental Protection Agency (November 30, 2023) (www.epa.gov/system/files/documents/2023-12/eo12866_oil-and-gas-nsps-eg-climate-review-2060-av16-final-rule-20231130.pdf).

[12] *EPA's Final Rule for Oil and Natural Gas Operations Will Sharply Reduce Methane and Other Harmful Pollution,* United States Environmental Protection Agency (December 2, 2023) (www.epa.gov/controlling-air-pollution-oil-and-natural-gas-operations/epas-final-rule-oil-and-natural-gas#:~:text=December%202%2C%202023%20%2D%2D%20EPA,time%2C%20from%20existing%20sources%20nationwide).

[13] *Supplementary Materials for the Regulatory Impact Analysis for the Final Rulemaking, "Standards of Performance for New, Reconstructed, and Modified Sources and Emissions Guidelines for Existing Sources: Oil and Natural Gas Sector Climate Review," EPA Report on the Social Cost of Greenhouse Gases: Estimates Incorporating Recent Scientific Advances,* United States Environmental Protection Agency (November 2023) (www.epa.gov/system/files/documents/2023-12/epa_scghg_2023_report_final.pdf).

CHAPTER 7   THE GLOBAL ESG REGULATORY PARADIGM

The social cost analysis weighs the economic estimates of reducing climate pollution versus taking no action and will likely underpin future climate regulatory action.

ESG Taxonomy

> *A [t]axonomy is a classification tool or system meant to help investors and companies make informed investment decisions on sustainable economic activities. It aims at establishing market clarity on what is robustly and consensually "sustainable" when it comes to environmental or social issues.*[14]
>
> —Natixis Corporate & Investment Banking

An ESG or green taxonomy generally refers to a classification system that defines which economic activities can be considered sustainable for investment purposes. Taxonomies are a tool of sustainable finance and help to provide clarity to investors, companies, policymakers, and other stakeholders by establishing a clear set of criteria for what constitutes an ESG-compliant investment or activity. A taxonomy typically includes thresholds and criteria for sustainable activities based on their substantial contributions to environmental objectives, such as climate change mitigation and adaptation, while also ensuring that these activities do not significantly hamper any of the objectives and that they meet minimum social safeguards.

Taxonomies are typically put forth by finance ministers or national bank supervisors in concert with technical working groups. Several countries and regions across the globe have begun developing and

---

[14] *The New Geography of Taxonomies*, Natixis (July 2023) (https://gsh.cib.natixis.com/api-website-feature/files/download/12776/the_new_geography_of_taxonomies_updated_july_2023.pdf).

CHAPTER 7   THE GLOBAL ESG REGULATORY PARADIGM

adopting their own ESG taxonomies to guide investments toward more sustainable practices. At this writing, 17 nations have taxonomies in place, and another 16 have taxonomies under development.[15] Figure 7-1 represents the expansion of green taxonomies around the world. These taxonomies have facilitated the development of sustainable finance including the growth of the green bond market which surpassed the USD$1 trillion mark in 2020.[16]

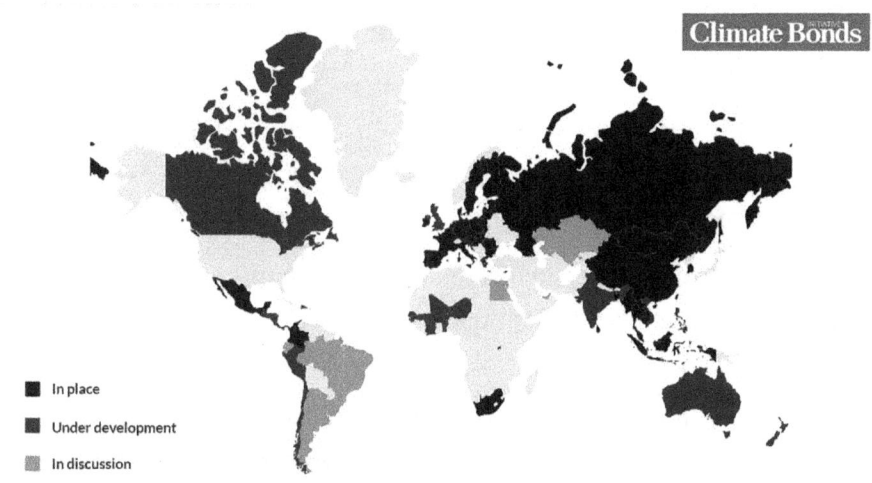

Figure 7-1. Climate Bonds Initiative Global Taxonomy Map[17]

---

[15] *The New Geography of Taxonomies*, Natixis (July 2023) (https://gsh.cib.natixis.com/api-website-feature/files/download/12776/the_new_geography_of_taxonomies_updated_july_2023.pdf).

[16] *FAQ: What is the EU Taxonomy and how will it work in practice?* European Commission (April 12, 2021) (https://finance.ec.europa.eu/system/files/2021-04/sustainable-finance-taxonomy-faq_en.pdf).

[17] *Taxonomy*, Climate Bonds Initiative (www.climatebonds.net/taxonomy).

CHAPTER 7   THE GLOBAL ESG REGULATORY PARADIGM

The following are examples of existing and developing taxonomies:

- **European Union:** The EU taxonomy for sustainable activities is one of the most advanced and comprehensive frameworks, introduced as part of the EU's Action Plan on Financing Sustainable Growth.[18]

- **Canada:** Canada has announced its intention to develop a taxonomy as part of its efforts to support sustainable finance and has established an expert group to work on it.[19]

- **United Kingdom:** Post-Brexit, the UK is working on its own green taxonomy, taking inspiration from the EU's framework but tailoring it to the UK market.[20]

- **Malaysia:** Malaysia has released a Principles-Based Sustainable and Responsible Investment Taxonomy for the Malaysian Capital Market.[21]

A number of countries also have climate finance or green bond standards. It's important to note that the development of ESG taxonomies is an ongoing process, and many other nations are considering or in the

---

[18] *EU Taxonomy Navigator*, European Commission (https://ec.europa.eu/sustainable-finance-taxonomy/).

[19] *Taxonomy Roadmap Report,* Sustainable Finance Action Council, Government of Canada (www.canada.ca/en/department-finance/programs/financial-sector-policy/sustainable-finance/sustainable-finance-action-council/taxonomy-roadmap-report.html).

[20] *Promoting the international interoperability of a UK Green Taxonomy,* Green Technical Advisory Group (February 2023) (www.greenfinanceinstitute.com/wp-content/uploads/2023/02/GFI-GTAG-INTERNATIONAL-INTEROPERABILITY-REPORT.pdf).

[21] *Principles-Based Sustainable and Responsible Investment taxonomy for the Malaysian Capital Market,* Securities Commission Malaysia (www.sc.com.my/api/documentms/download.ashx?id=a0ab5b0d-5d7d-4c66-8638-caec92c209c1).

early stages of creating their own expanded taxonomies. These taxonomies are expected to evolve as their use cases are broad, from policymaking to climate financial indexing to standards for sustainability-linked bonds and loans.

7.2 Key Global Regulatory Initiatives

A growing number of countries have enacted laws and regulations on a range of ESG and sustainability topics, including sustainable investing, sustainability claims, and transparency and enhanced disclosures of ESG and sustainability activities. The European Union has been at the forefront of ESG and sustainability regulation and has enacted some of the most advanced requirements in the world. The United Kingdom has been active in promulgating financial and non-financial ESG regulation. The United States has undertaken similar initiatives, as have other countries including Australia, Brazil, Canada, China, France, Germany, Hong Kong, India, Japan, the UAE, Singapore, and Switzerland.

Table 7-1 is a snapshot of the recent spate of global regulatory activity on ESG and sustainability issues by listed nation and supranational union, as is the case of the EU. It provides a horizon scan of what has been proposed and enacted but is by no means an exhaustive recitation of the pending and enacted statutes, regulations, executive orders, and listing rules. Because stock exchange rules generally only apply to listed companies, we've relegated all stock exchange rules irrespective of subject matter to the right-hand column of the chart. There are myriad legislative, regulatory, and executive actions (which are generally referred to with the catchall "regulatory" term) as well as stock exchange initiatives around the world. The scope is broad and in keeping with the range of environmental, social, and governance issues upon which companies are measured, evaluated, and scored.

CHAPTER 7 THE GLOBAL ESG REGULATORY PARADIGM

Table 7-1. *Global ESG Regulatory Activity Snapshot*

Nation	Climate Risk/ Disclosures	Sustainability, Supply Chain, & Human Rights	Greenwashing	Nature & Waste	Diversity	Taxonomy & Markets	Stock Exchange Rules
European Union	Non-Financial Reporting Directive (NFRD) Sustainable Finance Disclosure Regulation (SFDR)	European Green Deal Corporate Sustainability Reporting Directive (CSRD) Corporate Sustainability Due Diligence Directive (CSDDD) EU Conflict Minerals Regulation	Directive on Green Claims	Nature Restoration Law Sustainable Packaging Waste Regulation Circular Economy Action Plan Sustainable Products Regulation	EU Gender Equality Strategy EU Pay Transparency Directive	EU Taxonomy ECB Climate & Nature Plan CBAM ESG Rating Activities Regulation European Green Bond Standard Carbon Removal Certification	

CHAPTER 7   THE GLOBAL ESG REGULATORY PARADIGM

United Kingdom	Climate-Related Financial Disclosure Rules	Sustainability Disclosure Requirements (SDR)	Green Claims Code	Biodiversity Net Gain Regulation	Diversity and Inclusion on Boards Policy Statement	UK Draft Green Taxonomy	GHG Emissions Trading Scheme
		UK Modern Slavery Act	Green Agreements Guide		FCA Reporting Requirements	UK Transition Plan Task Force Disclosure Framework	
		Modern Slavery/ Human Trafficking Statement				ESG Ratings Regime	
		Mandatory ESG Reporting Resolution				Sustainable Finance Action Plan	
Brazil							

(continued)

225

CHAPTER 7 THE GLOBAL ESG REGULATORY PARADIGM

Table 7-1. (*continued*)

Nation	Climate Risk/ Disclosures	Sustainability, Supply Chain, & Human Rights	Greenwashing	Nature & Waste	Diversity	Taxonomy & Markets	Stock Exchange Rules
Hong Kong		Roadmap for ISSB Adoption				Green & Sustainable Finance Strategy; Hong Kong & Beijing MOU to Promote Green Finance	Enhanced Climate Disclosure Rule for Listed Issuers
Canada		Canadian Sustainability Disclosure Standards Proposal			Corporate Diversity Reporting Rules	Climate Investment Taxonomy; ESG-Related Investment Disclosure for Funds	Climate Related Disclosures for Listed Issuers

CHAPTER 7 THE GLOBAL ESG REGULATORY PARADIGM

Singapore	Climate-Related Reporting Rules	Taxonomy for Sustainable Finance	Environmental Risk Management for Asset Managers, Banks & Insurers	Code of Conduct for ESG Ratings & Data Providers
France	French Climate & Resilience Law	Corporate Sustainability Reporting	Guide to Environmental Claims	Anti-Waste and Circular Economy Law

(continued)

227

CHAPTER 7 THE GLOBAL ESG REGULATORY PARADIGM

Table 7-1. (*continued*)

Nation	Climate Risk/ Disclosures	Sustainability, Supply Chain, & Human Rights	Greenwashing	Nature & Waste	Diversity	Taxonomy & Markets	Stock Exchange Rules
Germany		Corporate Supply Chain Due Diligence Act					
	Climate Disclosures & Corporate Reporting	Revisions to Corporate Governance Code				Guidance on Financed Emissions	
Japan							
Australia	Climate-Related Financial Disclosure		ASIC Enhanced Enforcement of Greenwashing Claims			Sustainable Finance Strategy Prudential Practice Guide on Climate Change Financial Risks	

China	ESG-Related Amendments to the Securities Disclosure Rules	National Biodiversity Strategy & Plan	Hong Kong & Beijing MOU to Promote Green Finance
UAE			Climate & Environmental Risk Management -Sustainable Finance Task Force
Switzerland	Conflict Minerals & Metals and Child Labor Due Diligence Obligations		Transparency in Non-Financial Matters Reporting

(continued)

CHAPTER 7 THE GLOBAL ESG REGULATORY PARADIGM

Table 7-1. (continued)

Nation	Climate Risk/ Disclosures	Sustainability, Supply Chain, & Human Rights	Greenwashing	Nature & Waste	Diversity	Taxonomy & Markets	Stock Exchange Rules
United States	SEC Climate Disclosure Rule	Uyghur Forced Labor Act	SEC Name Rule			CFTC Voluntary Carbon Markets Guidance	NASDAQ Board Diversity Rule
	California Climate Corporate Accountability Act	California Supply Chains Transparency Act	FTC Green Guides			California Voluntary Carbon Market Disclosures Act	
	California GHG—Climate-Related Financial Risk	EPA Methane Rule				NY Banks Climate Guidance	

230

We'll cover a selection of those established and emerging regulations and initiatives. Companies should regularly monitor the continued advancement and maturity in the ESG and sustainability regulatory arena for updates that pertain to their business.

Comprehensive Sustainability Regulation in the European Union

The EU has set out an ambitious green plan for its member states, established a green taxonomy, and promulgated groundbreaking regulations aimed at promoting sustainable finance, human rights, sustainability disclosures, and due diligence. We'll review progress on a number of these fronts.

European Green Deal

The European Green Deal (the Green Deal) is a bold and ambitious vision initiated by the EU to combat climate change, promote sustainable development, and transition to a green economy. The Green Deal was presented by the European Commission in December 2019.[22] It was conceived in response to the pressing global concern of climate change and the need to transition toward a sustainable economic model. The Green Deal aims at transforming the EU into the world's first "climate-neutral continent" by 2050.[23]

---

[22] *The European Green Deal,* European Commission (https://commission.europa.eu/strategy-and-policy/priorities-2019-2024/european-green-deal_en).
[23] Id.

The Green Deal has a broad scope with several interlinked objectives. The overarching aim is to achieve climate neutrality by 2050, which means that the EU will strive to balance the amount of greenhouse gases it produces with the amount it removes from the atmosphere. The Green Deal aims to

- **Protect human life, animals, and plants by reducing pollution** and creating a pollution-free environment, ensuring clean air, water, and soil.

- **Ensure a just and inclusive transition** to a green economy. Recognizing it will impact different regions and sectors differently, the measures seek to ensure that the transition is socially fair and that no person or place is left behind.

- **Promote the development of clean, reliable, and affordable energy** where energy is sourced from renewable sources and used efficiently for a clean energy future.

- **Transform agriculture and rural regions** by including a "Farm to Fork" strategy to make food systems more sustainable and resilient.

- **Promote sustainable industry** in transforming the EU's industries and services, making them more sustainable and competitive.[24]

The Green Deal lays out a strategic roadmap with specific milestones to guide the EU's transition toward a green economy. Some of the key milestones include

---

[24] Id.

- **Climate neutrality by 2050:** This is a central goal of the Green Deal. The EU seeks to be net zero on net greenhouse gas emissions by 2050.

- **Fifty-five percent reduction in greenhouse gas emissions by 2030:** This is an interim target on the path to achieving climate neutrality. The EU aims to reduce its net greenhouse gas emissions by at least 55% by 2030, compared to 1990 levels.

- **Zero pollution:** The Deal aims to create a pollution-free environment by 2050, focusing on reducing air, water, and soil pollution.

- **Twenty-five percent of EU agriculture to be organic by 2030:** As part of the Farm to Fork strategy, the EU aims to increase the share of organic farming in the EU to 25% by 2030.[25]

The Green Deal is a comprehensive plan that requires a wide range of activities designed to achieve its ambitious goals. These activities span multiple sectors, including energy, transport, industry, agriculture, and finance. The sector-specific related goals include

- **Clean energy:** Energy production and use account for more than 75% of the EU's greenhouse gas emissions.[26] The Green Deal seeks to decarbonize the energy sector by promoting the use of renewable energy sources, enhancing energy efficiency, and modernizing energy infrastructure.

---

[25] Id.
[26] Id.

- **Sustainable industry:** The Green Deal aims to transform the EU's industries, making them more sustainable, efficient, and circular. This includes promoting the use of recycled materials, reducing waste, and encouraging the production of climate-neutral and circular products.

- **Buildings and renovations:** Buildings account for approximately 40% of the EU's energy consumption.[27] This includes measures to enhance the energy efficiency of buildings, promote sustainable construction methods, and reduce emissions from the building sector.

- **Sustainable mobility:** The Green Deal includes measures to promote sustainable mobility, including promoting the use of clean and efficient public and private transport and reducing emissions from cars, trains, ships, and aircraft.

- **Eliminating pollution:** The Green Deal includes a *Zero Pollution Action Plan* to achieve a pollution-free environment by 2050. This includes measures to reduce air, water, and soil pollution and to protect human health and the environment from hazardous chemicals.

- **Farm to fork:** The Farm to Fork strategy is a key component of the Green Deal. It aims to make the EU's food systems more sustainable, resilient, and fair. This includes measures to reduce the use of pesticides, promote organic farming, and ensure sustainable and healthy food for all EU citizens.

---

[27] Id.

CHAPTER 7   THE GLOBAL ESG REGULATORY PARADIGM

- **Preserving biodiversity:** The Green Deal includes measures to protect and restore the EU's biodiversity. This includes efforts to protect forests, restore degraded ecosystems, and create green spaces in cities.[28]

Despite being a relatively new and broad-scale initiative, the Green Deal has already achieved several notable milestones, including

- **Adoption of the European Climate Law:** In April 2021, the EU adopted the European Climate Law, which codifies the EU's 2050 climate neutrality target into law. The law also includes a legally binding target to reduce net greenhouse gas emissions by at least 55% by 2030, compared to 1990 levels.[29]

- **Launch of the just transition mechanism:** The EU has launched the Just Transition Mechanism, a EUR€100 billion funding mechanism designed to support regions and sectors most affected by the transition to a low-carbon economy.

- **Approval of the EU Biodiversity Strategy for 2030:** In May 2021, the EU approved the EU Biodiversity Strategy for 2030, which aims to protect and restore the EU's biodiversity and natural ecosystems.[30]

---

[28] Id.

[29] *Climate Action,* European Commission (https://climate.ec.europa.eu/eu-action/european-climate-law_en); Regulation (EU) 2021/1119 of the European Parliament and of the Council of 30 June 2021 establishing the framework for achieving climate neutrality and amending Regulations (EC) No 401/2009 and (EU) 2018/1999 ("European Climate Law") (June 30, 2021) (https://eur-lex.europa.eu/legal-content/EN/TXT/?uri=CELEX:32021R1119).

[30] Id.

CHAPTER 7 THE GLOBAL ESG REGULATORY PARADIGM

The European Green Deal is a long-term project that will shape the EU's economic and environmental policies for decades to come. It showcases the EU's commitment to leading the transition to a green and sustainable future. In the coming years, the EU will continue to implement the various activities and initiatives under the Green Deal, with a particular focus on achieving the 2030 targets. The transition to a green economy will require significant investment and may have significant social and economic impacts. The success of the Green Deal will depend on the EU's ability to mobilize the necessary financial resources, engage with stakeholders, and ensure that all member states are on board.

The European Commission (EC) updated its climate strategy with an impact assessment in 2024. The government set a new target for the reduction in GHG emissions of 90% by 2040 to achieve climate neutrality by 2050.[31] The strategy will focus on three technology pathways: (1) carbon capture for storage (CCS); (2) carbon capture for utilization (CCU); and (3) removal of carbon from the atmosphere for permanent storage. In its summary of the update, the EC said, "The Green Deal now needs to become an industrial decarbonization deal."[32]

Corporate Sustainability Reporting Directive (CSRD)

The European Union has been at the forefront of implementing sustainability reporting regulations to promote transparency and accountability in corporate environmental, social, and governance practices. The regulations aim to provide investors and stakeholders with reliable information on a company's sustainability performance and to support the transition toward a more sustainable economy. The Corporate Sustainability Reporting Directive (CSRD) is the cornerstone sustainability

---

[31] *Commission presents recommendation for 2040 emissions reduction target to set the path for climate neutrality by 2050,* European Commission (February 6, 2024) (https://ec.europa.eu/commission/presscorner/detail/en/ip_24_588).
[32] Id.

disclosure regulation, which was adopted by the European Commission (EC) in 2022 and entered force on January 5, 2023.[33] It amended and expanded upon the prior Non-Financial Reporting Directive (NFRD), which already required broad reporting.[34]

The CSRD requires comprehensive sustainability reporting requirements and obligations. This directive significantly expands the scope and content of existing nonfinancial reporting obligations, encompassing a broader range of entities and requiring more detailed reporting on ESG topics.[35] The CSRD applies to EU companies and non-EU companies with a substantial presence in the EU. There is a phased compliance timeline, and the first swath of companies must apply the new rules for the 2024 fiscal year with reports to be prepared in 2025.[36] The EC adopted its first set of European Sustainability Reporting Standards (ESRS), which were published on December 22, 2023.[37] Those standards establish a framework and methodology for reporting sustainability information, including environmental and social impacts, risks, and opportunities.

Under the CSRD, companies are required to report their sustainability impacts and to provide a comprehensive overview of their sustainability performance. The reporting obligations include

- **Cross-cutting reporting:** Companies must include a separate section in their management reports dedicated to general principles, strategy, governance, and materiality. Companies are required to conduct

---

[33] *Corporate sustainability reporting,* European Commission (https://finance.ec.europa.eu/capital-markets-union-and-financial-markets/company-reporting-and-auditing/company-reporting/corporate-sustainability-reporting_en).
[34] Id.
[35] Id.
[36] Id.
[37] Id.

materiality assessments to determine which sustainability topics are most relevant to their business and stakeholders.

- **Environmental reporting:** Companies are required to disclose their environmental impacts, including greenhouse gas (GHG) emissions, energy consumption, biodiversity, water usage, and waste management. Companies must align their reporting with established frameworks.

- **Social reporting:** Companies must report on social aspects, such as employee health and safety, labor practices, human rights, and diversity and inclusion. This includes information on workforce composition, employee training, and measures to ensure fair and ethical treatment.

- **Governance reporting:** Companies must disclose information on their corporate governance practices, risk management practices, internal controls, and business conduct. This aims to enhance transparency and accountability in corporate decision-making.

- **Materiality assessment:** Companies are required to conduct materiality assessments to determine which ESG topics are most relevant to their business and stakeholders. The CSRD incorporates a ***double materiality*** standard—financial materiality and impact materiality—which requires reporting on significant impacts from both an investor perspective and a societal stakeholder perspective.

CHAPTER 7   THE GLOBAL ESG REGULATORY PARADIGM

Companies in Scope for the CSRD

It is estimated that CSRD will apply to approximately 50,000 companies. In a nutshell, covered companies include EU companies and non-EU companies that meet certain criteria within the EU borders:

- All companies (EU and non-EU) with **EU-listed securities**, with limited exceptions.

- **"Large" unlisted EU companies**, defined as companies that meet two of the following: (1) balance sheet (EUR€25mm); (2) net revenue (EUR€50mm); and (3) workforce threshold (250 employees) in two consecutive years (European subsidiaries of US companies are required to report if threshold met), with specific rules for financial services companies.

- Non-EU headquartered global companies with European subsidiaries or branches that exceed a specific revenue threshold will be required to perform consolidated **"group-level"** reporting, with certain exceptions.

Compliance Timeline for the CSRD

The CSRD introduces a phased implementation approach to allow companies sufficient time to adapt to the new reporting requirements. The first compliance deadline is 2025 to report on FY2024 data for companies that are already subject to the Non-Financial Reporting Directive (NFRD) with a phased timeline for the remainder of in-scope companies. On February 14, 2024, the European Council and European Parliament announced a provisional agreement to delay reporting requirements for

CHAPTER 7 THE GLOBAL ESG REGULATORY PARADIGM

certain sectors and third-country companies by two years, which was adopted on April 30, 2024.[38] That agreement was formally endorsed and adopted by both the European Council and European Parliament. Under new regulatory regimes, it is not unusual for implementation timelines to be adjusted, particularly where there are newly established reporting standards.

EU member states must transpose the CSRD into their respective national laws by July 2024. In December 2023, France codified the CSRD into its national law.[39] This marked the first European Union member state to do so.

Corporate Sustainability Due Diligence Directive (CSDDD or CS3D)

The Corporate Sustainability Due Diligence Directive (CSDDD or CS3D) establishes a corporate due diligence duty. It focuses on ethical and sustainable business practices and will change the way businesses address and report on human rights and environmental impacts. The legislation is the first in the EU to require large, in-scope companies to adopt a climate transition plan aligned to their business strategies with the goal of limiting global warming to 1.5°C.[40]

---

[38] *Council and Parliament agree to delay sustainability reporting for certain sectors and third-country companies by two years,* Council of the European Union (February 7, 2024) (www.consilium.europa.eu/en/press/press-releases/2024/02/07/council-and-parliament-agree-to-delay-sustainability-reporting-for-certain-sectors-and-third-country-companies-by-two-years/). *Council adopts directive to delay reporting obligations for certain sectors and third country companies,* Council of the European Union (April 29, 2024) (https://www.consilium.europa.eu/en/press/press-releases/2024/04/29/council-adopts-directive-to-delay-reporting-obligations-for-certain-sectors-and-third-country-companies/).

[39] *Ordinance No. 2023-1142 of December 6, 2023 relating to the publication and certification of information regarding sustainability and the environmental, social and corporate governance obligations of commercial companies,* French Republic (www.legifrance.gouv.fr/jorf/id/JORFTEXT000048519395).

[40] Id.

CHAPTER 7   THE GLOBAL ESG REGULATORY PARADIGM

The CSDDD applies mandatory human rights and environmental due diligence obligations for EU and non-EU companies operating in the EU that meet specific thresholds for revenue and number of employees. A hallmark of the directive is that due diligence measures are required across the corporate value chain.

On December 14, 2023, the European Union announced that it had reached an agreement on the proposed text.[41] The provisional agreement was taken up by the European Council in early 2024. The text of the provisional agreement was intensely debated by the member states. Concessions were made, including the narrowing of the scope of coverage for the new directive. After machinations, the Council of the European Union approved a revised provisional agreement on March 15, 2024. The Legal Affairs Committee of the European Parliament approved the revised text of the legislation on March 19, 2024, and the European Union subsequently adopted the final directive.[42]

Key Requirements of the Agreement

- **Scope of Coverage:** The CSDDD contains a phased-in timeline. At the outset, it generally applies to

    (1) EU companies with more than 1,000 employees and a net annual global turnover of more than EUR€450 million for the prior financial year; and

---

[41] *Corporate sustainability due diligence: Council and Parliament strike deal to protect environment and human rights*, Council of the European Union (December 14, 2023) (www.consilium.europa.eu/en/press/press-releases/2023/12/14/corporate-sustainability-due-diligence-council-and-parliament-strike-deal-to-protect-environment-and-human-rights/).

[42] *First green light to new bill and firms' impact on human rights and environment*, European Parliament (March 19, 2024) (www.europarl.europa.eu/news/en/press-room/20240318IPR19415/first-green-light-to-new-bill-on-firms-impact-on-human-rights-and-environment), Council of the European Union. Corporate sustainability due diligence: Council gives its final approval (May 29, 2024) (https://www.consilium.europa.eu/en/press/press-releases/2024/05/24/corporate-sustainability-due-diligence-council-gives-its-final-approval/).

(2) non-EU companies that generate a net annual turnover of more than EUR€450 million in the EU for the prior financial year or, if that threshold is not met, but does meet certain thresholds for the ultimate parent company.

- **Corporate governance:** The directive requires integration of human rights and environmental impact into corporate governance.

- **Transition plans:** Businesses must prepare climate transition plans to support the Paris Agreement goal of limiting warming to 1.5°C. These include time-bound climate targets and key actions and investments to support reaching those targets.

- **Liability:** The directive includes the designation of a supervisory authority in charge of monitoring, investigating, and imposing civil fines on companies for non-compliance. The fines are specified to include penalty amounts correlated to a company's net worldwide turnover.[43]

The directive will require companies to thoroughly assess, mitigate, and report on potential and actual human rights and environmental impacts across their operations and supply chains.

Companies will need to fully integrate the principles of human rights, sustainability, and due diligence into their business strategies, operations, and relationships. This will necessitate a comprehensive review of existing practices, policies, and contracts, as well as the development of new procedures and risk management strategies.

---

[43] Id.

The CSDDD imposes administrative and civil penalties for non-compliance. It will significantly influence how businesses operate, both internally and across their value chains. The directive will enter force on the twentieth day following its publication in the EU *Official Journal*.

Sustainable Finance Disclosure Regulation (SFDR)

The European Union's Sustainable Finance Disclosure Regulation (SFDR) is a legislative measure enacted by the European Commission. The SFDR aims to promote transparency in the financial sector by requiring financial market participants and advisors to make sustainability and ESG disclosures.[44] The SFDR requires financial market participants and financial advisors to disclose their ESG policies, risks, impacts, and performance at both the entity and product levels.

The SFDR applies to financial market participants and financial advisors within the EU, those with EU shareholders, and those marketing to clients within the EU. The law mandates two levels of disclosure—at the entity level and the product level.

- At the **entity level**, organizations are required to disclose their policies on the integration of sustainability risks into investment decisions. They must be transparent about the process for evaluating sustainability risks and how consistent their policies are with those risks.[45]

---

[44] *Sustainability-related disclosure in the financial services sector*, European Commission (https://finance.ec.europa.eu/sustainable-finance/disclosures/sustainability-related-disclosure-financial-services-sector_en).

[45] *Regulation (EU) 2019/2088 of the European Parliament and the Council of 27 November 2019 on sustainability-related disclosures in the financial services sector*, Document 32019R2088, European Union Law (https://eur-lex.europa.eu/legal-content/EN/TXT/?uri=CELEX:32019R2088).

- At the **product level**, disclosures are required concerning the product's environmental or social characteristics and its sustainability objective. It covers a wide range of financial products, including mutual funds, insurance-based investment products, private and occupational pensions, and investment advice.[46]

The SFDR has been implemented in phases, and the first obligations went into effect in March 2021.[47] Part of these requirements includes the identification and disclosure of Principal Adverse Impacts (PAI), which must detail the negative impacts of an entity's investments on sustainability factors.[48] Compliance with the SFDR requires a company-wide effort to establish workflows for collecting, analyzing, and storing the necessary data and preparing the required disclosure forms.

The SFDR is part of the EU's broader sustainability agenda. In promoting transparency and accountability in private sector investment, the SFDR aligns with the objectives of the European Green Deal. The SFDR signifies a significant step toward achieving transparency in financial markets. This transparency is necessary to help investors compare available sustainable investment options accurately and understand the extent to which products integrate environmental, social, and governance considerations.

Nature Restoration Law

The European Parliament adopted a proposed Nature Restoration Law in 2024 and it will require adoption by the EU Council to enter into force.[49] The initial proposal was prepared by the European Commission in 2022. In

---

[46] Id.
[47] Id.
[48] Id.
[49] *Nature restoration: Parliament adopts law to restore 20% of EU's land and sea*, European Parliament (February 27, 2024) (www.europarl.europa.eu/news/en/press-room/20240223IPR18078/nature-restoration-parliament-adopts-law-to-restore-20-of-eu-s-land-and-sea).

the negotiation process, various aspects of the law were contested, and the final text reflects compromises made between different stakeholders. The consensus position aimed at restoring ecosystems, such as wetlands, forests, and coastlines, that have been degraded by human activity. The Nature Restoration Law is a component of the EU Biodiversity Strategy, which emphasizes the need for binding targets to restore damaged ecosystems in order to store carbon and mitigate the effects of climate change.

The proposed law sets an overarching restoration objective for the long-term recovery of nature of both land and sea areas, with binding restoration targets for specific habitats and species. It aims to cover at least 30% of the habitats by 2030, ramping up to 60% by 2040, and ultimately restoring 90% of ecosystems by 2050.[50] The specific targets include restoring and improving biodiverse habitats, reversing the decline of pollinator populations, and achieving an increasing trend for forest ecosystems, urban ecosystems, agricultural ecosystems, marine ecosystems, and river connectivity. Member states would be expected to submit National Restoration Plans to the Commission within two years of the Regulation coming into force, demonstrating how they will deliver on the targets. They would also be required to monitor and report on their progress. The EU was instrumental in the adoption of the goal to protect 30% of the world's land and sea resources under the UN Convention on Biological Diversity. As is often the case with initial legislative forays, if this proposed law is not successful in the first instance, we will see ongoing efforts towards enhancement and adoption.

Sustainability Disclosure Requirements in the United Kingdom

The United Kingdom's Financial Conduct Authority (FCA) has developed the Sustainability Disclosure Requirements (SDR) to promote transparency, consistency, and trust in the sustainable investment market. The FCA

---

[50] Id.

CHAPTER 7   THE GLOBAL ESG REGULATORY PARADIGM

published its highly anticipated Policy Statement setting out the final rules on the SDR and investment labels, in November 2023.[51] The regulation aims to prevent greenwashing, ensure fair marketing practices, and facilitate informed decision-making by investors. The SDR will require companies to provide investors with the information necessary to make informed decisions regarding sustainable investment products. It introduces sustainability-related product labels, requires product- and entity-level disclosures, and provides additional sustainable investing rules.[52]

The regulatory framework of the SDR is based on a series of consultation papers, policy statements, and regulatory requirements that have been released by the FCA. The FCA released consultation papers in 2021 and 2022, to seek feedback on the proposed SDR.[53] These documents laid out the proposed regulations, including the introduction of sustainable investment labels and related entity- and product-level disclosures. The FCA's policy statements provide further detail on the SDR, including the rules and requirements for firms. They provide a comprehensive overview of the SDR and its objectives and outline the regulatory responsibilities of firms.

The SDR introduces a series of regulatory requirements to which covered companies must adhere. These include the adoption of sustainable investment labels, adherence to naming and marketing rules, and the implementation of entity- and product-level disclosures. The FCA has laid out a timeline for the implementation of the SDR in its Policy Document (Figure 7-2).

---

[51] *Policy Statement PS23/16, Sustainability Disclosure Requirements (SDR) and investment labels*, United Kingdom Financial Conduct Authority (November 2023) (www.fca.org.uk/publication/policy/ps23-16.pdf).
[52] Id.
[53] *Sustainability Disclosure Requirements (SDR) and investments labels, Consultation Paper CP22/20****, United Kingdom Financial Conduct Authority (October 2022) (www.fca.org.uk/publication/consultation/cp22-20.pdf).

CHAPTER 7   THE GLOBAL ESG REGULATORY PARADIGM

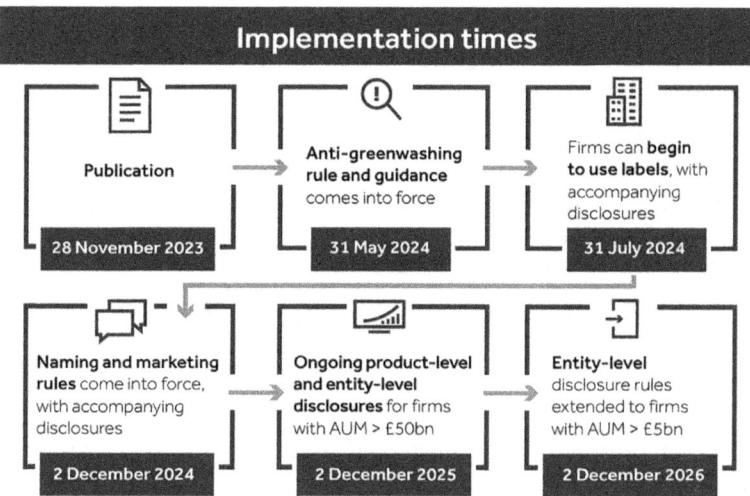

Figure 7-2. SDR Implementation, UK Financial Conduct Authority Policy Statement PS23/16[54]

One of the key provisions of the SDR is the introduction of sustainable investment labels.[55] These labels provide a clear distinction between products that are considered sustainable and those that are not. Each label is underpinned by specific criteria that firms must meet. This includes having a clear sustainability objective, following a sustainability-aligned investment policy, and demonstrating progress toward sustainability goals. The following are the four types of labels:

1. **Sustainable Focus:** For products maintaining a high standard of sustainability in their asset profile

---

[54] PS23/16: Sustainability Disclosure Requirements (SDR) and investment labels, United Kingdom Financial Conduct Authority (November 28, 2023) (www.fca.org.uk/publications/policy-statements/ps23-16-sustainability-disclosure-requirements-investment-labels).

[55] Policy Statement PS23/16, Sustainability Disclosure Requirements (SDR) and investment labels, United Kingdom Financial Conduct Authority (November 2023) (www.fca.org.uk/publication/policy/ps23-16.pdf).

247

2. **Sustainable Improvers**: For products aiming to deliver measurable sustainability improvements over time

3. **Sustainable Impact:** For products aiming to achieve positive, measurable real-world outcomes

4. **Sustainability Mixed Goals:** For products with a mix of assets that either focus on sustainability or aim to improve sustainability or sustainable impact [56]

The SDR also introduces entity- and product-level disclosure requirements:

- **Entity-level disclosures** require firms to disclose how they are managing sustainability-related risks and opportunities. This includes disclosing information on governance arrangements, risk management processes, and sustainability metrics and targets.

- **Product-level disclosures** aim to provide investors with a detailed understanding of a product's sustainability objectives and features. Firms will need to disclose information about a product's sustainability objective, investment policy, and stewardship approach.[57]

The SDR introduces naming and marketing rules to prevent greenwashing and promote fair marketing practices.[58] Products that qualify for a sustainable investment label may use sustainability-related terms in their names, and products that do not qualify for a label are

---

[56] Id.
[57] Id.
[58] Id.

CHAPTER 7   THE GLOBAL ESG REGULATORY PARADIGM

prohibited from using such terms. Similarly, marketing rules restrict the use of sustainability terms in marketing materials for products that do not qualify for a label. Firms are required to provide clear, fair, and non-misleading information about the sustainability characteristics of their products.

By introducing sustainable investment labels and requiring comprehensive disclosures, the SDR aims to enable investors to make informed decisions and drive the growth of sustainable investment. Compliance with these regulations will be a key responsibility for firms going forward, requiring a thorough understanding of the rules and a commitment to promoting sustainability in investment practices.

Stock Exchange and Securities Rules Requiring Sustainability Disclosures

China Stock Exchanges Sustainability Reporting Guidelines

The three primary stock exchanges in China, the Shanghai Stock Exchange (SSE), Shenzhen Stock (SZSE), and the Beijing Stock Exchange (BSE), jointly announced new sustainability reporting rules for listed companies. China, the world's second-largest economy and largest emitter of greenhouse gases, joins other markets in requiring listed companies to make sustainability disclosures. China adopted the EU's CSRD standard of double materiality and focused its rule on the core reporting components of governance, risks, and impacts of the company both internally on operations and externally on the environment and broader society.

CHAPTER 7   THE GLOBAL ESG REGULATORY PARADIGM

Brazil Securities and Exchange Commission (CVM) and Ministry of Finance Rules

> *We continue to hear strong support for the ISSB's Standards from regulators globally and I commend the Brazilian Ministry of Finance and Comissão de Valores Mobiliários for providing clarity to companies and investors in Brazil by setting out a clear roadmap towards mandatory adoption.*[59]
>
> —ISSB Chair Emmanuel Faber

The Securities and Exchange Commission (CVM) and Ministry of Finance of Brazil announced that listed companies will be required to report on sustainability and climate-related topics, beginning in 2026.[60] The disclosures will require adherence to the IFRS Foundation's International Sustainability Standards Board (ISSB) standards.

Brazil was the first country to adopt the two ISSB standards (which we'll cover in Chapter 8) and incorporate them into their regulatory framework. The securities rules follow the issuance of Brazil's Ecological Transformation Plan, which is meant to drive the energy transition within the country. That plan includes billions in public and private infrastructure investment.

---

[59] *Brazil adopts ISSB global baseline, as IFRS Foundation Trustees meet in Latin America,* IFRS (October 20, 2023) (www.ifrs.org/news-and-events/news/2023/10/brazil-adopts-issb-global-baseline/).

[60] Michael Kapoor, *Brazil Becomes First Country to Adopt Global ESG Reporting Rules,* Bloomberg Law (October 20, 2023) (https://news.bloomberglaw.com/esg/brazil-becomes-first-country-to-adopt-global-esg-reporting-rules).

CHAPTER 7   THE GLOBAL ESG REGULATORY PARADIGM

Hong Kong Exchange Listing Rules

The Stock Exchange of Hong Kong (SEHK) announced the implementation date of January 1, 2025, for its Listing Rules amendments on enhanced climate-related disclosures as part of its ESG framework.[61] Previously, the Green and Sustainable Finance Cross-Agency Streeting Group committed to key priorities to advance Hong Kong as a sustainable finance leader, including aligning with the ISSB standards and scaling capital deployment to support the net-zero transition.[62] The application of consistent frameworks and the provision of capital support will speed the carbon transition.

Climate-Focused Regulation, Legislation, Exchange Rules, and Tariffs

We'll review a selection of developing climate-specific initiatives around the globe. One topic of keen interest has been the calculation and reporting of an entity's Scope 1, Scope 2, and Scope 3 emissions. Scope 1 refers to a company's direct GHG emissions from sources owned and controlled. Scope 2 refers to a company's indirect GHG emissions from purchase or use in operations, such as electricity. Scope 3 refers to indirect emissions across the company's value chain, including upstream and downstream emissions. The World Resources Institute (WRI) has developed an informational graphic (Figure 7-3) that illustrates a company's full value chain emissions and depicts the specific activities that come under Scope 1, Scope 2, and Scope 3 greenhouse gas emissions.

---

[61] *Update on Consultation on Enhancement of Climate Disclosures under ESG Framework*, Stock Exchange of Hong Kong (November 3, 2023) (www.hkex.com.hk/News/Regulatory-Announcements/2023/231103news?sc_lang=en).

[62] *Cross-Agency Steering Group announces priorities to further strengthen Hong Kong's sustainable finance ecosystem*, Hong Kong Monetary Authority (August 7, 2023) (www.hkma.gov.hk/eng/news-and-media/press-releases/2023/08/20230807-3/).

CHAPTER 7   THE GLOBAL ESG REGULATORY PARADIGM

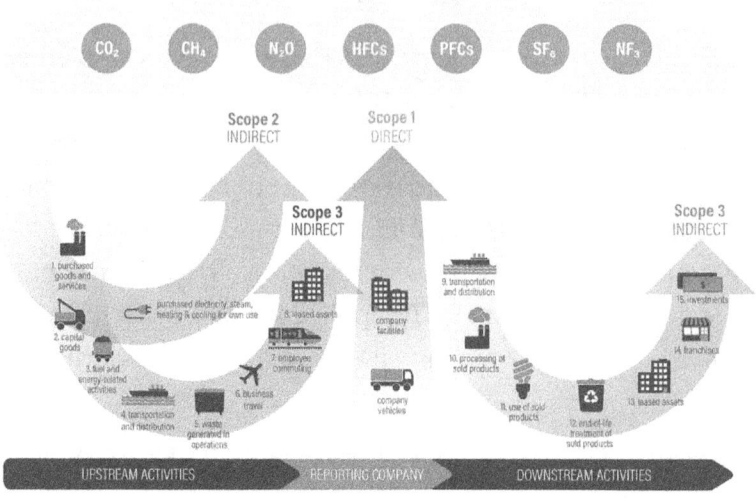

Figure 7-3. *WRI 3 Scopes Measure a Company's Full Emissions*[63]

Climate Risk and Investment in the United States
Climate Risk Disclosures and GHG Reporting
US Securities and Exchange Commission (SEC) Climate Disclosure Rule

The US Securities and Exchange Commission (SEC) promulgated its final Climate-Related Disclosure Rule on March 6, 2024.[64] In the first significant update to its initial 2010 material climate risk rule, the SEC has aimed to strike a middle ground while providing investors with a certain level of clarity on the climate risk profile of registered companies, including foreign issuers. The rulemaking significantly expands the climate reporting requirements for covered companies.

---

[63] Kyla Aiuto, Sarah Huckins, and Hannah Momblanco, *What Are Greenhouse Gas Accounting and Corporate Climate Disclosures? 6 Questions, Answered*, World Resources Institute (March 7, 2024) (www.wri.org/insights/ghg-accounting-corporate-climate-disclosures-explained).

[64] *SEC Adopts Rules to Enhance and Standardize Climate-Related Disclosures for Investors*, US Securities and Exchange Commission (March 6, 2024) (www.sec.gov/news/press-release/2024-31).

CHAPTER 7   THE GLOBAL ESG REGULATORY PARADIGM

The SEC Climate Disclosure Rule marks a significant step forward in corporate transparency, compelling publicly traded companies to disclose climate-related risks and their direct greenhouse gas emissions within their key public filings. This rule breaks new ground in the United States to provide stakeholders with a clearer understanding of climate risk impacts on businesses, toward more informed investment and voting decisions.

Set to take effect for financial years ending after December 31, 2025, for certain filers, the rule encompasses a comprehensive range of disclosures, including greenhouse gas emissions, risk management, and climate targets, aligning with global trends in ESG disclosures. The SEC received more than 24,000 comments on the proposed draft rule, amplifying the ongoing dialogue on sustainability and corporate responsibility. While many other jurisdictions and market-leading companies around the world have already committed to measuring and reporting on their full value chain emissions (Scope 1, 2, and 3), the agency walked back that requirement as had been proposed in the draft rule of March 21, 2022.[65]

Key Provisions of the Final Rule

The SEC Climate Disclosure Rule encompasses several key provisions designed to enhance transparency and provide stakeholders with critical information on how companies are addressing climate-related risks. These provisions include

1. **Greenhouse Gas Emissions Reporting**

    - Accelerated Filers (AFs) and Large Accelerated Filers (LAFs) are required to disclose Scope 1 and Scope 2 emissions data, if those emissions are material, with phased-in assurance requirements.

---

[65] *SEC Proposes Rules to Enhance and Standardize Climate-Related Disclosures for Investors*, US Securities and Exchange Commission (March 21, 2022) (www.sec.gov/news/press-release/2022-46).

CHAPTER 7 THE GLOBAL ESG REGULATORY PARADIGM

- Scope 3 emissions reporting, which pertains to indirect emissions in a company's value chain, has been eliminated from the requirement.

2. **Climate Risk Management, Governance, and Strategy**

    - Companies must disclose the material impact of climate risks (both short- and long-term risks) on their strategy, business model, and outlook.

    - Disclosure of management responsibility and board of directors oversight of material climate-related risks, including risk management processes, must be detailed.

    - Registrants must disclose climate targets and goals, if material to the business, operations, or financial condition.

    - Companies with plans addressing material transition risks must disclose details about those plans including scenario analysis and internal carbon prices if used, and quantify expenditures incurred.

3. **Financial Statement Disclosures**

    - In financial statement footnotes, companies must disclose impacts and material impacts on financial estimates and assumptions due to severe weather events.

    - Disclosures must include capitalized costs, expenditures, and losses related to carbon offsets and renewable energy credits (RECs) if they are a material component to achieving climate targets.

The rule also provides a safe harbor from liability for certain climate-related disclosures. By integrating these provisions, the SEC aims to provide investors with consistent, comparable, and decision-useful information, ensuring clear reporting requirements and fostering greater corporate accountability in the face of climate change. The rule, while scaled back from its original proposal, still represents a significant advancement in climate-related financial disclosure.

Compliance Deadlines and Phased Implementation
The SEC Climate Disclosure Rule establishes a staggered timeline for compliance, allowing companies of different sizes to adapt at a pace that aligns with their reporting capabilities. The phased implementation is as follows:

- **Large Accelerated Filers (LAFs):** These companies are at the forefront, with the first compliance period for fiscal years ending on or after December 31, 2025. This means LAFs will need to include the required disclosures in filings due by March 2026.

- **Accelerated Filers (AFs):** AFs will follow the LAFs, with their compliance starting for fiscal years ending on or after December 31, 2026, and disclosures due by March 2027.

- **Smaller Reporting Companies (SRCs), Emerging Growth Companies (EGCs), and Non-Accelerated Filers (NAFs):** These entities have an additional year to prepare, with compliance required for fiscal years ending on or after December 31, 2027, and disclosures due by March 2028.

CHAPTER 7   THE GLOBAL ESG REGULATORY PARADIGM

For **GHG emissions disclosures**, the rule introduces a phased-in approach for attestation: LAFs will not need to obtain limited assurance over their GHG emissions until fiscal years ending on or after December 31, 2029, and reasonable assurance is deferred until fiscal years ending on or after December 31, 2033.

There are **limited exceptions** for asset-backed issuers, smaller reporting companies, and emerging growth companies. The SEC considered but did not include a provision for substituted compliance for foreign private issuers in the final rule.

Comparison with Other Jurisdictions

As implementation begins to take shape, it is instructive to compare the SEC Climate Disclosure Rule with other jurisdictions that are also advancing their climate-related disclosure frameworks. The landscape of climate disclosure is becoming increasingly complex, with different regions adopting their own unique requirements.

In the European Union (EU), under the Corporate Sustainability Reporting Directive (CSRD), companies must report all greenhouse gas emissions, including Scopes 1, 2, and 3, highlighting the EU's comprehensive approach to climate transparency. It's important to note that the EU and other jurisdictions began with financial-related regulations and have now matured and moved into non-financial related and broader sustainability oversight, while the United States lags behind at the national level.

Within the United States, California got out ahead of the powers that be in Washington once again with the enactment of the Climate Corporate Data Accountability Act and the Climate-Related Financial Risk statute in 2023. The text upon passage of the California climate bills extended beyond direct emissions to the full value chain, mandating the disclosure of aggregate climate risks and emissions.

CHAPTER 7   THE GLOBAL ESG REGULATORY PARADIGM

The variations in climate disclosure norms underscore the challenge for multinational corporations, which will need to navigate and comply with multiple and potentially conflicting frameworks simultaneously. The SEC rule's reporting standards are built upon the gold-standard Greenhouse Gas Reporting Protocol (GHG Protocol) and the now-disbanded Task Force on Climate-related Financial Disclosures (TCFD) framework, which has essentially been subsumed by the International Sustainability Standards Board (ISSB). We'll cover those standards in Chapter 8. While progress has been made on international standards alignment particularly with the IFRS Sustainability Disclosure Standards developed by the ISSB, the global regulatory patchwork illustrates the continued need for harmonization.

Impact on Companies and the Broader Market

During the year the draft rule was considered and in the subsequent two years in which it was pending, the Chair of the Securities and Exchange Commission repeatedly stated that a primary driver for the rule was to provide consistent and decision-useful information to investors. Investors and financial services providers will benefit from the transparency, uniformity, and comparability of information that will result from the SEC Climate Disclosure Rule. The rule will provide stakeholders with a clearer picture of climate-related financial risks and strategies. It is also anticipated that this will lead to business operational improvements. This is a transformative period in corporate climate accountability.

The rule has incited legal challenges from certain states and industry groups, reflecting a contentious debate over the SEC's regulatory reach. It is not unusual for trade associations and states to file suits on administrative law and other grounds in an attempt to carve away or overturn a final agency action. The SEC voluntarily stayed the 886-page rule pending the outcome of consolidated challenges to the rule by the United States Court of Appeals for the Eighth Circuit and companies await a decision.

CHAPTER 7 THE GLOBAL ESG REGULATORY PARADIGM

Many global, market-leading companies have been preparing for compliance within the EU and have announced that they will be undertaking full value chain reporting. That will maintain consistency across jurisdictions and risk management, compliance, and audit function integrity for companies. The rule's ripple effect will be felt beyond registered companies as it will reach private company suppliers and vendors across supply chains, as well as asset managers and other market players.

California Climate Accountability Laws

California, the world's fifth-largest economy,[66] is leading the way in the United States on climate change regulation. The state legislature passed two groundbreaking bills—Senate Bill (SB) 253 (Climate Corporate Data Accountability Act)[67] and SB 261 (Greenhouse Gases: Climate-related Financial Risk)[68]—during its legislative session in 2023. These bills represent a turning point in corporate climate accountability and transparency at the state level, pushing businesses to align their practices with climate goals. California Governor Newsom signed both bills into law in October 2023.

California has been ahead of the federal government in enacting climate law and setting new climate policies, particularly in the energy and transportation sectors, to accomplish greenhouse gas emission reduction goals. Those two bills were part of a "Climate Accountability Package," which was a suite of bills that were introduced early in the 2023 California

---

[66] Matthew A. Winkler, *California Poised to Overtake Germany as World's No. 4 Economy,* Bloomberg (October 25, 2022) (www.bloomberg.com/opinion/articles/2022-10-24/california-poised-to-overtake-germany-as-world-s-no-4-economy).

[67] *SB-253 Climate Corporate Data Accountability Act,* California Legislative Information (October 7, 2023) (https://leginfo.legislature.ca.gov/faces/billNavClient.xhtml?bill_id=202320240SB253).

[68] *SB-261 Greenhouse gases: climate-related financial risk,* California Legislative Information (October 7, 2023) (https://leginfo.legislature.ca.gov/faces/billNavClient.xhtml?bill_id=202320240SB261).

CHAPTER 7   THE GLOBAL ESG REGULATORY PARADIGM

state legislative session. The package aimed to standardize disclosures and reporting and to improve transparency for corporate action to combat climate change.

SB 253 mandates all public and private business entities that do business in California and generate annual revenues over USD$1 billion to report their Scope 1 (direct), Scope 2 (purchase and consumption of energy), and Scope 3 (value chain) emissions. The bill stipulates that these companies will need to report their full carbon inventories, including Scope 3, which often accounts for a significant portion of an organization's climate impact. Covered companies must disclose their GHG emissions in conformance with the Greenhouse Gas Protocol. Those standards necessitate that companies take into account corporate structure changes, including divestments, mergers, and acquisitions.

SB 261 requires companies with revenues exceeding USD$500 million to disclose their climate-related financial risks in accordance with the Task Force on Climate-Related Financial Disclosures (TCFD) framework. This bill would obligate companies to disclose risk and mitigation measures related to climate change in their annual financial risk reports. It is estimated that 5,400 companies will be subject to SB 253 compliance and that 10,000 companies will be covered under SB 261.

There is a phased compliance timeline. For SB 253, companies must begin reporting their Scope 1 and 2 emissions in 2026 for the prior fiscal year and they must start reporting their Scope 3 emissions in 2027 for the prior fiscal year and annually thereafter. For SB 261, the first climate-related financial risk report must be prepared by December 31, 2024. Additionally, all reported emissions must be independently verified or assured by a third-party assurance provider.

The California Air Resources Board (CARB) is required to adopt regulations before January 1, 2025, to guide the disclosures. CARB is directed to structure their regulations to streamline reporting and meet relevant national and international reporting requirements. CARB is also authorized to adopt administrative penalties for non-filing, late filing, or other compliance failures. For SB 253, the total penalties that can be imposed on a reporting entity in a single reporting year are limited to USD$500,000, while SB 261 limits total annual penalties to USD$50,000.

In his signing statements accompanying both bills, Governor Newsom expressed concerns regarding the respective implementation times and the potential financial impacts to industry. In addition, he flagged concern over the uniform application of the GHG Protocol standard in SB 253.[69] Those concerns, along with the need to address the inclusion of the now-disbanded TCFD standard in the text of SB 261, will likely be reconciled in the 2024 California state legislative session.

The US Chamber of Commerce, the American Farm Bureau Federation, California Chamber of Commerce, Central Valley Business Federation, Los Angeles County Business Federation, and Western Growers Association filed a complaint against the State of California in the US District Court for the Central District of California challenging the two California climate accountability laws.[70] The lawsuit was filed on January 30, 2024, and its primary claims were brought on First Amendment, supremacy clause, and interstate commerce clause grounds.

Climate Investment

The Inflation Reduction Act (IRA) was predicted to catalyze the clean energy transition in the United States when it was enacted—and it has delivered on that expectation. The IRA has provided unprecedented investments in clean energy and infrastructure across industry sectors and

---

[69] *Governor Newsom Issues Legislative Update 10.7.2023*, Office of Governor Gavin Newsom (October 7, 2023) (www.gov.ca.gov/2023/10/07/governor-newsom-issues-legislative-update-10-7-23/).

[70] *Chamber of Commerce of the United States of America, et al., v. Liane M. Randolph, et al.*, Amended Complaint for Declaratory and Injunctive Relief, Case No. 2:24-cv-00801-KS, US District Court for the Central District of California (February 2, 2024) (www.uschamber.com/assets/documents/Amended-Complaint.pdf).

has spurred significant reductions in carbon emissions. A recent Princeton University research study[71] published in *Science* estimates that the IRA will cut US carbon emissions by 43%–48% (from 2005 levels) by 2035.[72]

The IRA, signed into law by President Biden on August 16, 2022, represented the largest investment in climate action in US history. The aim of the legislative clean energy and climate incentives was to stimulate the clean energy sector, to promote sustainable agriculture, and to enhance the resilience of communities to the impacts of climate change. The IRA provided tax incentives and credits for clean energy projects, which bolstered massive private investments in wind, solar, and other renewable energy technologies. The White House has prepared an IRA Guidebook to provide an overview of the available "clean energy, climate mitigation and resilience, agriculture, and conversation-related investment programs."[73]

The landmark legislation prompted a paradigm shift in the US energy landscape, igniting a clean energy revolution. In the past year and a half since passage, the country has witnessed an exponential surge in private investments in planned clean energy projects, with companies announcing over USD$270 billion worth of projects. In addition, for every single dollar the government allocated under the IRA, the private sector invested USD$5.47 alongside it, with a cumulative nearly quarter-trillion dollars flowing into the economy in just the first year after enactment of the IRA.[74]

---

[71] *New study evaluates the climate impact of the $400 billion Inflation Reduction Act*, Princeton University (July 12, 2023) (www.princeton.edu/news/2023/07/12/new-study-evaluates-climate-impact-ira).

[72] John Bistline, Geoffrey Blanford, et al. *Emissions and energy impacts of the Inflation Reduction Act,* Science (June 29, 2023) (www.science.org/doi/full/10.1126/science.adg3781).

[73] *Inflation Reduction Act Guidebook*, The White House (www.whitehouse.gov/cleanenergy/inflation-reduction-act-guidebook/).

[74] Syris Valentine, *The IRA has injected $240 billion into clean energy, The US still needs more.* Grist (March 12, 2024) (https://grist.org/economics/the-ira-has-injected-250-billion-into-clean-energy-it-might-not-be-enough/?utm_medium=social&utm_source=linktree&utm_campaign=the+ira+has+injected+%24240+billion+into+clean+energy.+the+us+still+needs+more).

CHAPTER 7   THE GLOBAL ESG REGULATORY PARADIGM

We have also witnessed a necessary change in the clean energy manufacturing sector. Upwards of 100 major clean energy manufacturing facilities have been announced in the past year, marking a significant increase compared to the rate of investment in the preceding years. It is estimated that the IRA has already created more than 170,000 clean energy jobs and accelerated adoption of hydrogen and other renewable sources, heralding a new era for renewable energy and manufacturing revitalization.[75]

The IRA has also made significant investments in forest restoration and conservation. It provided nearly USD$20 billion over five years for United States Department of Agriculture's (USDA) Natural Resources Conservation Service (NRCS) to address oversubscription in popular conservation programs. These programs aim to promote climate-smart agriculture, which integrates the dual goals of sustaining agricultural productivity and mitigating climate change. The IRA also allocated USD$5 billion in additional funding to the Forest Service for forest health treatments to protect communities from wildfires and other investments. This investment is consequential for carbon sequestration, for preserving biodiversity and ensuring the long-term sustainability of forest ecosystems.

With a spotlight on building climate resilience at the community level, the IRA provides funding for infrastructure upgrades, making communities more resilient to the impacts of climate change, such as drought, heat, and extreme weather. The Act has already delivered over USD$1 billion to communities to build resilience against the impacts of climate change. The IRA is not just about tackling climate change; it's also about driving

---

[75] *Fact Sheet: One Year In, President Biden's Inflation Reduction Act is Driving Historic Climate Action and Investing in America to Create Good Paying Jobs and Reduce Costs,* The White House (August 16, 2023) www.whitehouse.gov/briefing-room/statements-releases/2023/08/16/fact-sheet-one-year-in-president-bidens-inflation-reduction-act-is-driving-historic-climate-action-and-investing-in-america-to-create-good-paying-jobs-and-reduce-costs/).

CHAPTER 7   THE GLOBAL ESG REGULATORY PARADIGM

economic growth and job creation. The IRA is also fostering economic growth in rural communities. It provided nearly USD$11 billion for rural electric cooperatives to upgrade their energy systems and enhance the quality of life in rural communities. This investment is creating jobs and lowering energy costs in rural areas, contributing to their economic growth and development.

The IRA has made significant strides in advancing a clean economy in the United States. In 2023, 42GW of new renewable power-generating capacity was added to the grid, and 8.85% of total US energy demand and 25% of electricity demand was met by renewable energy.[76] While challenges remain, the progress made thus far is encouraging. As the IRA continues to be implemented and cabinet departments like the US Department of the Treasury's Internal Revenue Service (IRS), US Department of Energy (DOE), and USDA and agencies like EPA continue to roll out programs and guidelines, it holds the promise of driving further advancements in clean energy and infrastructure, climate-smart agriculture, and sustainable development, and shaping the transition toward a more sustainable and resilient future.

CASE STUDY: US DEPARTMENT OF ENERGY VOLUNTARY CARBON DIOXIDE REMOVAL CHALLENGE

The US Department of Energy (DOE) has been the lead agency implementing loan and other programs pursuant to the IRA. The agency announced a new Voluntary Carbon Dioxide Removal Challenge and launched a public–private partnership aimed "to catalyze carbon dioxide removal credit purchases

---

[76] *The 2024 Sustainable Energy in America Factbook*, BloombergNEF (February 28, 2024) (https://about.bnef.com/blog/the-2024-sustainable-energy-in-america-factbook/).

CHAPTER 7   THE GLOBAL ESG REGULATORY PARADIGM

and improve transparency of the carbon dioxide removal credit supply."[77] Organizations will purchase high-quality carbon dioxide removal credits through the program and DOE will support projects to remove, store, and utilize carbon from the atmosphere.[78]

Google was part of the DOE program announcement and committed to match the agency's initial outlay of USD$35 million in carbon dioxide removal (CDR) credits as part of the challenge.[79]

French Climate and Resilience Law

France enacted its Climate and Resilience Law in 2021.[80] This comprehensive legislation aims to accelerate the transition toward a carbon-neutral and resilient society by implementing new regulations and stringent limitations. It was passed after extensive deliberations and discussions with stakeholders from various sectors.

---

[77] *U.S. Department of Energy Announces Intent to Launch Voluntary Carbon Dioxide Removal Purchasing Challenge*, US Department of Energy (March 14, 2024) (www.energy.gov/fecm/articles/doe-helping-you-buy-good-carbon-dioxide-removal-credits#:~:text=To%20that%20end%2C%20we're,to%20have%20the%20greatest%20impact).

[78] *Notice of Intent Regarding Launching a Voluntary Carbon Dioxide Removal Purchasing Challenge; DOE Carbon Dioxide Removal Purchasing (CO2RP) Challenge*, Federal Register (March 14, 2024), 89 FR 18626 (www.federalregister.gov/documents/2024/03/14/2024-05269/notice-of-intent-regarding-launching-a-voluntary-carbon-dioxide-removal-purchasing-challenge-doe).

[79] Id.; *Our pledge to support carbon removal solutions, Company News*, Google (March 14, 2024) (https://blog.google/outreach-initiatives/sustainability/pledge-to-support-carbon-removal-solutions/).

[80] *Law of August 22, 2021 on the fight against climate change and strengthening resilience to its effects*, French Republic (August 24, 2021) (www.vie-publique.fr/loi/278460-loi-22-aout-2021-climat-et-resilience-convention-citoyenne-climat).

The law's scope extends beyond just reducing greenhouse gas emissions. It also focuses on the social and environmental implications of products and services throughout their life cycle. This holistic approach considers factors like biodiversity, water resources, and the overall environmental externalities of manufacturing systems. The law encompasses a broad spectrum of sustainability-related topics, including mandatory environmental labeling for certain products and services, as we discussed within the circular economy section of this chapter.

The Climate and Resilience Law bans the use of any language in advertising that suggests a product, service, or activity is carbon-neutral or has no negative impact on the climate. Exceptions will only be made for claims based on certifications that comply with recognized norms and standards at the French, European, and international levels.[81] It also imposes a ban on advertising related to the marketing or promotion of fossil fuels.[82]

European Union Carbon Border Adjustment Mechanism and Carbon Removal Certification System

Carbon Border Adjustment Mechanism

In addition to the comprehensive sustainability regulations discussed in this chapter, it is important for businesses to take note of the EU's Carbon Border Adjustment Mechanism (CBAM).[83] It is a landmark tool in the EU's green strategy, and it is poised to revolutionize the global carbon economy. The CBAM is a tariff designed for carbon-intensive products aimed at preventing carbon leakage, a phenomenon where production relocates to countries with less stringent environmental standards.

---

[81] Id.

[82] Id.

[83] *Carbon Border Adjustment Mechanism,* European Commission (https://taxation-customs.ec.europa.eu/carbon-border-adjustment-mechanism_en).

CHAPTER 7   THE GLOBAL ESG REGULATORY PARADIGM

It is also a supplementary measure to the EU ETS, as discussed in Chapter 4, as it imposes an adjusted charge on certain imports equal to that imposed on domestic goods under the ETS. By making imports subject to the same carbon price as domestic products, the CBAM aims to level the playing field and bolster the EU's climate policies. The CBAM is part of the EU's Fit for 55 Agenda under the Green Deal, which aims to reduce greenhouse gas (GHG) emissions by 55% by the year 2030.[84] To achieve this goal, the EU has developed a robust framework to ensure the equalization of carbon prices between domestic products and imports.

The CBAM came into effect in its transitional phase on October 1, 2023. During this phase, it applies only to imports of specific goods that account for a significant amount of GHG emissions, including cement, iron and steel, aluminum, fertilizers, electricity, and hydrogen. While no financial adjustment is required at this stage, EU importers must report on the volume of their imports and the GHG emissions from their production. The initial reporting phase covered data for the fourth quarter of 2023, with reports due to be submitted by January 31, 2024.

To support EU importers and non-EU companies in implementing the new rules, the European Commission launched a new CBAM transitional registry. This registry helps importers calculate and report the emissions associated with their products. The European Commission has made detailed guidance, training materials, webinars, sector-specific factsheets, and a step-by-step checklist available to businesses. The transitional period is meant to serve as a learning opportunity for all stakeholders involved. The European Commission will collect information on embedded emissions to refine the methodology to be used in the definitive period, which is slated to begin in 2026. Once that period begins, importers will purchase and surrender CBAM certificates according to the emissions profile of the imported goods.

---

[84] *European Green Deal, Fit for 55*, Council of the European Union (www.consilium.europa.eu/en/policies/green-deal/fit-for-55-the-eu-plan-for-a-green-transition/).

CHAPTER 7   THE GLOBAL ESG REGULATORY PARADIGM

The scope of the CBAM may also be expanded to include other goods produced in ETS sectors, pending review of the transitional period. The implementation of the CBAM has significant implications for global trade, particularly for businesses importing goods into the EU. With the CBAM in place, businesses must understand their emissions impact and make necessary adjustments to comply with the new regulations. Compliance requires the submission of detailed carbon emissions and energy consumption data that has undergone independent verification. Non-compliance with the measure exposes businesses to a range of potential sanctions, including monetary fines, product seizures, and market exclusion.

Carbon Removal Certification System

The European Parliament and Commission reached a provisional agreement on an EU-wide Carbon Removal Certification System, which is a framework designed to scale carbon removal activities and prevent greenwashing in the process.[85] The voluntary system is designed to support the climate neutrality by 2050 goal by reducing greenhouse gas emissions and compensating for residual emissions through carbon removals, which involves removing carbon dioxide (CO_2) from the Earth's atmosphere using natural and technological solutions.

The new framework covers various methods of carbon removal and provides that high-quality carbon removals must meet specific criteria to receive certification. These criteria include correct quantification, delivery of additional climate benefits, long-term carbon storage, prevention of carbon leaks, and contribution to sustainability. The system also includes requirements for third-party verification and certification to ensure environmental integrity and build public trust.

---

[85] *Commission welcomes political agreement on EU-wide certification scheme for carbon removals*, European Commission (February 20, 2024) (https://ec.europa.eu/commission/presscorner/detail/en/ip_24_885).

The certification process involves two main steps: (1) the establishment of high-level quality criteria by the EC under the proposed Regulation, and (2) the approval of detailed certification rules for the measurement, monitoring, reporting, and verification of carbon removals from both industrial and nature-based activities. Operators of carbon removal activities must apply to a recognized or approved public or private certification program and undergo regular verification and certification by independent bodies to ensure compliance with EU rules. The process results in the issuance of certificates of compliance and the recording of carbon removal units in public registries managed by certification schemes. The system aims to ensure transparency and environmental integrity and prevent negative impacts on biodiversity and ecosystems. It is intended to provide assurance about the quality of carbon removals and make the certification process reliable and trustworthy to combat greenwashing.

Singapore Climate-Related Reporting Rules

The Parliament of Singapore has announced it will introduce new, mandatory climate-related reporting rules for both listed and large non-listed companies.[86] The rules were proposed in 2023 by the Sustainability Reporting Advisory Committee, established by the Accounting and Corporate Regulatory Authority (ACRA) and Singapore Exchange Regulations (SGX RegCo), and expand upon the 2022 climate reporting rules for all issuers required by the Singapore Exchange.[87]

---

[86] *Climate reporting to help companies ride the green transition*, News, SGX Group (February 28, 2024) (www.sgxgroup.com/media-centre/20240228-climate-reporting-help-companies-ride-green-transition).

[87] Sustainability Reporting, Singapore Exchange (www.sgx.com/sustainable-finance/sustainability-reporting).

CHAPTER 7   THE GLOBAL ESG REGULATORY PARADIGM

The rules will be implemented in a phased approach to begin with the requirement that listed companies will first report in 2025 and followed by large, non-listed companies having at least SGD$1 billion in revenue and SGD$500 million in assets by 2027. Grants will be made available to reporting companies to assist with the costs of reporting via a grant program administered by the Singapore Economic Development Board (EDB) and Enterprise Singapore (EnterpriseSG).

The Singapore rules will require disclosure in accordance with the IFRS International Sustainability Standards Board (ISSB) framework. That trend is expected to continue as nations adopt and revise their sustainability reporting requirements.

Australia Climate-Related Financial Disclosures

The Australian Government has released and completed an exposure draft of new legislation, entitled "Treasury Laws Amendment Bill 2024: Climate related financial disclosure."[88] The Australian Treasury originally proposed a consultation paper on climate-related financial disclosure in 2022. If enacted, the new statute would require disclosures for large companies and registered companies, with the first requirements coming online for fiscal years commencing between July 1, 2024, and June 30, 2026.[89] The bill contains specific requirements for sustainability reporting and aligns to the IFRS standards. it is expected to be introduced to the Parliament of Australia later in 2024.

---

[88] *Climate-related financial disclosure: exposure draft legislation*, The Treasury, Australian Government (January 12, 2024-February 9, 2024) (https://treasury.gov.au/consultation/c2024-466491).
[89] Id.

269

CHAPTER 7   THE GLOBAL ESG REGULATORY PARADIGM

Executive Compensation

On August 25, 2022, the Securities and Exchange Commission (SEC) adopted final rules on the pay versus performance disclosure requirement pursuant to the Dodd-Frank Act.[90] According to a statement from SEC Chair Gary Gensler following its adoption, the pay versus performance rule "will strengthen the transparency and quality of executive compensation disclosure to investors."[91]

For clarity, SEC staff at the Division of Corporation Finance updated its Compliance & Disclosure Interpretations of Regulation S-K in February 2023, responding to certain practical and implementation questions about the new pay versus performance requirements.[92] Registrants were required to begin complying by including such compensation information in proxy and information statements for fiscal years ending on December 16, 2022, and thereafter. The rule standardizes issuer disclosures. The SEC wants registrants to align their pay versus performance disclosures with the agency's new rules, with an eye toward improving transparency and corporate accountability by providing investors with information that will help them assess registrants' executive compensation practices when, for example, exercising their right to cast advisory say-on-pay votes.

This pay versus performance rule makes it easier for investors and proxy advisors to conduct apples-to-apples comparisons when they analyze executive compensation. Public companies are under a new microscope to align compensation with market and industry

---

[90] *SEC Adopts Pay Versus Performance Disclosure Rules*, United States Securities and Exchange Commission (August 25, 2022) (www.sec.gov/news/press-release/2022-149).

[91] *Statement by Chair Gensler on Final Rule Regarding Pay Versus Performance*, Harvard Law School Forum on Corporate Governance (August 26, 2022) (https://corpgov.law.harvard.edu/2022/page/29/).

[92] *Compliance and Disclosure Interpretations*, United States Securities and Exchange Commission (www.sec.gov/corpfin/cfnew2023-02).

standards, as well as shareholder expectations. This level of disclosure requires companies to benchmark company performance and executive compensation and ultimately justify compensation decisions, often in investor relations meetings and during the annual shareholder meeting season.

Human Capital Management

A business's ability to attract, retain, and develop its workforce can significantly influence its success. In its Guidance on Human Capital Management Disclosures & Practices,[93] State Street Global Advisors cites J.S. Held research that "intangible assets including human capital comprise 90% of market value in the S&P 500."[94] Many of the leading ESG standards organizations (some of which have recently been consolidated) have long-included human capital reporting requirements, including the Global Reporting Initiative (GRI) and Sustainability Accounting Standards Board (SASB). Recognizing the importance of HCM metrics to the investor community and knowing that many companies were already measuring and reporting data, regulators worldwide have increasingly emphasized human capital disclosures in financial statements.

Understanding these developing regulatory requirements can help businesses understand and navigate the complex ESG landscape. In the United States, the US Securities and Exchange Commission has implemented specific regulations mandating these disclosures from publicly traded companies. The SEC introduced a landmark amendment to Regulation S-K, Items 101, 103, and 105, effectively modernizing HCM

---

[93] *Guidance on Human Capital Management Disclosures & Practices*, State Street Global Advisors (May 2023) (www.ssga.com/library-content/pdfs/asr-library/proxy-voting-human-capital-disclosure.pdf).
[94] *Intangible Asset Market Value Study*, Ocean Tomo, A Part of JS Held (https://oceantomo.com/intangible-asset-market-value-study/).

CHAPTER 7   THE GLOBAL ESG REGULATORY PARADIGM

disclosure regulations for publicly traded companies, which became effective on November 9, 2020.[95] This rule, known as the SEC Human Capital Disclosure Rule, mandates companies to provide more detailed information about their workforce management.

The SEC utilizes a principles-based approach, allowing companies some flexibility in reporting relevant issues and data points that directly relate to their business context. Key requirements of the SEC Human Capital Disclosure Rule include

- **Human capital objectives:** Companies are required to disclose information about how the company manages, develops, and retains its workforce.

- **Number of employees:** Companies are required to disclose the number of full-time, part-time, and seasonal employees.

- **Workplace health and safety:** Companies are required to disclose their policies and practices related to workplace health and safety.

- **Workforce diversity, equity, and inclusion (DEI):** Companies are required to disclose their DEI policies and practices, along with any related training and development programs.

- **Workforce composition:** Companies are required to disclose information on employee turnover rates and the number of contingent workers required.

---

[95] *Modernization of Regulation S-K Items 101, 103, and 105: A Small Entity Compliance Guide*, United States Securities and Exchange Commission (www.sec.gov/corpfin/modernization-regulation-s-k-compliance-guide#:~:text=On%20August%2026%2C%202020%2C%20the,make%20pursuant%20to%20Regulation%20S%2DK).

- **Additional necessary information:** Companies are directed to disclose any other information they deem necessary to the understanding of their workforce and any related risks or opportunities.[96]

The SEC disclosure rule cemented the importance of the "S" in ESG, underscoring the social aspect of running a successful business. The rule sets expectations for what companies should be doing on behalf of their workforces and emphasizes transparency and accountability in human capital management. The agency informally said it would begin to draft guidance to clarify the disclosure requirements in 2021.

In September 2023, the SEC's Investor Advisory Committee voted to recommend that the Commission write a rule requiring publicly traded companies to provide more relevant information about their workforce, including details on the category of workers (e.g., full-time or part-time employee, or contractor); turnover or comparable workforce data; total cost of the workforce; and workforce demographic data.[97]

Given the lack of maturity of the disclosure regulations, complying with the SEC's Human Capital Disclosure Rules can be challenging but is required. Companies should assess their current human capital disclosure practices and align them with the rules. This includes evaluating the company's mechanisms to collect, track, and report workforce data and reviewing the execution of workforce and human capital management policies. Companies must stay abreast of HCM regulations, ensuring their disclosures align with regulatory requirements and stakeholders' expectations.

---

[96] Id.

[97] *Draft Recommendation of the SEC Investor Advisory Committee's Investor-as-Owner Subcommittee regarding Human Capital Management Disclosures*, US Securities and Exchange Commission (September 14, 2023) (www.sec.gov/files/20230914-draft-recommendation-regarding-hcm.pdf).

CHAPTER 7   THE GLOBAL ESG REGULATORY PARADIGM

Supply Chain and Human Rights Regulation

Governments around the world are implementing regulations that require companies to investigate their supply chains for human rights abuses and other unethical practices. A number of countries have passed laws against human rights abuses. Some of the earlier laws, like the UK Modern Slavery Act, require transparency and reporting across supply chains but do not require issue remediation.[98]

As we've discussed, the proposed EU Corporate Sustainability Due Diligence Directive (CSDDD) establishes mandatory due diligence obligations to ensure responsible supply chains. The European Council and European Parliament also recently announced a provisional agreement on the regulation prohibiting products made with forced labor from entering the EU market. In announcing the provisional deal, the Council noted in its press release that "[r]oughly 27.6 million people are in forced labour around the world, in many industries and in every continent."[99]

A number of the EU member states have also enacted national human rights regulations that require remediation. The German Supply Chain Due Diligence Act requires companies to establish a risk management system to identify, mitigate, and prevent human rights and environmental risks in their supply chains.[100] The Dutch Child Labor Due Diligence Act

---

[98] *Modern Slavery Act*, Government of the United Kingdom (March 26, 2015) (www.legislation.gov.uk/ukpga/2015/30/enacted).

[99] *Council and Parliament strike a deal to ban products made with forced labour*, Council of the European Union (March 5, 2024) (www.consilium.europa.eu/en/press/press-releases/2024/03/05/council-and-parliament-strike-a-deal-to-ban-products-made-with-forced-labour/).

[100] *The German Act on Corporate Due Diligence Obligations in Supply Chains*, Federal Ministry for Economic Cooperation and Development, Federal Republic of Germany (www.bmz.de/resource/blob/154774/lieferkettengesetz-faktenpapier-partnerlaender-eng-bf.pdf).

CHAPTER 7   THE GLOBAL ESG REGULATORY PARADIGM

requires companies to perform due diligence and identify and address child labor across their supply chains.[101]

The Duty of Vigilance Law was enacted in France in 2017.[102] It is a comprehensive statute that requires large companies to both identify and prevent human rights and environmental abuses across their entire supply chains.[103] Under the law, companies are required to publish annual "vigilance plans" that identify risk management measures and are subject to consequences for compliance failures.

In the United States, the California Transparency in Supply Chains Act, enacted in 2015, requires retail sellers or manufacturers doing business in the State of California with annual worldwide gross receipts in excess of USD$100 million to disclose their efforts to eradicate forced labor and human trafficking from their supply chains.[104] Forced labor abuses and violations of fair labor standards led the US Congress to enact and President Biden to sign the Uyghur Forced Labor Protection Act (UFLPA) into law on December 23, 2021.[105] The law is aimed at prohibiting

---

[101] Anneleos Hoff, *Dutch child labour due diligence law: a step towards a mandatory human rights due diligence*, Oxford Human Rights Hub (June 10, 2019) (https://ohrh.law.ox.ac.uk/dutch-child-labour-due-diligence-law-a-step-towards-mandatory-human-rights-due-diligence/). See also *Staatsbland van het Koninkrijk der Nederlanden, 2019, 401*, Ministerie van Buitenlandse Zaken (November 13, 2019) (https://zoek.officielebekendmakingen.nl/stb-2019-401.html).

[102] *Law no. 2017-399 of March 27, 2017 relating to the duty of vigilance of parents companies and ordering companies*, French Republic (March 28, 2017) (www.legifrance.gouv.fr/jorf/id/JORFTEXT000034290626/).

[103] *Law no. 2017-399 of March 27, 2017 relating to the duty of vigilance of parents companies and ordering companies*, French Republic (March 28, 2017) (www.legifrance.gouv.fr/jorf/id/JORFTEXT000034290626/).

[104] *The California Transparency in Supply Chains Act*, State of California Department of Justice (https://oag.ca.gov/SB657).

[105] *Public Law 117-78-Dec. 23, 2021*, 117th Congress (www.congress.gov/117/plaws/publ78/PLAW-117publ78.pdf).

CHAPTER 7   THE GLOBAL ESG REGULATORY PARADIGM

the importation to the United States of goods manufactured with forced labor in the People's Republic of China, particularly from the Xinjiang region.[106]

Circular Economy and Greenwashing Regulation
Circular Economy Regulation

The European Union enacted a circular economy action plan in March 2020.[107] It has been described as one of the building blocks of the European Green Deal, which we discussed earlier in this chapter.[108]

The European Parliament and Council reached an agreement pertaining to a new regulation to make consumer goods more durable on December 5, 2023.[109] The Ecodesign and Sustainable Products Regulation proposal outlines a framework for new requirements on product durability, energy and resource efficient, recycled content, remanufacturing and recycling, and carbon footprint, among other circular topics.[110] The announcement of the agreement included a statement by Kadri Simson, European Commissioner for Energy, that the circular regulation was reaping "billions of euros in consumer savings."[111]

---

[106] Id.

[107] *Circular economy action plan*, European Commission (https://environment.ec.europa.eu/strategy/circular-economy-action-plan_en).

[108] Id.

[109] *Commission welcomes provisional agreement for more sustainable, repairable and circular products*, European Commission (December 5, 2023) (https://ec.europa.eu/commission/presscorner/detail/en/ip_23_6257).

[110] *Ecodesign for Sustainable Products Regulation*, European Commission (https://commission.europa.eu/energy-climate-change-environment/standards-tools-and-labels/products-labelling-rules-and-requirements/sustainable-products/ecodesign-sustainable-products-regulation_en).

[111] *Commission welcomes provisional agreement for more sustainable, repairable and circular products*, European Commission (December 5, 2023) (https://ec.europa.eu/commission/presscorner/detail/en/ip_23_6257).

CHAPTER 7   THE GLOBAL ESG REGULATORY PARADIGM

The European Parliament voted 514-20 to adopt a set of rules to reduce waste from both the textiles and the food sectors on March 13, 2024.[112] The vote amends relevant provisions of the EU Waste Framework. The new rules include Extended Producer Responsibility (EPR) requirements, which are characteristic of other manufacturing industries. Those provisions include requiring textile producers and fashion companies to pay for the collection and recycling of clothing and footwear and setting binding food waste reduction targets among EU member states. In its press release announcing the adoption, Parliament noted that less than 1% of textiles are recycled worldwide, and every year, the EU generates 60 million metric tons of food waste and 12.6 million metric tons of textile waste.[113]

European Union member state France has also taken a bold step forward in minimizing waste and promoting a sustainable, circular economy. With the adoption of the Anti-Waste and Circular Economy Law in 2020, the nation has taken a systems-solution approach to tackle global challenges like climate change, biodiversity loss, waste, and pollution.[114] France focused on the link between waste generation and pollution of the natural environment, which threatens biodiversity.

The French law attempts to eliminate waste beginning in the design phase and to transform the entire product cycle into a circular model. The law encourages businesses across various sectors to eliminate waste and adopt more circular practices. It takes a holistic view of design, production, consumption, and turning end of use into reuse. It encourages durable and recyclable products and promotes smarter use. For example, it aims to phase out all single-use plastic packaging by 2040.

---

[112] *MEPs call for tougher EU rules to reduce textiles and food waste*, News, European Parliament (March 13, 2024) (www.europarl.europa.eu/news/en/press-room/20240308IPR19011/meps-call-for-tougher-eu-rules-to-reduce-textiles-and-food-waste).

[113] Id.

[114] *Anti-Waste Circular Economy Act: Measures in Place and Coming*, French Republic (www.service-public.fr/particuliers/actualites/A16390?lang=en).

CHAPTER 7   THE GLOBAL ESG REGULATORY PARADIGM

There are a number of localities worldwide that have recently undertaken waste reduction and more comprehensive circular economic initiatives. Amsterdam, the capital city of the Netherlands, is one such example. The city became the first in the world to commit to being a 100% circular economy by 2050, and it has released a Circular Strategy setting out more than 70 actions the city plans to carry out pursuant to that goal from 2023 to 2026.[115]

Greenwashing Regulation

As greenwashing has become more prevalent, regulators worldwide have begun to take a tougher stance on the practice. From consumer protection laws to advertising standards, a multitude of legal frameworks are being applied to squash greenwashing.

European Union

Green Claims

> *It is time to put an end to greenwashing. Our position ends the proliferation of misleading green claims that have deceived consumers for far too long. We will ensure businesses have the right tools to embrace genuine sustainability practices. European consumers want to make sustainable choices; all those offering products or services must guarantee their green claims are scientifically verified.*[116]
>
> —Cyrus Engerer,
> European Parliament Environment
> Committee Rapporteur

---

[115] *Policy: Circular economy*, City of Amsterdam (www.amsterdam.nl/en/policy/sustainability/circular-economy/).

[116] *Parliament wants to improve consumer protection against misleading claims*, News, European Parliament (March 12, 2024) (www.europarl.europa.eu/news/en/press-room/20240308IPR19001/parliament-wants-to-improve-consumer-protection-against-misleading-claims#:~:text=The%20green%20claims%20directive%20would,having%20%E2%80%9Cbio%20based%20content%E2%80%9D).

CHAPTER 7   THE GLOBAL ESG REGULATORY PARADIGM

The European Union has been proactive in its fight against greenwashing. In September 2023, the European Council and European Parliament reached a provisional agreement on new rules to ban misleading advertisements,[117] and it was approved in November 2023.[118] This move follows on the European Commission's proposed Green Claims Directive, which was published in March 2023. The Green Claims Directive is designed to prevent greenwashing and to improve reliability of information provided to consumers.[119]

Two studies conducted by the European Commission on a sample of 150 environmental claims found that 53.3% provided vague, misleading, or unfounded information, and those results were cited in the text of the Green Claims Directive proposal.[120] The new rules aim to harmonize the minimum requirements for making and substantiating environmental claims. This also serves a competitive purpose by requiring businesses to play by the same marketing rules. The rules seek to establish transparent environmental labeling schemes that are regularly reviewed and independently verified. The European Parliament voted to approve the set of new rules by a vote of 467-65 on March 12, 2024.[121]

---

[117] *EU to ban greenwashing and improve consumer information on product durability*, European Parliament (September 19, 2023) (www.europarl.europa.eu/news/en/press-room/20230918IPR05412/eu-to-ban-greenwashing-and-improve-consumer-information-on-product-durability).

[118] *Parliament backs new rules for sustainable, durable products and no greenwashing*, European Parliament (November 5, 2023) (www.europarl.europa.eu/news/en/press-room/20230505IPR85011/parliament-backs-new-rules-for-sustainable-durable-products-and-no-greenwashing).

[119] *Directive of the European Parliament and of the Council*, 2023/0085 (COD), European Commission, Brussels (March 23, 2023) (https://eur-lex.europa.eu/legal-content/EN/TXT/PDF/?uri=CELEX:52023PC0166).

[120] Id.

[121] Id.

CHAPTER 7   THE GLOBAL ESG REGULATORY PARADIGM

Green Bond Standard

The European Council adopted a voluntary standard on the use of a European green bond label on October 24, 2023.[122] The standard aligns with the EU taxonomy and aims to prevent greenwashing in sustainable finance.[123] Organizations are required to disclose substantive information about the bond when using the label for marketing purposes. The regulation also establishes a registration system and a supervisory framework for the independent entities that conduct external reviews of European green bonds. All proceeds of European green bonds must be invested in economic activities pursuant to the EU taxonomy for sustainable activities.

France

In addition to legislation, France has also established guidelines to combat greenwashing. The French General Department of Competition, Consumer Affairs and Fraud Control updated its *Practical Guide to Environmental Claims*, which provides a framework for acceptable claims and identifies those prohibited or strictly regulated.[124] This guidance is a complement to and classification of the environmental claim provisions in the Circular Economy Law and to the Climate and Resilience Law. France has also imposed legal sanctions to deter companies from resorting to greenwashing.

---

[122] *European Green Bonds: Council adopts new regulation to promote sustainable finance*, European Council (October 24, 2023) (www.consilium.europa.eu/en/press/press-releases/2023/10/24/european-green-bonds-council-adopts-new-regulation-to-promote-sustainable-finance/).

[123] *European Green Bonds: Council adopts new regulation to promote sustainable finance*, European Council (October 24, 2023) (www.consilium.europa.eu/en/press/press-releases/2023/10/24/european-green-bonds-council-adopts-new-regulation-to-promote-sustainable-finance/).

[124] Practical Guide to Environmental Claims, French Republic (2023) (www.economie.gouv.fr/files/files/directions_services/dgccrf/documentation/publications/publications_externes/bro-guide-cnc-VF.pdf?v=1685109324).

CHAPTER 7 THE GLOBAL ESG REGULATORY PARADIGM

United Kingdom

The UK has also been making strides in addressing greenwashing. The UK's regulatory framework for environmental claims is primarily guided by the CMA's Green Claims Code.[125] The code emphasizes that businesses must ensure their claims are truthful and accurate, clear and unambiguous, and substantiated. The UK Advertising Standards Authority (ASA) has investigated marketing claims by many companies. For example, the ASA banned a television advertisement for Persil, a brand owned by Unilever, based on the claim that the product was "kinder to our planet."[126]

United States

There has been a patchwork of activity on greenwashing in specific sectors within the United States. The SEC has adopted a rule to crack down on greenwashing practices in investment funds. The "Names Rule" aims to quelch deceptive and misleading marketing practices, particularly related to funds with ESG-related names.[127] As we'll discuss in Section 7.6, the US Federal Trade Commission has undertaken to update its *Green Guides for the Use of Environmental Marketing Claims*, which pertain to the consumer market.[128] At the state level, California enacted AB 1305, the

---

[125] *Green claims code: making environmental claims*, Government of the United Kingdom (www.gov.uk/government/publications/green-claims-code-making-environmental-claims).

[126] *Persil advert banned for misleading green claims*, BBC (August 30, 2022) (www.bbc.com/news/business-62726666).

[127] *SEC Adopts Rule Enhancements to Prevent Misleading r Deceptive Investment Fund Names*, US Securities and Exchange Commission (September 20, 2023) (www.sec.gov/news/press-release/2023-188).

[128] *Federal Trade Commission Extends Public Comment Period on Potential Updates to its Green Guides for the Use of Environmental Marketing Claims*, US Federal Trade Commission (January 31, 2023) (www.ftc.gov/news-events/news/press-releases/2023/01/federal-trade-commission-extends-public-comment-period-potential-updates-its-green-guides-use).

Voluntary Carbon Market Disclosures Business Regulations Act (VCMDA), in 2023, which became effective on January 1, 2024.[129] The legislation, as codified, applies to certain businesses operating in California that make *net-zero*, *carbon-neutral*, and similar emissions reduction claims.

Government and Stock Exchange Board Diversity Rules

Governments and stock exchanges have begun to understand the value of diversity on boards of directors and have taken steps to promote that diversity. The European Union and the United Kingdom have passed significant mandates for diversity on boards. The mandate model was enacted in California, but that ultimately failed as a result of judicial challenges. The Republic of Korea is another example of a nation that has enacted a board diversity requirement in an effort to increase the representation of women on corporate boards.

Select exchanges around the world have set gender diversity rules for boards of listed companies. Those rules typically apply a mandate model for a minimum representation of women. A major listing initiative in the United States by NASDAQ that focuses on transparency, rather than a mandate, has so far been successful and has survived challenges to roll it back from diversity opponents.

European Union

The European Union has taken a decisive step to ensure gender balance in corporate leadership roles. The *EU Gender Equality Strategy 2020-2025* sets out objectives and action steps toward achieving gender equality

---

[129] *AB-1305 Voluntary carbon market disclosures*, California Legislative Information (October 9, 2023) (https://leginfo.legislature.ca.gov/faces/billTextClient.xhtml?bill_id=202320240AB1305).

across the continent.¹³⁰ The EU's *Women on Boards Directive* is designed to improve the representation of women on the boards of listed companies.¹³¹ Under the directive, larger listed companies registered in an EU member state with more than 250 employees must strive to meet the goal of women representing 40% of their board among non-executive directors or 33% among all directors by June 30, 2026, with certain exceptions.¹³² Companies must also report their progress against these targets annually on their company websites and to local authorities.

Over the years, an increasing number of EU member states have introduced statutory quotas on their company boards to promote gender balance. Despite these efforts, progress has been slow and uneven as women continue to be underrepresented on corporate boards.

The Directive went into effect on December 27, 2022, and all EU member states have until December 28, 2024, to transpose its provisions into national legislation.¹³³

United Kingdom

The UK Financial Conduct Authority (FCA) implemented new diversity reporting requirements for premium and standard listed companies, beginning in 2022.¹³⁴ Covered companies must include a statement in

---

[130] *EU action to promote gender balance in decision-making*, European Commission (https://commission.europa.eu/strategy-and-policy/policies/justice-and-fundamental-rights/gender-equality/equality-between-women-and-men-decision-making/eu-action-promote-gender-balance-decision-making_en).

[131] *Directive (EU) 2022/2381 of the European Parliament and of the Council of 23 November 2022 on improving the gender balance among directors of listed companies and related measures*, Official Journal of the European Union https://eur-lex.europa.eu/legal-content/EN/TXT/PDF/?uri=CELEX:32022L2381).

[132] Id.

[133] Id.

[134] *PS22/3: Diversity and inclusion on company boards and executive management*, UK Financial Conduct Authority (April 20, 2022) (www.fca.org.uk/publications/policy-statements/ps22-3-diversity-inclusion-company-boards-executive-managment).

CHAPTER 7   THE GLOBAL ESG REGULATORY PARADIGM

their annual reports as to whether they have achieved 40% women on their boards and at least one director from a diverse ethnic background, and they must also disclose data on the gender and ethnic diversity of their boards and executive management positions. This disclosure model is similar to the *comply or explain* approach taken by NASDAQ.

A 2022 study of FTSE350 companies (the Financial Times Stock Exchange is a share index of the top companies listed on the London Stock Exchange with the highest market) found that corporations with more than 25% of women on their board's executive committees equated to a profit margin of 16%—which was more than ten times higher than those with no female board members. The researchers extrapolated that number and reported that if companies were to have women make up a quarter of all board members, then the UK economy would receive a boost of an additional GBP£67 billion (equivalent of USD$54 billion)—representing a 2.5% bump to the GDP—and each company would realize an extra GBP£900 million (equivalent of USD$1.1 billion) in pre-tax profit. The report concluded that equivalency meant "[e]ach year, the UK is losing the equivalent of more than the [defense] budget, the entire schools budget, and triple the police budget, because of gender imbalance at the top of our companies."[135]

Republic of Korea

The Republic of Korea amended the Financial Investment Services and Capital Markets Act in 2020, to include board diversity requirements. According to MSCI, women held 12.8% of director seats on corporate boards in South Korea in 2022.[136] The revised law, which went into effect in 2022,

---

[135] *Women Count 2022: The Role, Value, and Number of Female Executives in the FTSE 350, Strength Through Diversity*, The Pipeline (May 2022) (https://d2e9sck8yc9g7g.cloudfront.net/wp-content/uploads/2019/09/06103843/Women-Count-2022-FINAL.pdf).

[136] Tanya Matanda, Carrie Wang, and Olga Emelianova, *Women on Board 2022 Progress Report*, MSCI (www.msci.com/documents/10199/36771346/Women_on_Boards_Progress_Report_2022.pdf).

requires Korean companies with assets in excess of KRW₩ 2 trillion to have at least one woman on their boards of directors.[137] That threshold is anticipated to cover nearly 100 of the largest companies in Korea, many of which are global household brands.

United States

The US Court of Appeals for the Fifth Circuit recently upheld the NASDAQ Board Diversity Rule.[138] This rule, aimed at public disclosure and transparency, stipulates that companies listed on the NASDAQ stock exchange should have a diverse board of directors or provide a compelling explanation for any deviations from that guideline. With the rule, the exchange set a precedent for board diversity in the United States.

The rule expects that NASDAQ-listed companies should have two diverse board members in terms of gender and demographic background. The rule takes into account the public company board nominating and governance process and allows for a transition period. NASDAQ companies should have at least one director who identifies as female, an underrepresented race or ethnicity, or an LGBTQ+ member by the end of 2023. By the end of 2026, companies are generally expected to have two diverse directors to satisfy the rule, with flexibility provided for Smaller Reporting Companies and Foreign Issuers.

In a prior judicial test of board diversity rules, two California state laws mandating board diversity were overturned by the courts on constitutional grounds.[139] The NASDAQ rule did not follow the mandate model; it

---

[137] Financial Investment Services and Capital Markets Act (https://elaw.klri.re.kr/kor_service/lawView.do?hseq=57344&lang=ENG).

[138] *NASDAQ's Board Diversity Rule: What Companies Should Know*, NASDAQ (Last Updated February 28, 2023) (https://listingcenter.nasdaq.com/assets/Board%20Diversity%20Disclosure%20Five%20Things.pdf).

[139] Ufonobong Umanah and Andrew Ramonas, *California Board Diversity Law Violates Federal* Constitution, Bloomberg Law (May 17, 2023) (https://news.bloomberglaw.com/us-law-week/california-board-diversity-law-violates-federal-constitution).

established a disclosure and reporting—essentially, a comply or explain—model. If a company fails to meet the diversity requirement, they are obliged to explain why. Furthermore, companies must annually disclose how their board members identify in terms of gender, race, and sexual orientation utilizing the NASDAQ Board Diversity Matrix.[140]

Conservative groups filed a lawsuit against the US Securities and Exchange Commission (SEC), which had previously approved the NASDAQ rule in August 2021.[141] The petitioners argued that the rule contravened the US Constitution's prohibition of discriminatory laws and free speech restrictions. They contended that these government restrictions extended to NASDAQ, as the SEC could penalize the exchange for failing to enforce the rule. They also alleged violations of the Securities Exchange Act (Exchange Act) and the Administrative Procedure Act (APA).

The Fifth Circuit upheld the NASDAQ diversity rule and issued a panel opinion in October 2023.[142] The court determined that the SEC acted within its statutory authority and was correct in considering the importance of diversity data to investors in making investment decisions. The court further noted the utility of the rule to provide standardized information on board diversity in the marketplace. The unanimous panel of three judges ruled that the constitutional challenges fell short

---

[140] *Board Diversity Matrix Instructions and Templates*, NASDAQ (Last Updated June 6, 2023) (https://listingcenter.nasdaq.com/assets/Board%20Diversity%20 Disclosure%20Matrix.pdf).

[141] *Self-Regulatory Organizations; The Nasdaq Stock Market LLC; Order Approving Proposed Rule Changes, as Modified by Amendments No. 1, to Adopt Listing Rules Related to Board Diversity and to Offer Certain Listed Companies Access to a Complimentary Board Recruiting Service*, Securities and Exchange Commission (Release No. 34-92590; File Nos. SR-NASDAQ-2020-081; SR-NASDAQ-2020-082) (August 6, 2021) (www.sec.gov/files/rules/sro/nasdaq/2021/34-92590.pdf).

[142] *Alliance for Fair Board Recruitment v. U.S. Securities and Exchange Commission*, ACLU (October 18, 2023) (www.aclu.org/cases/alliance-for-fair-board-recruitment-v-u-s-securities-and-exchange-commission).

CHAPTER 7   THE GLOBAL ESG REGULATORY PARADIGM

as NASDAQ is regulated by the SEC but is not a state actor, and that the petitioners gave no reason to conclude that the SEC's approval of the rule violated the Exchange Act or the APA.

In approving the rule, the SEC found substantial evidence to support the view that information on board diversity would inform market behavior. The SEC determined that the rule would provide "widely available, consistent, and comparable information that would contribute to investors' investment and voting decisions."[143] The SEC findings were based on industry demand and various letters from institutional investors. The court pointed out that investors deem board diversity information important to their investment decisions. As of February 2024, the Fifth Circuit agreed to rehear the challenge *en banc* as the two plaintiff groups had petitioned for a review of the decision.

The validation of the NASDAQ Board Diversity Rule has several important implications:

1. **Reinforced Importance of Board Diversity**
   The court's decision underlines the significance of board diversity in the corporate world. It emphasizes that promoting inclusivity within corporate boards is not just an ideal, but a necessary criterion that public companies must strive to meet.

2. **Emphasis on Transparency**
   The ruling features the need for transparency in disclosing board diversity statistics. The court supported the rule's provision requiring companies to disclose the demographic makeup of their boards annually.

---

[143] *Statement on NASDAQ's Diversity Proposals – A Positive First Step for Investors*, United States Securities and Exchange Commission (August 6, 2021) (www.sec.gov/news/public-statement/statement-nasdaq-diversity-080621).

3. **Enhanced Accountability**
   The NASDAQ Diversity Rule holds companies accountable for their board composition. Companies that fail to meet the diversity standards must provide an explanation to investors and other stakeholders, which could catalyze corporate action.

4. **Affirmed Value of Diversity Data for Investors**
   The court's decision affirms the importance of diversity data for investors. It is a clear indication that investors value this information and that it impacts their investment and voting decisions.

Regulatory Efforts to Address the Gender Pay Gap

Governments have begun taking moderate steps to address gender pay inequality. The models run the gamut from requesting pay transparency to requiring equal pay. The approach taken by the United Kingdom has been one of transparency for companies with a threshold number of employees. The UK requires organizations with more than 250 employees to report on their gender pay gaps.[144]

As discussed, the European Commission has adopted the *Gender Equality Strategy* to outline the actions required to close the EU's gender pay gap. The *EU Pay Transparency Directive* aims to enforce gender pay

---

[144] *Statutory guidance: Who needs to report,* United Kingdom Government Equalities Office (www.gov.uk/government/publications/gender-pay-gap-reporting-guidance-for-employers/who-needs-to-report#:~:text=Any%20employer%20with%20250%20or,year%20of%20your%20snapshot%20date).

equality across EU member states.[145] The pay gap between men and women in the EU stood at 12.7% in 2021, with minimal change across the prior decade.[146] The directive requires large employers to report on their gender pay gaps and to take measures to eliminate them. The primary sustainability financial disclosure regulation in the EU—the Sustainable Finance Disclosure Regulation (SFDR)—specifically identifies the "unadjusted gender pay gap" as an indicator for investors to take into account.[147]

Spain has been a leader in the EU on DEI. The Spanish government introduced two Royal Decrees to address gender equality in 2020: one to regulate gender equality plans and the second on equal pay to address the gender pay gap. These regulations represented a strengthening of diversity regulations first passed years earlier. More than 1,300 private and public companies have signed the Spanish Diversity Charter, which was launched in 2009 to promote equal opportunities and DEI in the workplace.[148]

---

[145] *Pay transparency in the EU*, Council of the European Union (www.consilium.europa.eu/en/policies/pay-transparency/).

[146] *The gender pay gap situation in the EU*, European Commission (https://commission.europa.eu/strategy-and-policy/policies/justice-and-fundamental-rights/gender-equality/equal-pay/gender-pay-gap-situation-eu_en).

[147] *Regulation (EU) 2019/2088 of the European Parliament and of the Council of 27 November 2019 on sustainability-related disclosures in the financial sector*, European Union Law (November 27, 2019) (https://eur-lex.europa.eu/legal-content/EN/TXT/?uri=CELEX:32019R2088).

[148] Spanish Diversity Charter, EU Commission (https://commission.europa.eu/strategy-and-policy/policies/justice-and-fundamental-rights/combatting-discrimination/tackling-discrimination/diversity-and-inclusion-initiatives/diversity-charters-eu-country/spanish-diversity-charter_en).

CHAPTER 7   THE GLOBAL ESG REGULATORY PARADIGM

Regulatory and Legislative Actions on Antitrust

Antitrust and competition laws are designed to ensure fair competition and prevent market dominance. They prohibit agreements and practices that restrict free trade and competition between businesses. This can include agreements by competitive businesses to fix prices or divide markets. These laws are designed to protect consumers by promoting fair competition, which can lead to lower prices, improved services, and more choice.

Antitrust has become a concern for companies collaborating on the ESG front. ESG commitments often take the form of pledges and sometimes involve interactions between competitors, often under the rubric of industry efforts. It is common practice for companies to work together in trade associations, business groups, and other efforts to establish standards and best practices, and to coordinate on other issues, particularly in emerging fields.

While ESG initiatives are aimed at environmental sustainability and social good, they could potentially violate antitrust laws when they lead to restrictions on competition. An increased collaboration between competitors has brought antitrust laws to the forefront. Depending on the nature of the agreement and the extent of collaboration, these initiatives could be seen as coordinated behavior, group boycotts, or price-fixing cartels.

United Kingdom

The Competition and Markets Authority (CMA) in the UK published the Green Agreements Guidance in October 2023, which provides clarity on the intersection of ESG and antitrust laws.[149] This guidance

---

[149] *Draft guidance on environmental sustainability agreements,* HM Government, United Kingdom (October 12, 2023) (www.gov.uk/government/consultations/draft-guidance-on-environmental-sustainability-agreements).

supplements the CMA Guidance on Horizontal Agreements and applies to "environmental sustainability agreements," which are agreements between competitors aimed at preventing, reducing, or mitigating adverse economic impact on the environment.[150] The CMA adopts a more tolerant approach to climate change agreements, allowing for the consideration of benefits to all UK citizens, regardless of whether they also buy products or services in the relevant market affected by the agreement. This permissive approach is a key element in addressing negative externalities and market failures arising from firms' production costs not reflecting the societal costs of greenhouse gas emissions.

United States

ESG commitments have been under fire from a growing anti-ESG movement led by state attorneys general and federal lawmakers. In letters sent to asset managers and insurance companies, state attorneys general warned that their ESG activity might violate antitrust laws.[151] Members of the US House Judiciary Committee have also raised concerns about investment fund ESG policies and issued subpoenas to multiple firms.[152] With the growing scrutiny of ESG initiatives, companies should ensure their policies and practices align with antitrust laws to avoid potential legal pitfalls.

---

[150] Id.

[151] Tommy Wilkes, Alexander Hübner, and Tom Sims, *Insurers flee climate alliance after ESG backlash in the U.S.*, Reuters (May 26, 2023) (www.reuters.com/business/allianz-decides-leave-net-zero-insurance-alliance-2023-05-25/).

[152] *House panel subpoenas Vanguard, Arjuna in ESG 'collusion' probe*, US House of Representatives Judiciary Committee (December 11, 2203) (https://judiciary.house.gov/media/in-the-news/house-panel-subpoenas-vanguard-arjuna-esg-collusion-probe#:~:text=House%20panel%20subpoenas%20Vanguard%2C%20Arjuna%20in%20ESG%20'collusion'%20probe,-December%2011%2C%202023&text=WASHINGTON%20%E2%80%94%20A%20House%20panel%20subpoenaed,governance%20policies%20violate%20antitrust%20laws).

CHAPTER 7 THE GLOBAL ESG REGULATORY PARADIGM

7.3 Distinguishing Regulation, Legislation, and Executive Action

We've noted that while they are collectively termed "regulatory activity" as an accepted shorthand, legislation, regulation, and executive action are distinct methods through which laws are created and enforced. While they share the common goal of governing society, they differ in terms of the bodies responsible for their creation and the force of law they possess. Each carries a different level of authority, originates from different branches of government, and serves different functions within the legal framework. Respective powers and associated duties are typically found in the relevant state or national constitution.

Legislation refers to the process of creating laws by the legislative branch of a government, such as a parliament or congress. It involves the introduction, debate, amendment, and voting on a bill, which, once passed, becomes a statute and is codified in the legislative code. Legislators, who are elected representatives, propose and shape legislation based on the needs and interests of their constituents. Legislation is generally broad and comprehensive, covering a wide range of issues and applying to all individuals within a jurisdiction. Legislation is subject to judicial review.

Regulation involves the promulgation, implementation, and enforcement of regulations by executive branch departments or administrative agencies or bodies pursuant to a legislative directive. Regulations are a set of rules issued by government agencies that are created to enforce and specify the requirements of legislation. They fill in the details and the practical measures needed for the practical application of statutes and must not go beyond the scope of the legislation pursuant to which they are promulgated. Regulations are also subject to judicial review.

CHAPTER 7   THE GLOBAL ESG REGULATORY PARADIGM

Executive action refers to actions taken by the executive branch of a government, typically by the head of state or government, such as a president or prime minister. These actions are unilateral, derived from the executive's inherent powers or delegated authority, and are aimed at implementing and enforcing laws. Executive actions can take various forms, including executive orders, proclamations, directives, and memoranda. While they do not have the same level of permanence as legislation or regulation, they carry the force of law within the executive branch and can have significant impacts on policy and governance.

7.4 Concept of Materiality
Financial Materiality

Financial materiality was an aspect of first mover ESG regulatory disclosures. The concept of financial materiality formed the crux of the first wave of ESG disclosures, shaping the dialogue between corporations and investors about sustainability issues. Financial materiality generally refers to issues that are reasonably likely to impact a company's financial condition or operating performance. This tracks, as the expectation with ESG is that companies should provide clear and comprehensive information about their risks relative to impact on financial performance.

With regulators like the SEC promulgating climate disclosure rules and the financial materiality of climate-related risks and actions increasing, companies are urged to enhance their sustainability disclosure strategy and prepare for future reporting requirements. In the United States, there has been intense debate over whether ESG information should be considered *material* for purposes of securities laws. Regardless of the outcome of the strict materiality debate, changes in the practices of investors are expanding the amount of ESG information they consider to be material. That is driving the scope of information considered to be material in practice.

CHAPTER 7   THE GLOBAL ESG REGULATORY PARADIGM

Non-financial Materiality and Double Materiality

Contrasting with financial materiality is the concept of impact materiality, which focuses on how a company's activities impact the world around it. This approach is informed by the concept of sustainability, meeting the needs of today without compromising the needs of the future. There is an ongoing shift to require broader information, with investors and regulators alike recognizing the importance of ESG in determining long-term value creation.

The Global Reporting Initiative (GRI), the widely used global sustainability reporting framework, discusses materiality as a filter of an organization's significant economic, environmental, and social impacts that substantively influence the assessments and decisions of stakeholders, and notes the two main directions of materiality are financial materiality and impact materiality, which together make up the concept of double materiality.[153] Many companies have conducted materiality assessments pursuant to the GRI approach and have made related ESG disclosures.

As discussed, the European Union's Corporate Sustainability Reporting Directive (CSRD) integrates the concepts of financial materiality and impact materiality in a structurally significant way, guiding companies in determining their reporting requirements under the EU's European Sustainability Reporting Standards (ESRS). If not already utilizing a double materiality assessment, covered companies will need to adapt their existing approaches to comply with the ESRS, including conducting a full value chain analysis and documentation of the double materiality process.

---

[153] *The GRI Perspective, The materiality madness: why definitions matter*, Global Reporting Initiative (February 22, 2022) (www.globalreporting.org/media/r2oojx53/gri-perspective-the-materiality-madness.pdf).

7.5 Disclosure and Reporting Models

Regulatory disclosure and reporting requirements vary across regions, reflecting the unique challenges and priorities of each jurisdiction. Some jurisdictions have implemented mandatory reporting requirements, while others still rely on voluntary frameworks or a combination of both. As an example of a mandatory reporting regime, the European Union's CSRD amended and expanded its Non-Financial Reporting Directive (NFRD) which already required large companies to disclose information on environmental, social, and employee matters, human rights, and anti-corruption and bribery issues.

There is a growing demand for transparency and accountability across every sector. When it comes to regulatory disclosure and reporting, there are two main approaches::

- **Mandatory reporting** refers to regulations that require companies to disclose specific information, often based on industry or company size. These regulations are enforceable by law, and non-compliance may result in penalties. They are also beginning to prescribe the use of specific reporting standards.

- **Voluntary reporting** allows companies to choose whether or not to disclose information. While voluntary reporting may not be legally binding, many companies choose to participate in voluntary reporting frameworks to demonstrate their commitment to transparency and sustainability.

In the United States, the Securities and Exchange Commission (SEC) does require disclosures on material climate risk and related information, and human capital management for publicly traded companies. The new climate accountability laws in California will represent a change and require disclosures for covered companies once they come into effect. Aside from those requirements, reporting on ESG factors in the United States has been primarily voluntary and it is driven by stakeholder expectation and utilizes voluntary reporting standards.

While regulatory disclosure and reporting frameworks have made significant progress in promoting transparency and accountability, they are not without challenges and limitations. One of the key challenges is the lack of harmonization and consistency across frameworks. Companies often must navigate multiple reporting requirements, leading to increased complexity and resource constraints. Another challenge is the difficulty in quantifying and measuring certain ESG components. For example, assessing the social impact of a company's activities or measuring the long-term financial implications of climate change can be complex and subjective. The availability and reliability of data also pose challenges, as companies may struggle to collect and report accurate information. Then there is the looming risk of greenwashing, where companies engage in misleading or superficial reporting to create a positive image without making meaningful changes.

This highlights the importance of independent verification and assurance to support the credibility and reliability of reported information. Numerous tools and resources are available to support companies in their regulatory disclosure and reporting efforts. These tools provide guidance, templates, frameworks, and indicators that help streamline the reporting process and ensure compliance with relevant regulations. In addition to these frameworks, there are also software solutions and platforms that automate data collection, analysis, and reporting. These tools help companies streamline their reporting processes, improve data accuracy, and enhance the efficiency of their reporting efforts.

Understanding the various types of regulatory disclosure and reporting models for climate, ESG, and sustainability is imperative for companies seeking to demonstrate their commitment to sustainability, attract responsible investors, and mitigate risks associated with climate change and ESG concerns. By following the right reporting practices and utilizing the available tools and resources, organizations can foster trust among stakeholders and drive positive change in their industries.

As the global focus on climate change and sustainability intensifies, regulatory frameworks will continue to evolve, demanding greater transparency and accountability. It is essential for companies to stay informed about the latest reporting requirements and best practices to effectively navigate the diverse landscape of regulatory disclosure and reporting models for climate, ESG, and sustainability globally.

7.6 Rise of Enforcement, Litigation, and Shareholder Challenges

Regulatory Enforcement

Regulators such as the US Securities and Exchange Commission (SEC) are taking a more proactive approach to enforcing ESG-related disclosures and practices. In March 2021, the SEC introduced its Climate and ESG Task Force. This task force, within the Division of Enforcement, was formed to identify potential misconduct, misstatements, and material gaps in issuer disclosures of ESG and climate risk. The SEC has leveraged the antifraud, reporting, and internal controls provisions of the Securities Act and the Exchange Act to bring forth several ESG-related enforcement actions.

The Climate and ESG Task Force brought a high-profile enforcement action against Vale S.A., a publicly traded Brazilian mining company. Following a 2019 mine collapse that resulted in numerous deaths and substantial environmental damage, the SEC alleged that Vale had manipulated its safety audits and misled investors about its mines meeting safety

CHAPTER 7   THE GLOBAL ESG REGULATORY PARADIGM

standards.[154] In its press release on the enforcement action, the SEC specifically noted Vale's ESG disclosures including its annual Sustainability Reports in reporting that the company "allegedly manipulated those disclosures."[155] The action resulted in Vale paying a USD$55.9 million settlement for the charges of making false and misleading disclosures about the safety of its dams in 2023.[156] In its announcement of the settlement, the SEC noted that the agency's "action against Vale illustrates the interplay between the company's sustainability reports and its obligations under federal securities laws."[157]

The task force also brought an enforcement action against BNY Mellon Investment Adviser, Inc. for alleged misstatements regarding its ESG investments during a three-year period from July 2018 to September 2021.[158] The SEC claimed that while BNY represented that the investments underwent an ESG quality review, some investments did not have an associated quality review score at the time they were made.[159] BNY Mellon Investment Adviser, Inc. paid USD$1.5 million to settle the charges. In its press announcement on the settlement, the SEC said, "[i]nvestors are increasingly focused on ESG considerations when making investment decisions," and "the Commission will hold investment advisers accountable when they do not accurately describe their incorporation of ESG factors into their investment selection process."[160]

---

[154] *SEC Charges Brazilian Mining Company with Misleading Investors about Safety Prior to Deadly Dam Collapse*, US Securities and Exchange Commission (April 28, 2022) (www.sec.gov/news/press-release/2022-72).

[155] Id.

[156] *Brazilian Mining Company to Pay $55.9 Million to Settle Charges Related to Misleading Disclosures Prior to Deadly Dam Collapse*, US Securities and Exchange Commission (March 28, 2023) (www.sec.gov/news/press-release/2023-63).

[157] Id.

[158] *SEC Charges BNY Mellon Investment Advisers for Misstatements and Omissions Concerning ESG Considerations*, US Securities and Exchange Commission (May 23, 2022) (www.sec.gov/news/press-release/2022-86).

[159] Id.

[160] Id.

CHAPTER 7   THE GLOBAL ESG REGULATORY PARADIGM

The SEC charged Goldman Sachs Asset Management, L.P. for failures involving two mutual funds and one separately managed account that were marketed as ESG investments from April 2017 to February 2020.[161] Goldman Sachs Asset Management paid a USD$4 million penalty in settlement of the charges.[162] The SEC stated in its announcement that the "action reinforces that investment advisers must develop and adhere to their policies and procedures over their investment processes, including ESG research, to ensure investors receive the advisory services they would expect to receive from an ESG investment."[163]

In September 2021, the SEC issued a *Sample Letter to Companies Regarding Climate Change Disclosures*.[164] The SEC's sample letter was part of a broader initiative aimed at ensuring that public companies are transparent about their climate-related impacts. The letter reiterates the disclosure matters discussed in the 2010 Climate Change Guidance and outlines nine key factors that companies must consider when crafting their climate change disclosures.[165] The nine factors range from general considerations to specific risk factors and financial implications.

- **General factors:** The general factor emphasizes the need for consistency in climate-related disclosures across all corporate platforms. If a company's corporate social responsibility report provides more detailed information on its climate impacts than its SEC filings, the company may need to justify this discrepancy.

---

[161] *SEC Charges Goldman Sachs Asset Management for Failing to Follow its Policies and Procedures Involving ESG Investments*, US Securities and Exchange Commission (November 22, 2022) (www.sec.gov/news/press-release/2022-209).

[162] Id.

[163] Id.

[164] *Sample Letter to Companies Regarding Climate Change Disclosures*, US Securities and Exchange Commission (www.sec.gov/corpfin/sample-letter-climate-change-disclosures).

[165] Id.

- **Risk factors:** The risk factors section calls on companies to detail the climate-related risks they face. These can include regulatory changes, market trends affecting business opportunities, credit risks, technological changes, and the potential for climate-related litigation.

- **Management's Discussion and Analysis of Financial Condition and Results of Operations (MD&A):** The MD&A section requires companies to discuss a wide range of climate-related topics. These include pending or existing legislation, capital expenditures on climate-related projects, indirect consequences of climate regulation on business trends, and the purchase or sale of carbon credits or offsets.[166]

Since releasing the sample letter, the SEC has actively engaged with companies to enhance their climate change disclosures. This engagement has taken various forms, including issuing comment letters to companies and encouraging them to review their disclosure practices. As the SEC continues to refine its approach to climate change disclosures, companies should stay abreast of these developments and adjust their disclosure practices accordingly.

The SEC is not the only regulator focusing on ESG-related enforcement in the United States. The US Federal Trade Commission has the authority to initiate enforcement actions based on deceptive or unfair marketing practices, including for environmental marketing claims.[167] The agency has exercised that authority against global retailers, including in a publicized

---

[166] Id.

[167] *Green Guides*, US Federal Trade Commission (www.ftc.gov/legal-library/browse/rules/green-guides).

action against Kohl's and Walmart for eco product claims, which was settled for USD$5.5 million in 2022.[168] The Federal Trade Commission (FTC) is in the process of updating its *Green Guides for the Use of Environmental Marketing Claims*, which were first issued in 1992 and subsequently updated in 1996, 1998, and 2012.[169]

The enforcement actions shine a light on the importance of a high level of diligence in the setting, reviewing, conducting, and disclosing of ESG claims, commitments, metrics, and activities. It is expected that enforcement will likely increase as enhanced and additional ESG laws and regulations are passed and implemented.

Climate Litigation

The total number of climate change court cases worldwide has more than doubled since 2017 and that number continues to grow, according to the *Global Climate Litigation Report: 2023 Status Review*, published jointly by the UN Environment Programme (UNEP) and the Sabin Center for Climate Change Law at Columbia University.[170] In calendar year 2022, there were 2,180 climate-related cases filed in 65 jurisdictions.[171] The majority of those

---

[168] *$5.5 million total FTC settlements with Kohl's and Walmart challenge "bamboo" and eco claims, shed light on Penalty Offense enforcement*, US Federal Trade Commission (April 8, 2022) (www.ftc.gov/business-guidance/blog/2022/04/55-million-total-ftc-settlements-kohls-and-walmart-challenge-bamboo-and-eco-claims-shed-light).

[169] *Federal Trade Commission Extends Public Comment Period on Potential Updates to its Green Guides for the Use of Environmental Marketing Claims*, US Federal Trade Commission (January 31, 2023) (www.ftc.gov/news-events/news/press-releases/2023/01/federal-trade-commission-extends-public-comment-period-potential-updates-its-green-guides-use).

[170] *Global Climate Litigation Report: 2023 Status Review*, United Nations Environment Programme (July 27, 2023) (www.unep.org/resources/report/global-climate-litigation-report-2023-status-review).

[171] Id.

CHAPTER 7   THE GLOBAL ESG REGULATORY PARADIGM

cases were brought in the United States and approximately 17% were filed in developing nations, including Small Island Developing States which are most vulnerable to climate change.[172]

Climate litigation has historically been a peculiar area of environmental law in the United States, as most of the cases are brought under the common law nuisance doctrine. Environmental law is rooted in regulations and laws such as the Clean Air Act, the Clean Water Act, and the Endangered Species Act in the United States, among others. However, the lack of regulation and law on climate or GHG emissions has led plaintiffs to resort to the nuisance doctrine, which predates all the primary environmental laws.

In a nuisance claim, individuals or groups file an action against activities that they deem harmful to the public or interfere with the enjoyment of their property. Climate litigation, therefore, represents a shift from traditional environmental law, focusing on the broader impacts of human activity on the global climate. The two main types of climate litigation cases are public actions, where governments or regulatory bodies are involved, and private plaintiff actions, where individuals or groups bring cases against entities deemed responsible for climate change.

Public actions typically involve governments or regulatory bodies being taken to court for failing to meet their obligations. Public actions in climate litigation often involve lawsuits brought against governments for their insufficient efforts to mitigate climate change. These lawsuits generally argue that the government's inaction or inadequate action infringes on the plaintiffs' rights to life, health, and a clean environment.

Private plaintiff actions usually involve individuals or groups, often represented by non-governmental organizations, suing corporations or other entities for their contribution to climate change. The plaintiffs in

---

[172] *Climate litigation more than doubles in five years, now a key tool in delivering climate justice*, United Nations Environment Programme (July 27, 2023) (www.unep.org/news-and-stories/press-release/climate-litigation-more-doubles-five-years-now-key-tool-delivering).

these cases often argue that the defendants are responsible for significant greenhouse gas emissions, contributing to climate change, and causing harm to the plaintiffs. Many of these cases revolve around allegations of negligence, nuisance, and failure to warn the public about the potential harm caused by the defendants' products.

The climate litigation era began in the early 2000s, with a significant increase in cases observed in the past decade. These early cases laid the groundwork for the use of the legal system as a tool to combat climate change. Over time, the scope of climate litigation expanded to encompass a wide range of legal claims, including constitutional law, administrative law, consumer protection, the public trust doctrine, and human rights.

In the United States, one of the earliest and most significant cases was *Massachusetts v. Environmental Protection Agency (EPA)*, which was ultimately decided by the US Supreme Court in 2007.[173] This lawsuit was brought by several states against the EPA after the agency declined to regulate carbon dioxide and other greenhouse gas emissions under the Clean Air Act (CAA). In a 5-4 decision, the Supreme Court ruled in favor of the states, finding that carbon dioxide and other greenhouse gases were indeed harmful and should be regulated by EPA. This decision shaped the trajectory of climate action in the United States. US courts have made a series of key rulings in climate litigation, many of which have set important precedents for future cases.

Globally, courts have made landmark rulings in climate litigation cases, setting important precedents and driving climate change governance reform. One of the most high-profile climate cases, *Milieudefensie et al. v. Royal Dutch Shell*,[174] was brought by a group of seven nonprofit organizations against Shell, one of the biggest global oil

---

[173] *Massachusetts et al. v. Environmental Protection Agency et al.*, No. 05-1120, US Supreme Court (April 2, 2007) (https://tile.loc.gov/storage-services/service/ll/usrep/usrep549/usrep549497/usrep549497.pdf).

[174] Milieudefensie et al. v. Royal Dutch Shell plc., Hague District Court, File number 90046903 (April 5, 2019) (https://climatecasechart.com/non-us-case/milieudefensie-et-al-v-royal-dutch-shell-plc/).

CHAPTER 7   THE GLOBAL ESG REGULATORY PARADIGM

companies, referred to as "majors," for violations of human rights law and tort law and heard in the Hague District Court in 2021. Shell had previously announced a net-zero plan in place to reach carbon neutrality by 2050. The court found that Shell had an obligation to reduce its carbon emissions on a faster timetable and ordered the company to reduce its carbon emissions from both its operations and its products, from 2019 levels by 45% by 2030. That formula aligns with the Paris Agreement threshold goals, as discussed in Chapter 3. Shell appealed the decision in 2022.

The number of climate litigation cases has been steadily increasing. The field of climate litigation has seen several key trends emerge over the past decade, including an increase in cases and a shift toward human impact-based arguments. This trend in activist judicial strategy is expected to continue, reflecting the growing awareness and urgency of climate change issues.

The prediction is that judicial challenges on climate will continue rising in the coming years, driven by increasing public awareness of climate change and the urgency of the issue. This trend is likely to be seen both in the United States and globally. Future climate litigation is likely to involve innovative legal strategies, including the use of the public trust doctrine, arguments based on human rights, and claims related to corporate responsibility for climate change. The growing body of case law and emerging trends indicate that climate litigation will remain a potent tool for those groups mobilizing to hold governments and corporations accountable for their actions and inactions.

Greenwashing Litigation

As we discussed in Chapter 6, greenwashing is a deceptive practice where companies overstate, exaggerate, or outright mislead consumers about the environmental benefits of their products, services, or corporate practices. While the term was coined in the 1980s, the practice has gained traction in the past decade due to increasing consumer demand for environmentally friendly products and services. Greenwashing has become a hot-button issue in the corporate, legal, and regulatory spheres. This has led to a surge in greenwashing lawsuits aimed at preventing such misleading practices across the globe.

Concurrent with the increased regulatory focus, there has been a significant rise in the number of greenwashing lawsuits across the globe. These cases often involve allegations that companies have misled the public and investors about their environmental commitments or the sustainability of their products. As more greenwashing cases are being litigated, court decisions are shaping the legal landscape.

In the United States, more than a dozen large class action lawsuits have been filed each year since 2020.[175] Many of these involve the sustainability-related claims made by high-profile consumer companies like Nike, REI, and Colgate. While few have been successful to date, greenwashing claims are setting important precedents regarding what constitutes deceptive environmental marketing. The risk of legal challenge is not only shaping the way companies market their products, but also how they approach sustainability. The risks serve as guideposts for companies striving to avoid greenwashing, be good actors, and stay on the right side of the law.

The impact of greenwashing is far-reaching, affecting not just consumers but also investors and the environment. Misleading environmental claims can lead to consumers making purchases based on false information, investors making misinformed decisions, and genuine environmental efforts being overshadowed. Looking ahead, it is expected that both greenwashing litigation and regulation will continue to expand.

---

[175] *By the Numbers: Greenwashing Class-Action Lawsuits*, Truth in Advertising (October 30, 2023) (https://truthinadvertising.org/articles/by-the-numbers-greenwashing-class-action-lawsuits/#:~:text=Since%20TINA.org%20began%20tracking,have%20been%20filed%20each%20year).

CHAPTER 7   THE GLOBAL ESG REGULATORY PARADIGM

Shareholder Litigation

Shareholder lawsuits on ESG issues typically involve allegations of securities fraud, breach of fiduciary duty, failure to disclose risk, and misrepresentation of claims against a company and its board of directors based on its ESG statements and disclosures. Shareholder actions allege derivative harm, such as depreciated stock value. The challenges span a range of ESG topics, including human rights, green marketing, carbon emissions, DEI, and anti-ESG issues.

More than a dozen companies have faced shareholder lawsuits based on their DEI claims in the past few years. The basis for these actions includes failure to comply with anti-discrimination laws, false statements in public materials, and a lack of representation on the board of directors and in leadership positions while claiming support for DEI. Two case studies are illustrative.

CASE STUDY: WELLS FARGO EMPLOYMENT PRACTICES

An asset management company and a firefighters' pension fund jointly brought a shareholder derivative action against Wells Fargo, one of the largest US banks, in federal court in San Francisco, which is the location of the bank's corporate headquarters.[176] The shareholder pleading included claims of breach of fiduciary duties, unjust enrichment, and violation of securities laws.[177] Allegations included that the company boasted about its commitment to diversity in its hiring practices in public statements while it was actually conducting sham interviews of non-white and female job candidates to inflate its diversity candidate interview numbers.

---

[176] *SEB Investment Management AB, and West Palm Beach Firefighters' Pension Fund, et al. v. Scharf et al.*, US District Court for the Northern District of California (3:22-cv-03811-TLT) (January 31, 2023).

[177] *Amy Cook v. Steven D. Black, et al.*, United States District Court Case No. 3:23-cv-04934-JCS (September 26, 2023) (https://dockets.justia.com/docket/california/candce/3:2023cv04934/418638).

The New York Times first reported on the alleged sham interviewing scheme at Wells Fargo on May 19, 2022.[178] The article lays out details from conversations with seven current and former Wells Fargo employees. When the interview scheme came to light, the Wells Fargo stock price dropped more than 10% in the two following days, "wiping out more than $17 billion of market value."[179]

The court ultimately dismissed the case against the bank, finding that plaintiffs could not prove that a sham interviewing process occurred, that it was widespread, or that it was known to the CEO or two key DEI executives.[180]

CASE STUDY: SHELL'S BOARD OF DIRECTORS DUTY TO ADDRESS CLIMATE RISK

In February 2023, ClientEarth, a Shell shareholder, filed a lawsuit against Shell's Board of Directors in the United Kingdom under the Companies Act of 2006.[181] The derivative suit alleged that Shell's board had failed in their duties of reasonable care, skill, and diligence to properly address climate risk and had failed to comply with the *Milieudefensie* ruling. In part, the plaintiff shareholder claimed that the board of directors had failed to properly protect the company from risk exposure and a decline in long-term value. The High Court ultimately dismissed the case for failure to make a prima facie case.

---

[178] Emily Flitter, *At Wells Fargo, a Quest to Increase Diversity Leads to Fake Job Interviews,* The New York Times (May 19, 2022) (www.nytimes.com/2022/05/19/business/wells-fargo-fake-interviews.html).

[179] Jonathan Stempel, *Wells Fargo defeats shareholder lawsuit over fake job interviews,* Reuters (August 21, 2023) (www.reuters.com/business/finance/wells-fargo-defeats-shareholder-lawsuit-over-fake-job-interviews-2023-08-21/).

[180] Id.

[181] *In the High Court of Justice Business and Property Courts of England and Wales Insolvency and Companies List (ChD), Derivative Claim,* Case No. BL-2023-000215 (July 24, 2023) (www.judiciary.uk/wp-content/uploads/2023/07/ClientEarth-v-Shell-judgment-240723.pdf).

While these two cases were not fruitful for the shareholders in court, they demonstrate the pattern of shareholders seeking change on ESG issues. Shareholders are taking action in various ways, including through activism in the proxy season and through judicial action. Battling a shareholder suit can have significant implications for a company's financial performance, reputation, and long-term sustainability. It can also create a public profile of the company that shapes the viewpoints of the public and regulators. The cases are also a drain on corporate resources and influence future public relations behavior.

7.7 Navigating Regulations and Emerging ESG Risk

Regulatory efforts are reshaping the ESG landscape and will continue to do so in the short term. Global regulations are progressing, necessitating that businesses disclose financial and non-financial facets of their operations to stakeholders. These requirements include a wide range of ESG aspects like human rights, climate change activities, and supply chain due diligence. As new ESG protocols and reporting expectations are enumerated, the specific guidelines vary by geography.

Companies should understand the requirements of the jurisdictions in which they operate and assess their relative regulatory risk. Non-compliance can lead to administrative, civil, and criminal penalties and damage to a company's social license to operate. Table 7-2 provides guidance in how companies can navigate the changing regulatory environment.

CHAPTER 7   THE GLOBAL ESG REGULATORY PARADIGM

Table 7-2. Best Practices for Keeping Pace with Envolving Global Regulations

1. Ensuring **readiness of legal, finance, and compliance teams** by assessing resources and addressing gaps.
2. Identifying and **mapping regulations** applicable to the company and reconciling any conflicting **jurisdictional requirements.**
3. Reviewing **regulatory requirements** coming online and conducting a **gap assessment.**
4. Revising **corporate governance materials and processes** as needed.
5. Evaluating **board oversight** and **management responsibilities.**
6. Assessing the **current state of reporting for voluntary and mandatory disclosures**, including for data collection and management, for legal, finance, and leadership review, in the control environment, and in the audit function.
7. Revising **risk management practices** as necessary.
8. Implementing a system to **monitor, track, and update** colleagues of **regulatory developments.**
9. Assigning a team to **work with partners and suppliers** on compliance.
10. Developing a **matrixed plan** that encompasses the above and continuously revising it **to comply with developing regulatory requirements.**

As stakeholder expectations and engagement have taken center stage, ESG is increasingly becoming a top priority for organizations worldwide. Companies should integrate ESG considerations into their core business strategies and their risk management frameworks. The shift from voluntary to mandatory ESG reporting is a major trend in sustainability disclosures. It necessitates adherence to reporting requirements, cogent data collection protocols, strong risk, compliance, and audit practices, and both cross-functional collaboration and exemplary leadership from management teams and boards of directors. We'll go into more detail on specific emerging risks for which companies should develop plans in Chapter 10.

CHAPTER TAKEAWAYS

- Global regulations governing ESG are rapidly developing
- Regulators are expanding their scope of coverage for ESG factors
- Regulators are stepping up their enforcement regimes for ESG issues
- Climate litigation, greenwashing claims, and shareholder suits are increasing worldwide
- Instituting a monitoring and tracking system to keep up with ESG regulatory and enforcement updates is a business imperative
- Companies should assess their ESG governance, disclosures, and reporting processes, as well as their legal, compliance, risk, and audit coverage, and address any gaps

CHAPTER 8

The ESG Standards and Frameworks

ESG frameworks and standards provide a structured approach to ESG reporting. They offer guidelines for organizations to evaluate their ESG-related practices and disclose the associated business risks and opportunities. By utilizing these frameworks and standards, businesses can prioritize consistency, comparability, and reliability in their ESG disclosures, which can lower the risk profile and enable enhanced regulatory compliance.

8.1 Progression of Standards and Frameworks

The standards and frameworks were typically developed by independent organizations. These frameworks not only set the metrics and qualitative elements that a company should disclose but also provide the format and reporting frequency. Some of these frameworks are voluntary, while others are now mandated by governments, as we're seeing governments and stock exchanges move toward adopting specific approaches.

ESG standards and frameworks are designed to be applicable across a wide range of industries. Some frameworks provide sector-specific guidance. For example, the Sustainability Accounting Standards Board

CHAPTER 8   THE ESG STANDARDS AND FRAMEWORKS

(SASB) developed a set of standards for 77 industries, and the Global Reporting Initiative (GRI) offers sector-specific standards in addition to its universal standards.

The earliest sustainability standards were launched more than 25 years ago. A number of the standards organizations agreed to align and consolidate in the past couple of years, in an effort to reduce complexity, improve reporting comparability, and enhance the quality of disclosure. Prior to this alignment, there were about eight primary frameworks. Understanding the background and the development of the standards is important to put the recent calibration into context. We provide summaries of a select number as follows:

- **International Financial Reporting Standards Foundation (IFRS)**

    The IFRS Foundation's Sustainability Disclosure Standards are developed by the International Sustainability Standards Board (ISSB).[1] These new standards aim to provide a unified set of disclosure standards for reporting ESG data to investors, consolidating the existing SASB Standards and Climate Disclosure Standards Board (CDSB) Framework.[2] IFRS was established in 2001 and ISSB was announced at UNFCCC COP26 in 2021.[3]

    SASB and the International Integrated Reporting Council (IIRC) merged to form the Value Reporting Foundation, which was subsequently absorbed by the International Financial Reporting Standards

---

[1] *About the International Sustainability Standards Board*, IFRS Foundation (www.ifrs.org/groups/international-sustainability-standards-board/).
[2] Id.
[3] Id.

CHAPTER 8   THE ESG STANDARDS AND FRAMEWORKS

(IFRS) Foundation, leading to the creation of the International Sustainability Standards Board (ISSB) to develop a single set of global sustainability standards.[4]

- **Sustainability Accounting Standards Board (SASB)**

    SASB was developed in 2011 and the SASB standards provided sector-specific guidelines on a broad range of ESG topics.[5] These standards enabled organizations to disclose financially material sustainability information, emphasizing topics considered material for issuers in specific industries. SASB ultimately merged into ISSB.[6]

- **Task Force on Climate-related Financial Disclosures (TCFD)**

    TCFD provided both general and sector-specific guidance on climate-related topics alone.[7] It helped organizations articulate how ESG performance could materially impact future financial performance and value creation. It was created in 2015 and became the primary standard for climate-related disclosures used by companies. TCFD disbanded after its consolidation with ISSB.[8]

---

[4] *Consolidations organisations (CDSB & VRF)*, IFRS Foundation (www.ifrs.org/about-us/consolidated-organisations/).
[5] *About us*, SASB Standards (https://sasb.org/about/).
[6] Id.
[7] *About*, Task Force on Climate-related Financial Disclosures (www.fsb-tcfd.org/about/).
[8] Id.

CHAPTER 8   THE ESG STANDARDS AND FRAMEWORKS

- **Global Reporting Initiative (GRI)**

    GRI was created in 1997.[9] The GRI Standards utilize a modular approach to guide organizations in reporting on a comprehensive range of sustainability issues.[10] These standards assess materiality based on the impacts made by issuers on the economy, environment, and society.[11]

- **Carbon Disclosure Project (CDP)**

    CDP was established in 2000.[12] CDP offers a platform for companies to disclose environmental information to their stakeholders through different questionnaires on climate change, water security, and deforestation.[13]

- **Equator Principles**

    The Equator Principles were developed in 2003, as a risk management framework for financial institutions to assess environmental and social risk in the project finance space.[14] The principles were initially designed as due diligence standards and influenced by the International Finance Corporation (IFC) policy frameworks. The IFC standards and

---

[9] *About GRI*, Global Reporting Initiative (www.globalreporting.org/about-gri/).
[10] *The global standards for sustainability impacts*, Global Reporting Initiative (www.globalreporting.org/standards/).
[11] Id.
[12] *About us*, CDP (www.cdp.net/en).
[13] Id.
[14] *About the Equator Principles*, Equator Principles (https://equator-principles.com/about-the-equator-principles/).

CHAPTER 8   THE ESG STANDARDS AND FRAMEWORKS

other World Bank guidelines have subsumed some of the work grounding project finance best practices as many of the large banks are no longer signatories to the principles.[15]

- **European Financial Reporting Advisory Group (EFRAG) Sustainability Reporting Board**

    The EFRAG Sustainability Reporting Board develops the European Sustainability Reporting Standards (ESRS).[16] These standards will be enforceable by the European Commission.[17] We discussed ESRS in Chapter 7 and it's important to note the mandatory reporting model.

- **Taskforce on Nature-related Financial Disclosures (TNFD)**

    TNFD is the newest set of original standards, having launched in 2021. It is a "market-led, science-based and government-backed initiative providing [organizations] with the tools to act on evolving nature-related issues."[18]

---

[15] Simon Jesspo, Isla Binnie and Ross Kerber, *Leading U.S. banks leave ESG project finance group* (March 6, 2024) (www.reuters.com/business/finance/jpmorgan-citi-wells-boa-are-no-longer-signatories-equator-principles-website-2024-03-05/).

[16] *Sustainability Reporting Standards*, European Financial Reporting Advisory Group (https://efrag.org/About/Governance/40/EFRAG-Sustainability-Reporting-Board?AspxAutoDetectCookieSupport=1).

[17] Id.

[18] *About us*, Taskforce on Nature-related Financial Disclosures (https://tnfd.global/about/).

315

CHAPTER 8   THE ESG STANDARDS AND FRAMEWORKS

The International Finance Corporation (IFC) has developed an illustration that tracks the advent and convergence of the global reporting standards from 1997 to 2024 (Figure 8-1).

Figure 8-1. *IFC Pathway to ESG Disclosure Going Mainstream*[19]

---

[19] *Understanding the Global Reporting Frameworks*, International Finance Corporation (www.ifcbeyondthebalancesheet.org/understanding-global-reporting-frameworks).

8.2 Examination of Key Standards and Frameworks

There are currently four primary and independent (i.e., nongovernmental) global standards that have emerged following the alignment in 2024. Those are (1) the International Sustainability Board's IFRS standards, (2) GRI, (3) CDP, and (4) Taskforce on Nature-related Financial Disclosures (TNFD) standards. We will examine each standard in detail.

International Sustainability Standards Board (ISSB)

The International Sustainability Standards Board (ISSB) was announced at UNFCCC COP26 in Glasgow in November 2021. This marked the beginning of a new era in sustainability disclosures. The decision to establish the ISSB was spurred by mounting demands from the market for a unified, globally recognized standard for sustainability disclosures.

In developing its standards, ISSB leverages the work of several predecessor standards and frameworks, including the Climate Disclosure Standards Board (CDSB), the Task Force on Climate-related Financial Disclosures (TCFD), the Value Reporting Foundation's Integrated Reporting Framework, industry-based SASB Standards, and the World Economic Forum's Stakeholder Capitalism Metrics.[20] The ISSB has become the standard-bearer in the realm of corporate ESG reporting, and its consolidation of other standards should lead to more comprehensive, comparable, and reliable reporting.

---

[20] Id.

CHAPTER 8   THE ESG STANDARDS AND FRAMEWORKS

The ISSB has taken a pragmatic and investor-centric approach in developing these standards. It aims to create standards that are cost-effective, decision-useful, and informed by market realities.[21] ISSB's mission is focused on four key objectives:

1. "To develop standards for a global baseline of sustainability disclosures.
2. To meet the information needs of investors.
3. To enable companies to offer comprehensive sustainability information to global capital markets.
4. To facilitate interoperability with jurisdiction-specific disclosures and disclosures aimed at broader stakeholder groups."[22]

The ISSB has broad international backing in the development of global sustainability standards. Its efforts are supported by "the G7, the G20, the International Organization of Securities Commissions Oversight (IOSCO), the Financial Stability Board, and Finance Ministers and Central Bank Governors from over 40 jurisdictions."[23]

ISSB's Inaugural Standards: IFRS S1 and IFRS S2

In June 2023, the ISSB issued its inaugural standards, IFRS S1 and IFRS S2, which were formally rolled out in January 2024.[24] That marked the start of a new phase for global sustainability standards. IFRS S1 focuses

---

[21] *About the International Sustainability Standards Board*, IFRS (www.ifrs.org/groups/international-sustainability-standards-board/).
[22] *International Sustainability Standards Board*, IFRS Foundation (www.ifrs.org/groups/international-sustainability-standards-board/).
[23] Id.
[24] *ISSB issues inaugural global sustainability disclosure standards*, IFRS Foundation (www.ifrs.org/news-and-events/news/2023/06/issb-issues-ifrs-s1-ifrs-s2/).

CHAPTER 8   THE ESG STANDARDS AND FRAMEWORKS

on general disclosure requirements, enabling companies to communicate the sustainability-related risks and opportunities they face in the short, medium, and long term.[25] IFRS S2 zeroes in on specific climate-related disclosures.[26] Both standards are based on the prior work of the Task Force on Climate-related Financial Disclosures (TCFD).

The organization aims to provide confidence in the new standards and to illustrate the benefits of aligning standards and developing a global baseline. To that end, to accompany their release, the ISSB listed ten characteristics of the new standards (Table 8-1).

Table 8-1. *Ten Characteristics of the New ISSB Standards*[27]

1. Global disclosure standards: ISSB Standards allow companies and investors to [standardize] on a single, global baseline of sustainability disclosures for the capital markets, with any additional jurisdictional requirements being built on top of this global baseline.
2. International support: The ISSB's work has received strong support from investors, companies, policy makers, market regulators and others from around the world, including the International Organization of Securities Commissions Oversight (IOSCO), the Financial Stability Board, the G20 and the G7 Leaders.

(continued)

---

[25] *IFRS S1 General Requirements for Disclosure of Sustainability-related Financial Information*, IFRS Foundation (www.ifrs.org/issued-standards/ifrs-sustainability-standards-navigator/ifrs-s1-general-requirements/).

[26] *IFRS S2 Climate-related Disclosures*, IFRS Foundation (www.ifrs.org/issued-standards/ifrs-sustainability-standards-navigator/ifrs-s2-climate-related-disclosures/).

[27] *Ten things to know about the first ISSB Standards*, ISSB (June 27, 2023) (www.ifrs.org/news-and-events/news/2023/06/ten-things-to-know-about-the-first-issb-fstandards/?utm_medium=email&utm_source=website-follows-alert&utm_campaign=imme).

Table 8-1. (*continued*)

3. **Disclosure of decision-useful, material information:** Focusing exclusively on capital markets means that ISSB Standards only require information that is material, proportionate and decision-useful to investors. Moreover, by beginning with climate, companies can phase-in their sustainability disclosures.
4. **Building on and consolidating existing initiatives:** IFRS S1 and IFRS S2 are built on and consolidate the TCFD recommendations, SASB Standards, CDSB Framework, Integrated Reporting Framework and World Economic Forum metrics to streamline sustainability disclosures. Consolidation will help companies to benefit from their investments they've already made in sustainability disclosures while reducing the "alphabet soup" of sustainability disclosures.
5. **Reducing duplicative reporting:** The baseline approach provides a way to achieve global comparability for financial markets, and allow jurisdictions to further develop additional requirements if needed to meet public policy or broader stakeholder needs. This approach helps to reduce duplicative reporting for companies subject to multiple jurisdictional requirements.
6. **Helping companies communicate worldwide cost-effectively:** ISSB Standards have been designed to provide reliable information to investors; helping companies to communicate how they identify and manage the sustainability-related risks and opportunities they face over the short, medium and longer term.
7. **Connections with financial statements:** The information required by the ISSB Standards is designed to be provided alongside financial statements as part of the same reporting package. ISSB Standards have been developed to work with any accounting requirements, but they are built on the concepts underpinning IFRS Accounting Standards, already required for use by more than 140 jurisdictions.

(*continued*)

CHAPTER 8   THE ESG STANDARDS AND FRAMEWORKS

Table 8-1. (*continued*)

8. **Developed through rigorous consultation:** ISSB Standards have been developed using the same inclusive, transparent due process used to develop IFRS Accounting Standards—with more than 1,400 responses to the ISSB's proposals. All ISSB papers, feedback and technical decision-making are available to view online.
9. **Interoperability with broader sustainability reporting:** The ISSB's partnership with the Global Reporting Initiative enables the ISSB to build its requirements to be interoperable with GRI standards, helping to reduce the disclosure burden for companies using both ISSB and GRI Standards for reporting.
10. **A partnership for capacity building:** The ISSB's responsibilities do not stop at standard setting. At COP27, the ISSB announced plans for a capacity building partnership programme, helping to establish the necessary resources for high quality, consistent reporting across developed and emerging economies.[28]

The ISSB has made considerable strides in a relatively short time frame. The consolidated standards will modernize how companies disclose sustainability-related information. The organization continues to consult with stakeholders to prioritize next steps, which are expected to further streamline the standards universe.

Companies, investors, and regulators worldwide seem to be embracing these standards, recognizing their potential in promoting transparency and comparability of corporate sustainability disclosures. The ISSB has also committed to working with stakeholders to support the adoption of its standards, launching capacity-building initiatives to enable effective implementation.

---

[28] *Ten things to know about the first ISSB Standards*, ISSB (June 27, 2023) (www.ifrs.org/news-and-events/news/2023/06/ten-things-to-know-about-the-first-issb-fstandards/?utm_medium=email&utm_source=website-follows-alert&utm_campaign=imme).

CHAPTER 8   THE ESG STANDARDS AND FRAMEWORKS

Global Reporting Initiative (GRI)

GRI was initially developed as a collaboration between Ceres and the Tellus Institute in coordination with the United Nations Environment Programme (UNEP) in 1997.[29] GRI has gained significant traction among businesses over the decades. According to a KPMG survey in 2022, 78% of the world's 250 largest companies utilize GRI in their sustainability and ESG reporting, and GRI "remains the dominant global standard for sustainability reporting."[30]

The GRI standards cover a broad range of economic, environmental, and social topics, and have been adopted by organizations across the globe to report their sustainability performance and impacts. The approach is to deliver a detailed framework for reporting on various ESG issues. The standards are reviewed every three years by the Global Sustainability Standards Board (GSSB), an independent body created by GRI.[31] The standards are divided into three main sections:

- **Universal Standards:** These standards provide a foundation for all GRI reporting, covering core sustainability issues related to an organization's impact on the economy, society, and the environment.

- **Sector Standards:** These standards offer additional guidance for organizations in specific sectors, including high-impact sectors.

---

[29] *Our mission and history*, Global Reporting Initiative (www.globalreporting.org/about-gri/mission-history/).

[30] *KPMG 2022 U.S. CEO Outlook*, KPMG (https://kpmg.com/us/en/home/insights/2022/08/us-ceo-outlook-2022.html).

[31] *Global Sustainability Standards Board*, Global Reporting Initiative (www.globalreporting.org/about-gri/governance/global-sustainability-standards-board/).

- **Topic Standards:** These standards provide disclosures relevant to topical areas, such as waste, occupational health and safety, and biodiversity.[32]

GRI offers a standardized approach to sustainability reporting that enhances accountability and fosters trust with stakeholders. It allows companies to link their GRI reports to their disclosures for other business-focused sustainability reporting guidelines for uniformity of sustainability disclosures. The approach helps companies understand the scope of their environmental, social, and governance impacts, which can guide better decision-making and resource allocation.

CDP (Formerly Carbon Disclosure Project)

Since its founding in 2000, CDP, which was established as the Carbon Disclosure Project, has emerged as a leading global environmental disclosure system.[33] CDP is an international nonprofit organization with offices and local partners across 50 countries around the globe. Funding comes from a diverse range of sources, including government and philanthropic grants, membership fees, administrative fees, sponsorships, and data licensing.[34]

With its focus on climate change, water security, and deforestation, CDP drove a new era of corporate transparency that is transforming how businesses and governments operate. The primary aim of CDP was to make environmental reporting and risk management a business norm, thereby encouraging a sustainable economy. It was founded with the support of the investor community as a questionnaire answered by

---

[32] *The global standards for sustainability impacts*, Global Reporting Initiative (www.globalreporting.org/standards/).
[33] *What we do*, CDP (www.cdp.net/en/info/about-us/what-we-do).
[34] *Who we are*, CDP (www.cdp.net/en/info/about-us).

CHAPTER 8   THE ESG STANDARDS AND FRAMEWORKS

reporting companies, and its reach has grown exponentially. Today, more than 23,000 companies accounting for more than two-thirds of the world's market capitalization report through CDP.[35]

CDP announced that it will incorporate the ISSB climate disclosure standard into the CDP disclosure standard, beginning in 2024.[36] It is important to highlight that CDP and GRI, two of the original sustainability framework providers, "work together to align best practice and avoid duplication of disclosure effort to ease the reporting burden for the thousands of companies that report through CDP and the GRI [standards]."[37]

CDP operates the world's largest environmental disclosure system, covering thousands of companies, cities, states, and regions.[38] These disclosure areas include climate change, water security, and deforestation. This system not only provides a platform for organizations to report their emissions but also stimulates a more sustainable economy by encouraging transparency and accountability.

Taskforce on Nature-Related Financial Disclosures (TNFD)

The Taskforce on Nature-related Financial Disclosures (TNFD) is a pioneering initiative developed to facilitate the incorporation of nature into businesses and financial decision-making. Established in June 2021, the TNFD aims to support the global shift in financial flows from

---

[35] Id.

[36] *CDP and environmental disclosure standards and frameworks*, CDP (www.cdp.net/en/guidance/environmental-disclosure-standards-and-frameworks).

[37] Id.

[38] *Accelerating the Rate of Change: CDP Strategy 2012-2025*, CDP (https://cdn.cdp.net/cdp-production/comfy/cms/files/files/000/005/094/original/CDP_STRATEGY_2021-2025.pdf).

CHAPTER 8   THE ESG STANDARDS AND FRAMEWORKS

nature-negative to nature-positive outcomes.[39] The objective of the TNFD is to reduce the unprecedented rate of nature and biodiversity loss that poses significant risks to businesses, financial institutions, and the foundations of society.[40]

TNFD is funded by the governments of Australia, France, Germany, the Netherlands, Norway, Switzerland, and the United Kingdom, as well as the United Nations Development Programme (UNEP), the Global Environment Facility (GEF), the Children's Investment Fund Foundation (CIFF), and the Macdoch Foundation.[41] The Taskforce consists of 40 members representing financial institutions, corporations, and market service providers with more than USD$20 trillion in assets.[42]

TNFD is working to develop a robust risk management and disclosure framework to enable organizations to identify, evaluate, manage, and report nature-related risks and opportunities. The Taskforce unveiled its first beta iteration of the disclosure framework in March 2022 and the final version in September 2023.[43] The TNFD Recommendations are also designed to be in alignment with the Kunming-Montreal Global Biodiversity Framework, which we covered in Chapter 4. This alignment is key to facilitating global consistency in environmental disclosure standards.

The framework follows the TCFD model, and it promotes the integration of nature-related considerations into strategic and capital allocation decision-making processes. It is centered around four key disclosure pillars, mirroring the structure of the TCFD. These pillars include

---

[39] *Our mission*, Taskforce on Nature-related Financial Disclosures (https://tnfd.global/about/#mission).

[40] Id.

[41] *About us*, Taskforce on Nature-related Financial Disclosures (https://tnfd.global/about/#funded).

[42] Id.

[43] *Our history*, Taskforce on Nature-related Financial Disclosures (https://tnfd.global/about/history/).

325

CHAPTER 8   THE ESG STANDARDS AND FRAMEWORKS

1. Governance
2. Strategy
3. Risk Management
4. Metrics and Targets[44]

The TNFD framework is voluntary, and its adoption depends on the willingness of organizations to disclose their nature-related risks and opportunities. Given the growing regulatory risk and stakeholder demand for transparency, it is anticipated that many organizations will likely embrace the framework. It's a first mover framework for the common assessment, disclosure, and management of nature-related risks and impacts.

On January 16, 2024, TNFD announced that 320 organizations from more than 46 countries such as public companies and financial institutions, including asset managers representing USD$14 trillion AUM, had committed to begin making nature-related disclosures in line with the TNFD Recommendations issued in September 2023.[45] Nature-related financial disclosures will continue to mature as investors, corporations, and regulators join other stakeholders in embracing the value and impact of natural capital and biodiversity.

---

[44] Id.

[45] *320 companies and financial institutions to start TNFD nature-related corporate reporting*, Taskforce on Nature-related Financial Disclosures (January 16, 2024) (https://tnfd.global/320-companies-and-financial-institutions-to-start-tnfd-nature-related-corporate-reporting/#:~:text=DAVOS%2C%20SWITZERLAND%20%E2%80%93%2016%20January%202024,published%20in%20September%20last%20year).

8.3 Carbon Accounting and the GHG Protocol

Carbon Accounting

Carbon accounting is an accounting methodology designed to measure, track, and report an organization's carbon dioxide and other GHG emissions, also known as its carbon footprint. It operates somewhat like financial accounting, but instead of tracking income and expenses, it measures the direct and indirect emissions produced by a company. The concept of carbon accounting has evolved over the past few decades. As previously discussed, initial efforts focused on national-level GHG accounting with the UNFCCC requirement that countries report their annual emissions.

With climate change exacerbating and stakeholders increasingly pressuring companies to improve their sustainability, carbon accounting practices should be part of the corporate strategy. Carbon accounting provides a tangible measure of a company's carbon emissions and a snapshot of operational efficiency. There are several accepted standards and guidelines that govern carbon accounting and provide companies with a structured approach to measuring and reporting their GHG emissions.

The challenges associated with carbon accounting include the veracity of data and collection procedures, as tracking and gathering accurate data from various internal and external sources is difficult, and the resources required to establish and manage the end-to-end process. To overcome these challenges, companies can leverage technology and digital tools for efficient data collection, analysis, and reporting. Automation can help streamline the carbon accounting process, reducing time and resource constraints. Companies can invest in training and capacity building to ensure that decision-makers understand the strategic importance of carbon accounting. This will help integrate carbon accounting into the organization's overall business strategy, driving sustainability and corporate responsibility.

Greenhouse Gas (GHG) Protocol

The Greenhouse Gas (GHG) Protocol is a standardized framework and set of tools to measure and manage greenhouse gas emissions.[46] It has earned widespread adoption and is an integral part of many sustainability certifications and reporting systems. The GHG Protocol was established in 1998, born out of a partnership between the World Resources Institute (WRI) and the World Business Council for Sustainable Development (WBCSD).[47]

The impetus behind its inception was to address the urgent need for an internationally standardized mechanism that corporations could use to account for and report GHG emissions. The GHG Protocol is applied to measure and manage emissions from private and public sector operations, products, cities, and policies.

The GHG Protocol also offers guidance, tools, and training to various entities, including businesses, governments, and nongovernmental organizations. This facilitates the measurement, management, and reporting of climate-warming emissions. The GHG Protocol serves as a foundation for numerous sustainability certifications and reporting systems, making it a globally recognized framework.[48]

---

[46] *Standards*, Greenhouse Gas Protocol (https://ghgprotocol.org/standards).

[47] *Greenhouse Gas Protocol*, World Resources Institute (www.wri.org/initiatives/greenhouse-gas-protocol).

[48] *Calculation Tools and Guidance*, Greenhouse Gas Protocol (https://ghgprotocol.org/calculation-tools-and-guidance).

8.4 Incorporation of Standards into Legislation and Regulation

We've seen the incorporation of and prescription for the use of a number of voluntary reporting standards in legislative and regulatory vehicles. For example, the GHG Protocol has been incorporated into legislative and regulatory language mandating its use for greenhouse gas measurement and reporting, including California SB 253: The Climate Corporate Data Accountability Act, signed into law in 2023. Another example is the Task Force on Climate-related Financial Disclosures (TCFD) standard, which had been similarly incorporated into California SB 261: The Climate-Related Financial Risk Act, signed into law in 2023. Given the disbanding of TCFD and the alignment with the IFRS Standards, it is expected that those requirements will be amended. A number of jurisdictions have already begun incorporating the IFRS Standards into their regulatory lexicon. We covered those measures in detail in Chapter 7.

In addition to the incorporation of accepted voluntary standards into the text of legislation and regulations, some governmental bodies have structured their own sustainability standards. One leading example is EFRAG, as we've covered in previous sections. With the alignment and consolidation of standards, it is expected that requiring the adoption and use of these voluntary models will become more commonplace. That follows the ongoing shift from voluntary to mandatory ESG reporting. Companies will have to continue to navigate the updates and comply with the relevant jurisdictional requirements.

CHAPTER 8   THE ESG STANDARDS AND FRAMEWORKS

8.5 Implications for Compliance and Risk Management

The burgeoning regulatory environment has created a groundswell for risk, compliance, legal, and audit teams within companies. The shift from voluntary to mandatory reporting and the regulations that require reporting have added focus and scrutiny for the second and third lines of defense. In a recent study conducted by Deloitte of 300 senior finance, legal, and sustainability leaders, 57% reported that data availability and quality remain their greatest challenges for ESG disclosure, and 89% said that their company will likely enhance its ESG controls.[49]

Establishing robust compliance and risk management systems and protocols is becoming increasingly important in the new standards and regulatory environment. We've discussed the importance of integrity in the data collection and management process. Companies should carefully vet any ESG comments, forward-looking statements, reports, and disclosures to ensure that they are factually accurate, do not omit any material information, and align with required standards and frameworks.

CHAPTER TAKEAWAYS

- There is an ongoing consolidation and alignment of the historic global ESG standards

- The importance of natural capital and biodiversity conservation has led to the creation of new nature-related risk standards

---

[49] *US Public Companies Prepare for Increasing Demand for High Quality ESG Disclosures*, Deloitte (March 14, 2022) (www2.deloitte.com/us/en/pages/about-deloitte/articles/press-releases/us-public-companies-prepare-for-increasing-demand-for-high-quality-esg-disclosures.htmlwww.deloitte.com/us/en/pages/about-deloitte/articles/press-releases/us-public-companies-prepare-for-increasing-demand-for-high-quality-esg-disclosures.html).

CHAPTER 8   THE ESG STANDARDS AND FRAMEWORKS

- Regulators are increasingly mandating adherence to specific standards frameworks in legislation and regulation
- Companies should stay abreast of the standards developments and apply relevant standards to their reporting regimes
- Businesses should be attuned to and incorporate nature-related risks into their strategies

CHAPTER 9

The ESG Ratings Providers and Indices

In addition to frameworks and standards, ESG ratings are part of the ESG reporting paradigm. ESG ratings are produced by third-party research firms and credit ratings agencies (CRAs) based on proprietary methodologies. They assess a company's ESG performance and provide scores that are used by market participants in capital allocation decisions.

9.1 Role of ESG Ratings

> *ESG ratings provide a synthetic indicator of an issuer's ESG characteristics or exposure to ESG risks. ESG ratings are typically produced by ESG information providers and are non-credit products. They are synthetic indicators of the ESG characteristics or exposure to ESG risks of an issuer of equity or debt instruments.*[1]
>
> —UN Principles for Responsible Investment

---

[1] *ESG in credit ratings and ESG ratings,* UN Principles for Responsible Investment (March 5, 2023) (www.unpri.org/credit-risk-and-ratings/esg-in-credit-ratings-and-esg-ratings/11071.article).

© Kristyn Noeth 2024
K. Noeth, *The ESG and Sustainability Deskbook for Business,*
https://doi.org/10.1007/979-8-8688-0261-4_9

CHAPTER 9   THE ESG RATINGS PROVIDERS AND INDICES

ESG ratings are a quantifiable measure of a company's sustainability and societal impact based on ESG criteria. Over the years, ESG issues have grown in significance, and with it, the necessity to evaluate companies based on their ESG performance. Consequently, companies are paying more attention to their ESG performance, and many have introduced ESG scorecards to evaluate their own performance, independent of external appraisals. Given the expanding importance of ESG investing, transparency around how ESG scores are computed and what they measure is necessary.

An ESG score is meant to be an objective evaluation of an organization's performance against environmental, social, and governance criteria. The score can be used to validate a company's ESG efforts, enable industry benchmarking, manage progress, attract investors, and identify potential areas of concern and risk from an ESG perspective. Ratings providers undertake to collect relevant information about the companies. The sources are both self-reported data submitted by companies and data collected from third parties. A company's ESG score is based on the company's initiatives, and the data used to rate the company can come from a variety of sources, including publicly available information, corporate ESG disclosures, and interviews with company management.

The scores are used in different ways by different stakeholders. For investors, ESG scores provide a quantifiable measure that can help to inform and screen investment decisions. Companies also use ESG scores to evaluate and benchmark their own performance with those of their competitors, and to identify areas for improvement as well as to inform strategic moves in the market.

CHAPTER 9   THE ESG RATINGS PROVIDERS AND INDICES

Regulators are also interested in ESG scores as they are useful in assessing whether ESG information might be material to a company's financial performance and in evaluating ESG claims made by a company. ESG scores measure a company's performance in relation to environmental, social, and governance issues. We've covered the three ESG pillars and their key factors throughout the book, and those generally correspond to the criteria that ratings providers measure.

Scores are calculated using a multistep process that varies slightly depending on the ratings agency. Scores are assigned to each rating factor and are combined using a defined weighting mechanism to arrive at a composite ESG score. The final score is then classified according to a rating scheme similar in many cases to the credit ratings agency scale, which is used to assess the credit risk of an issuer of debt instruments (e.g., AAA score (Leader), CCC score (Laggards)[2]).

As an example, MSCI, a prominent ESG ratings provider, considers around 35 key ESG issues as the basis for its rating (Figure 9-1). Data is then collected across each of the issues identified in the ESG Ratings Key Issue Framework, as well as potentially additional criteria relevant to a company's specific industry, to create an ESG score.

---

[2] *How does MSCI ESG Ratings work?* MSCI (www.msci.com/our-solutions/esg-investing/esg-ratings).

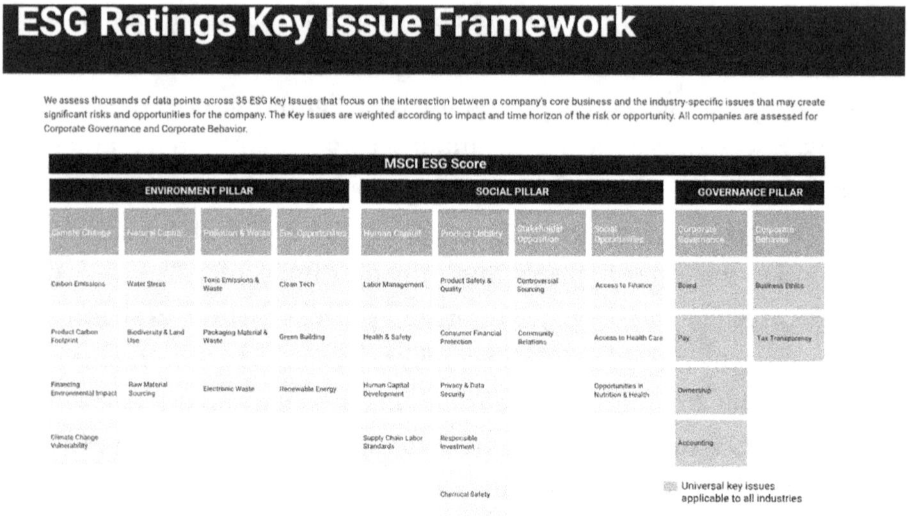

CHAPTER 9   THE ESG RATINGS PROVIDERS AND INDICES

ESG Ratings Key Issue Framework

We assess thousands of data points across 35 ESG Key Issues that focus on the intersection between a company's core business and the industry-specific issues that may create significant risks and opportunities for the company. The Key Issues are weighted according to impact and time horizon of the risk or opportunity. All companies are assessed for Corporate Governance and Corporate Behavior.

MSCI ESG Score			
ENVIRONMENT PILLAR	SOCIAL PILLAR		GOVERNANCE PILLAR
Climate Change / Natural Capital / Pollution & Waste / Env. Opportunities	Human Capital / Product Liability / Stakeholder Opposition / Social Opportunities		Corporate Governance / Corporate Behavior
Carbon Emissions / Water Stress / Toxic Emissions & Waste / Clean Tech	Labor Management / Product Safety & Quality / Controversial Sourcing		Board / Business Ethics
Product Carbon Footprint / Biodiversity & Land Use / Packaging Material & Waste / Green Building	Health & Safety / Consumer Financial Protection / Community Relations	Access to Health Care	Pay / Tax Transparency
Financing Environmental Impact / Raw Material Sourcing / Electronic Waste / Renewable Energy	Human Capital Development / Privacy & Data Security	Opportunities in Nutrition & Health	Ownership
Climate Change Vulnerability	Supply Chain Labor Standards / Responsible Investment		Accounting
	Chemical Safety		

■ Universal key issues applicable to all industries

Figure 9-1. *MSCI ESG Ratings Key Issue Framework*[3]

9.2 ESG Ratings Providers

MSCI and Sustainalytics are considered two of the biggest ESG ratings providers, mainly due to the scope of their wide coverage. Many of the providers offer a range of data products under the ESG umbrella.

---

[3] *ESG Ratings Key Issue Framework*, MSCI (www.msci.com/our-solutions/esg-investing/esg-ratings/esg-ratings-key-issue-framework). Reproduced by permission of MSCI ESG Research LLC ©2024. All rights reserved. No further reproduction or dissemination is permitted. Although MSCI's information providers, including without limitation, MSCI ESG Research LLC and its affiliates (the "ESG Parties"), obtain information from sources they consider reliable, none of the ESG Parties warrants or guarantees the originality, accuracy and/or completeness, of any data herein and expressly disclaim all express or implied warranties, including those of merchantability and fitness for a particular purpose. None of the ESG Parties shall have any liability for any errors or omissions in connection with any data herein, or any liability for any direct, indirect, special, punitive, consequential or any other damages (including lost profits) even if notified of the possibility of such damages.

CHAPTER 9   THE ESG RATINGS PROVIDERS AND INDICES

The following is a list of some of the primary ESG ratings and analytics providers, along with basic information on their platforms and coverage:

- **MSCI ESG ratings:** MSCI publishes ratings and research on over 8,500 companies and also rates equity and fixed income securities, loans, mutual funds, ETFs, and countries.[4]

- **Sustainalytics:** Sustainalytics, a Morningstar subsidiary, offers ESG ratings on more than 20,000 companies and 172 countries.[5]

- **Bloomberg ESG:** Bloomberg offers ESG data for more than 15,000 companies in over 100 countries, organized into 2,000 fields.[6]

---

[4] *ESG Ratings*, MSCI (www.msci.com/our-solutions/esg-investing/esg-ratings).

[5] *Company ESG Risk Ratings*, Morningstar Sustainalytics (www.sustainalytics.com/esg-ratings?utm_source=google&utm_medium=paid+search&utm_campaign=scs_genericads_global_2305_publicratingspage_en&utm_content=scs_ga_global_230518_publicratingsgoogleads_en&utm_term=esg%20reporting&utm_campaign=Shared+-+ESG+Risk+Rating+Public+Page+-+July+2020&utm_source=adwords&utm_medium=ppc&hsa_acc=4619360780&hsa_cam=10594802130&hsa_grp=104743607957&hsa_ad=459189391085&hsa_src=g&hsa_tgt=aud-2009601961397:kwd-2211793615&hsa_kw=esg%20reporting&hsa_mt=b&hsa_net=adwords&hsa_ver=3&gad_source=1&gclid=CjOKCQiA1rSsBhDHARIsANB4EJbUgh5EEPFkIyHHKbow5Ufy8DVONPqQL94c6_ssWaa_ORwv4P7o6HOaAvWwEALw_wcB).

[6] *ESG and Sustainable Finance Solutions*, Bloomberg (www.bloomberg.com/professional/solution/sustainable-finance/?utm_medium=Adwords_SEM&utm_source=pdsrch&utm_content=AMER_ESGAwareness_2023&utm_campaign=720598&tactic=720598&gad_source=1&gclid=CjOKCQiA1rSsBhDHARIsANB4EJYPvH4ioNjdjS7Y7q2vo8dRLZp8oSCJeH4O0gRdWaAUZnoLXO1B37gaApYmEALw_wcB).

337

CHAPTER 9   THE ESG RATINGS PROVIDERS AND INDICES

- **FTSE Russell ESG scores:** The ratings include 7,200 securities in 47 developed and emerging markets.[7] It is owned by the London Stock Exchange Group.

- **ISS (Institutional Shareholder Services) ESG:** ISS ESG rates 7,800 companies, 26,500 funds, and 830 sovereign issuers.[8]

- **S&P Global:** The S&P Global ESG data set includes more than 10,000 companies.[9]

- **Moody's ESG:** Moody's ESGView, part of Moody's Corporation, provides ESG ratings for over 10,000 entities.[10]

- **LSEG Data & Analytics (formerly Refinitiv):** LSEG Data & Analytics calculates ESG scores for more than 15,000 companies.[11] It is also owned by the London Stock Exchange Group.

---

[7] *ESG Scores*, FTSE Russell (www.ftserussell.com/financial-data/sustainability-and-esg-data/esg-ratings).
[8] *ISS ESG*, ISS (www.issgovernance.com/esg/).
[9] *ESG Scores*, S&P Global (www.spglobal.com/esg/solutions/data-intelligence-esg-scores?utm_source=google&utm_medium=cpc&utm_campaign=Brand_ESG_Search&utm_term=s%26p%20global%20esg%20scores&utm_content=534418150272&gclid=CjOKCQiA1rSsBhDHARIsANB4EJYOuTGCUbkXILeAFM6YSIHijKn7rf45S3bnNwqCkUlRpDGLIkNuKxsaAgzLEALw_wcB).
[10] *Moody's ESGView*, Moody's (www.moodys.com/web/en/us/esgview.html).
[11] *LSEG ESG Scores*, LSEG Data & Analytics (www.lseg.com/en/data-analytics/sustainable-finance/esg-scores).

CHAPTER 9   THE ESG RATINGS PROVIDERS AND INDICES

9.3 ESG Indices and Relevancy in the Market

An index tracks a group of securities designed to represent a particular market.[12] ESG indices differ from traditional broad market indexes in that they incorporate ESG criteria into the security selection. They are performance benchmarks and are drivers of sustainable investing. In addition, these indices also serve as a benchmark for asset allocation in wealth management.

Several ESG ratings providers are also index providers. Many of the same criticisms of ESG ratings apply to ESG indexes, which we'll examine in the next section. Popular ESG indices include those compiled by MSCI, FTSE Russell, Bloomberg, and S&P Global. An interesting development is the recent launch of two biodiversity indices by S&P Dow Jones: (1) the S&P 500 Biodiversity Index and (2) the S&P Global LargeMedCap Biodiversity Index. In its announcement, the company notes that its "research shows that 85 percent of the world's largest companies have a significant dependency on nature and biodiversity."[13] There are also ESG data companies, such as RepRisk, that identify ESG risks by compiling comprehensive data sources mapped against the UN Global Compact principles, the UN Sustainable Development Goals, and specific standards sets.[14]

---

[12] *Market Indices*, US Securities and Exchange Commission (www.sec.gov/answers/indices.htm).

[13] *S&P Dow Jones Indices Launches Biodiversity-Focused Benchmarks*, S&P Down Jones Indices (February 27, 2024) (www.spglobal.com/spdji/en/index-launches/article/sp-dow-jones-indices-launches-biodiversity-focused-benchmarks/#:~:text=%22The%20S&P%20Biodiversity%20Indices%20are,resilient%20and%20ecologically%20conscious%20investment).

[14] *Delivering the transparency that drives better decisions*, RepRisk (www.reprisk.com/about).

CHAPTER 9 THE ESG RATINGS PROVIDERS AND INDICES

9.4 Criticism of Ratings Providers

ESG ratings agencies have received a wave of criticism, with questions being raised about their methodologies, credibility, and overall effectiveness.[15] The as yet unregulated nature of the ESG ratings market has led to a host of concerns that ratings can potentially mislead consumers, investors, or other market participants, resulting in uninformed decisions regarding ESG-related investments, risks, impacts, and opportunities. Investors, corporations, and regulators alike rely on ESG ratings providers in navigating the complex world of sustainability. These agencies sift through masses of data, providing insights into a company's environmental, social, and governance performance. Their ability to accurately analyze and benchmark is key in investment decisions and corporate strategy.

As the influence of ratings expands, so does the scrutiny over ratings procedures and assessments. Critics argue that "significant shortcomings exist in [the] objectives, methodologies, and incentives" of ESG ratings agencies, "which detract from the informativeness of their assessments."[16] This is compounded by the looming and irreversible impacts of climate change as climate risk is an area of primary importance in municipal, insurance, construction, and other markets. At present, climate risk is not necessarily accurately quantified or done so with detailed specificity in ESG ratings.

---

[15] Boffo, R. and R. Patalano, *ESG Investing: Practices, Progress and Challenges*, OECD Paris (2020) (www.oecd.org/finance/ESG-Investing-Practices-Progress-and-Challenges.pdf).

[16] David F. Larcker, Lukasz Pomorski, et al. *ESG Ratings: A Compass without Direction*, Rock Center for Corporate Governance at Stanford University (August 4, 2022) (https://papers.ssrn.com/sol3/Delivery.cfm/SSRN_ID4179647_code1199479.pdf?abstractid=4179647&mirid=1&type=2).

CHAPTER 9   THE ESG RATINGS PROVIDERS AND INDICES

One of the most prominent criticisms against ESG ratings providers is the potential for conflicts of interest. For example, some of the ratings providers offer consulting services to the companies they rate, advising them on how to improve their scores. This practice raises questions about the impartiality of these agencies. Critics argue that regulatory measures should be put in place to prevent such conflicts.

Another issue lies in the inconsistency of ESG ratings across different providers. Each provider uses its own unique methodology to collect, analyze, and score data. As a result, the same company can receive vastly different ratings from different agencies, leading to confusion among investors and companies alike. It's important to note that while credit ratings typically correlate between credit ratings agencies, ESG ratings do not correlate between ESG ratings providers and there is widespread disparity.[17] A study of ESG ratings issued by six different providers (MSCI, Sustainalytics, Moody's ESG, Refinitiv, S&P Global, and KLD) found the correlation between ratings ranged from 0.36 to 0.71.[18]

The lack of transparency around the ratings methodologies is another major bone of contention. Critics argue that the secretive nature of these methodologies undermines the credibility of the ratings. There is an ongoing call for greater transparency in how the ratings are calculated and what they aim to measure.

Data limitations pose another major challenge for ESG ratings agencies. Companies self-report the bulk of the data and not all companies disclose the same level of information, leading to gaps in the data. In some instances, ratings providers seek third-party data such as from local water utilities that only provide estimated or even more problematic,

---

[17] Beth Stackpole, *ESG ratings: Don't throw the baby out with the bath water*, MIT (February 23, 2023) (https://mitsloan.mit.edu/ideas-made-to-matter/esg-ratings-dont-throw-baby-out-bath-water).

[18] Florian Berg, Julian F Kölbel, and Roberto Rigobon, *Aggregate Confusion: The Divergence of ESG Ratings*, Oxford Academic (May 23, 2022) (https://academic.oup.com/rof/article/26/6/1315/6590670?login=false).

regional site usage data. These input limitations can result in inaccurate or incomplete ratings. Additionally, aggregating ESG scores into a single metric rather than relative weighting taking into account industry and other factors may not provide the most accurate profile.[19] Even when ratings agencies provide that they normalize ratings by industry, an industry sector bias could remain.

The cost of obtaining ESG ratings is typically flagged as another concern. ESG ratings can be expensive, with costs often passed on to investors in the form of higher management fees. A recent ERM research study found that 33 institutional investors spent an average of $487,000 per year on external ESG ratings, data, and consultants.[20] Critics argue that this raises barriers to entry for smaller investors and companies, limiting the potential for widespread adoption of ESG practices.

9.5 Potential Regulation and Oversight of Ratings Providers

The future of ESG rating agencies hinges on regulatory action to establish standardised reporting for companies and maintain a competitive market that promotes quality and efficiency. Emphasising transparency, addressing conflicts of

---

[19] *Greater ESG rating consistency could encourage sustainable investments*, Institute for Energy Economics and Financial Analysis (October 11, 2022) (https://ieefa.org/resources/greater-esg-rating-consistency-could-encourage-sustainable-investments).

[20] *Costs and Benefits of Climate-Related Disclosure Activities by Corporate Issues and Institutional Investors*, The SustainAbility Institute by ERM (May 17, 2022) (www.sustainability.com/globalassets/sustainability.com/thinking/pdfs/2022/costs-and-benefits-of-climate-related-disclosure-activities-by-corporate-issuers-and-institutional-investors-17-may-22.pdf).

interest and refining methodologies will help ESG ratings better serve investors and society at large in a rapidly changing landscape.[21]

—Julian F Kölbel, Florian Berg,
and Professor Roberto Rigobon

As regulators across the globe begin to scrutinize ESG ratings agencies more closely, these providers will need to navigate the changing regulatory landscape. New rules and regulations could have significant implications for their business models and methodologies. Regulatory oversight could help streamline ESG ratings, enhance their credibility, and address conflicts of interest and other issues. Regulatory efforts to standardize ESG disclosures and reporting requirements would potentially lend more structure and consistency to the ratings system.

The comparison of credit ratings agencies (CRAs) to ESG ratings providers is informative. CRAs are regulated entities, and they must go through a certification process to operate.[22] There is transparency around their methodologies and CRAs also must conduct their assessments and activities in accordance with an adopted set of principles of conduct. This is a primary reason why credit ratings are comparable across providers. Some of the ESG ratings providers are also CRAs and some are not.

---

[21] Florian Berg, Julian F Kölbel, and Roberto Rigobon, *Rating the ESG rating agencies*, Financial Times (July 3, 2023) (www.ft.com/content/e9eaa11a-31e0-4f60-9a65-b6883546e8da).

[22] *ESG in credit ratings and ESG ratings*, UN Principles for Responsible Investment (March 5, 2023) (www.unpri.org/credit-risk-and-ratings/esg-in-credit-ratings-and-esg-ratings/11071.article).

CHAPTER 9   THE ESG RATINGS PROVIDERS AND INDICES

The European Union (EU) has taken a significant step to regulate providers of ESG ratings. The European Parliament and Council have agreed to a provisional agreement to oversee ESG ratings.[23] The proposal came after a lengthy consultation process with market participants.[24] This new regulatory oversight is designed to enhance the transparency and integrity of ESG ratings, which are increasingly driving investment decisions and risk management strategies.

The proposed regulation will provide rules governing ESG ratings providers across the EU. The European Commission noted the importance of maintaining a variety of approaches in the EU ESG ratings market; thus, it does not intend to require harmonization of the methodologies used by ESG ratings providers. The proposed regulation is intended to

1. "Enhance the integrity, transparency, responsibility, good governance, and independence of ESG rating activities.

2. Contribute to the transparency and quality of ESG ratings.

3. Mitigate the risk of greenwashing in the financial field.

---

[23] *Environmental, social and governance (ESG) ratings: Council and Parliament reach agreement*, Council of the European Union (February 5, 2024) (www.consilium.europa.eu/en/press/press-releases/2024/02/05/environmental-social-and-governance-esg-ratings-council-and-parliament-reach-agreement/); *Proposal for a Regulation of the European Parliament and of the Council on the transparency and integrity of Environmental, Social and Governance (ESG) ratings activities*, European Commission, COM/2023/314 final, Strasbourg, June 13, 2023 (https://eur-lex.europa.eu/legal-content/EN/TXT/?uri=CELEX%3A52023PC0314).

[24] *Targeted consultation on the functioning of the ESG ratings market in the European Union and on the consideration of ESG factors in credit ratings*, European Commission (https://finance.ec.europa.eu/regulation-and-supervision/consultations/finance-2022-esg-ratings_en).

344

CHAPTER 9 THE ESG RATINGS PROVIDERS AND INDICES

4. Increase comparability and reliability of ESG ratings."[25]

The proposed regulation applies broadly to ESG ratings issued by providers operating in the EU that either publicly disclose ratings or distribute ratings to regulated financial firms within the EU or to public authorities in the member states. The proposal defines an ESG rating as

> *an opinion, a score, or a combination of both, regarding an entity, a financial instrument, a financial product, or an undertaking's ESG profile or characteristics or exposure to ESG risks or the impact on people, society, and the environment, that are based on an established methodology and defined ranking system of rating categories and that are provided to third parties, irrespective of whether such ESG rating is explicitly labelled as "rating" or "ESG score."*[26]

Under the proposed regulation, any person intending to provide ESG rating activities in the EU must be authorized to do so by the European Securities Markets Authority (ESMA). ESMA is granted powers under the proposed regulation, including the authority to request information, carry out on-site inspections, withdraw or suspend the use of ESG ratings, impose fines and penalty payments, and issue public notices.[27] ESMA will also supervise ESG ratings providers' compliance with the regulation as it does for CRAs.[28]

---

[25] Id.

[26] Id.

[27] Id.

[28] *Credit Ratings Agencies*, European Securities Markets Authority (ESMA) (www.esma.europa.eu/esmas-activities/investors-and-issuers/credit-rating-agencies).

CHAPTER 9   THE ESG RATINGS PROVIDERS AND INDICES

The proposed regulatory approach promotes integrity and reliability and requires certain practices of ESG ratings providers. These practices include preventing conflicts of interest, requiring competence and independence of rating analysts, maintaining record-keeping, implementing complaints mechanisms, and refraining from outsourcing where it would risk undermining the quality of the provider's internal control policies and procedures.[29] For ESG ratings providers, the draft proposal would require compliance with stringent measures to prevent conflicts of interest, the necessity for significant organizational changes, and the burden of increased supervision and potential fines.

The proposal must be negotiated by the European Parliament and EU member states. The timeline for passing the proposal is in the near term as EU elections are scheduled for June 2024.[30] Other jurisdictions, including the UK Financial Conduct Authority, the Securities and Exchange Board of India, and the Financial Services Agency of Japan, are undertaking reviews of ESG ratings with considerations for regulation.[31]

---

[29] Id.

[30] John Ainger and Frances Schwartzkopff, *EU Puts ESG Rating Firms on Notice as Major Overhaul Planned*, Bloomberg (June 8, 2023) (www.bloomberg.com/news/articles/2023-06-08/eu-seeks-major-reform-of-esg-ratings-firms-in-sweeping-overhaul).

[31] Hazel Ilango, *An unregulated ESG ratings system reveals its flaws*, Institute for Energy Economic and Financial Analysis (May 2, 2023) (https://ieefa.org/resources/unregulated-esg-rating-system-reveals-its-flaws); *Britain says it will regulate ESG ratings later in 2024*, Reuters (March 6, 2024) (www.reuters.com/world/uk/britains-finance-ministry-regulate-esg-ratings-2024-03-06/#:~:text=LONDON%2C%20March%206%20(Reuters),but%20gave%20no%20specific%20timeline).

CHAPTER TAKEAWAYS

- ESG ratings are used to assess a company's investment risk profile

- There is no consistent ESG ratings methodology or oversight of the ratings system

- The first foray into regulating the ratings system is underway in the EU, which is likely to have global implications

- Businesses should evaluate the accuracy and integrity of their ESG data collection and reporting processes

- Companies should understand their ESG ratings profile and how that information is used in the market

CHAPTER 10

The Role of the Corporation

An effective ESG strategy is a crucial aspect of a company's long-term success. Adopting ESG strategies can help companies attract and retain talent, meet regulatory requirements, appeal to a broader investor or shareholder base, and bolster reputation among customers, communities, and society. We'll review corporative purpose, structures, activities, programs, and innovation in this chapter.

10.1 Focus on Emerging ESG Trends and Risks

There is a tremendous amount of activity on the ESG front, for which companies will have to plan and strategize. We've pointed out several emerging trends and risks for businesses throughout the book. Table 10-1 is an outline of some of the key emerging trends and risks as a ready reference tool.

CHAPTER 10   THE ROLE OF THE CORPORATION

Table 10-1. *List of Key Emerging ESG Trends and Risks for Businesses*

1. Developments in **standards alignment** and the new **global regulatory frontier** should be monitored and tracked. In a short timeline, the new ESG requirements will come into effect, and regulatory agencies will turn their focus to **enforcement**.
2. Companies must assess **regulatory, governance, finance, and legal capabilities**; **adjust compliance, risk, and audit plans**; and **address knowledge and skills gaps at all levels** of the organization, including leadership ranks and boards of directors, to prepare for the new paradigm. Companies should also better integrate ESG into their **operations and financial planning**.
3. The regulatory press on climate-risk disclosures and GHG accounting is causing businesses to verify underlying **data collection** processes, to advance **climate transition plans**, and to view and report emissions across their complete **value chains**. It also places **heightened scrutiny** on disclosures and **transparency** in reporting.
4. Consumer expectations and regulatory pressure on **greenwashing** should spur responsible businesses to address their marketing practices and their environmental claims for veracity.
5. The importance of **protecting biodiversity and the value of natural capital** has come into sharper focus. Businesses will be expected to better measure and evaluate their ecosystem impacts and to practice **responsible sourcing and resource use** in their supply chains.
6. Current world affairs have renewed a focus on **human rights and fair labor standards.** Businesses will be expected to perform better **monitoring and due diligence** to protect human rights and promote fair labor in their supply chains.

(continued)

Table 10-1. (*continued*)

7. The interconnected aspects of ESG, such as human health and the disparate geographical impacts of climate change, will also bring the *just transition* into view as companies must better *bridge the connection between environmental and social issues*.

8. There is an ongoing *expansion of sustainable finance and investment* that will open opportunities, foster innovation, and bring viable solutions to market. As we noted in Chapter 2, ESG assets are projected to exceed USD$50 trillion by 2025—representing one-third of total global AUM.[1]

10.2 Fiduciary Duties and ESG
Fundamentals of the Fiduciary Duties

Corporate directors and officers of both private and publicly held corporations in the United States provide specific expertise and guide strategic direction while ensuring the operational efficiency of the corporation. Directors and officers have fiduciary duties to the corporation they serve, and those duties are fundamental to their role in and obligations to the corporation. Fiduciary duties refer to the legal obligations imposed upon corporate directors and officers, requiring them to act in the best interest of the corporation.

Fiduciary duties also serve as a guiding principle for directors and officers in carrying out their responsibilities. The legal basis for fiduciary duties is rooted in corporate law, specifically the statutes governing the operation of corporations. These laws outline the roles and responsibilities

---

[1] Bloomberg Intelligence. *ESG assets may hit $53 trillion by 2025, a third of global AUM* (February 23, 2021) (www.bloomberg.com/professional/blog/esg-assets-may-hit-53-trillion-by-2025-a-third-of-global-aum/).

CHAPTER 10   THE ROLE OF THE CORPORATION

of corporate directors and officers, explicitly mentioning their fiduciary obligations. In addition, court decisions interpreting those statues have shaped the understanding and enforcement of fiduciary duties.

Fiduciary duties generally encompass three primary obligations: the duty of care, the duty of loyalty, and the duty of good faith. Each duty carries unique responsibilities and expectations.

Duty of Care

The duty of care refers to the obligation of corporate directors and officers to exercise a degree of diligence and prudence that an ordinary individual would under similar circumstances. This duty requires them to make informed decisions after reasonable inquiry and careful consideration. The duty of care implies that directors and officers must remain informed about the corporation's affairs and make decisions based on thorough analysis and understanding. For example, before approving a significant corporate investment, they should review relevant financial information, understand the potential risks and benefits, and consult with experts if necessary.

Duty of Loyalty

The duty of loyalty mandates that officers and directors place the corporation's interests above their own. This duty prohibits them from engaging in self-dealing or taking advantage of corporate opportunities for personal gain. The duty of loyalty means that directors and officers must avoid situations that could potentially conflict with the corporation's interests. For instance, they should not engage in business transactions with the corporation for personal profit or use corporate assets for personal advantage.

A seminal case before the Delaware Court of Chancery, *In re Caremark International Inc. v. Derivative Litigation*, further refined a director's duties of care and loyalty to include a duty of oversight.[2] This is often referred to as the *Caremark duty*, and in response to the decision, corporations strengthened their compliance programs and their board oversight, reporting, and communication procedures. Subsequent cases have expanded the duty of oversight to officers as well as directors.[3]

Duty of Good Faith

The duty of good faith requires directors and officers to act honestly and with a sincere intention to serve the corporation's best interests. This duty complements the duties of care and loyalty, reinforcing the expectation for ethical conduct and integrity. In practice, the duty of good faith serves as a moral compass for directors and officers. It mandates that they perform their duties with sincerity and honesty. For instance, if they discover mismanagement or unethical behavior within the corporation, they should take appropriate action.

Business Judgment Rule

When discussing fiduciary duties, it's essential to consider the business judgment rule. This rule gives directors and officers the freedom to make business decisions without fearing legal repercussions, provided they act in good faith, based on reasonable information, and in the corporation's best interests. Courts routinely review directors' and officers' decisions in the context of the business judgment rule.

---

[2] *In re Caremark International Inc. Derivative Litigation*, 698 A.2d 959 (Del. Ch. 1996).
[3] *In re McDonald's Corp. Stockholder Derivative Litigation*, 2023 WL 387292, C.A. No. 2021-0324-JTL (Del. Ch. Jan. 26, 2023).

CHAPTER 10   THE ROLE OF THE CORPORATION

Fiduciary Duties and ESG Application

The rise of ESG has brought a range of challenges and considerations for corporate officers and directors. As a result, ESG has been transformative in how we think of fiduciary duties in the corporate world. Given the heightened scrutiny of corporate directors and officers with respect to their action or inaction on ESG issues, it has become imperative for leaders to understand the intersection of ESG with their established fiduciary duties.

There is ongoing debate on how ESG impacts director and officer fiduciary responsibilities, with the anti-ESG camp seeking to remove ESG considerations from the boardroom. Companies aim to balance conflicting demands from regulators, investors, shareholders, and public officials. That said, amid the shifting landscape, being reactive is no longer a choice; companies must proactively incorporate ESG into their business strategy and operations.

Corporate directors, officers, and leadership must ensure compliance with evolving reporting requirements around ESG issues. Fiduciaries should assess how ESG needs vary based on their industry, business, and investment strategy, among other factors. They should also press management to make the case explicitly and credibly for why particular ESG components align with their respective business plans, investment goals, and long-term strategies. Management and boards alike must be able to explain how ESG considerations help optimize shareholder or investor assets and value and minimize identifiable material risks to the enterprise.

10.3 Rise of Corporate Purpose and the Aligned Corporation

The corporate world has experienced a shift from shareholder capitalism to a more balanced approach that incorporates environmental, social, and governance priorities. Corporate purpose goes beyond profit generation

CHAPTER 10   THE ROLE OF THE CORPORATION

and encompasses a company's values, mission, and broader societal impact. The rise of corporate purpose has generated discussion, as has the concept of mission-aligned corporations.

Purpose-driven businesses focus on creating shared value that benefits both the company and its stakeholders. They can be said to outperform their competitors in the market. An oft-cited research study of 500,000 survey responses from a sample of US companies found that businesses that exhibit both high purpose and clarity around that purpose have higher stock market performance.[4] Purpose plays a central role in guiding business strategy, and it influences decision-making processes. That can guide a company, particularly when it must navigate downturns and manage crises.

A 2021 Deloitte report surveyed more than 200 C-suite executives and conducted in-depth interviews with leaders across industries to understand how their roles and companies are prioritizing purpose.[5] They learned that 79% of the executives surveyed reported their company had a clear and defined purpose strategy integrated within the core business strategy.[6] That same percentage reported that purpose supports talent recruitment, engagement, and retention.[7] The approach garners better results and more engaged employees, as well as customer loyalty and a better public standing. Generally, companies must do a better job to measure and show impact derived from purpose beyond the talent calculus.

---

[4] Gartenberg, Claudine, Andrea Prat, and George Serafeim. "Corporate Purpose and Financial Performance." Organization Science 30, no. 1 (January–February 2019): 1–18.

[5] Shira Beery, John Mennel, and Kwasi Mitchell, *C Suite insights: How purpose delivers value in every function and for the enterprise*, Deloitte (www2.deloitte.com/content/dam/Deloitte/us/Documents/about-deloitte/us-state-of-purpose.pdf).

[6] Id.

[7] Id.

Leaders have a critical role in championing corporate purpose and modeling a purpose-driven culture. Mission-aligned companies have embedded their mission, purpose, and values within their strategies and operations. They have an established objective to balance economics with environmental, social, and governance factors. As we'll discuss in the next section, with respect to Public Benefit Corporations and Certified B Corporations, those goals are not mutually exclusive. This mission alignment may be described as creating shared value. It's a more circular viewpoint of the interconnectedness between business and society.

10.4 Alternative Corporate Structures

Public Benefit Corporations (PBCs)

Public Benefit Corporations (PBCs) stand out as an innovative model that harmonizes financial profit with societal benefit. This new breed of corporations strives to cultivate a more sustainable and inclusive economy by balancing the interests of shareholders with those of broader stakeholders, including employees, communities, and the environment. A PBC is a legally recognized corporate entity that seeks to generate profit while also making a positive impact on society. Unlike traditional corporations, PBCs are obligated to consider the interests of all stakeholders, not just shareholders, and to apply that rubric in their decision-making process. This relatively new business model emanates from the concept of stakeholder capitalism, aiming to balance economic prosperity with social responsibility.

CHAPTER 10   THE ROLE OF THE CORPORATION

The PBC entity structure emerged in response to the growing call for corporate accountability and social responsibility. The first US state to pass benefit corporation legislation was Maryland in 2010.[8] Today, nearly 40 have enacted legislation allowing for the creation of Public Benefit Corporations. Internationally, many countries including Canada, Ecuador, Italy, and Colombia have enacted similar legislation. The incorporation of or transformation of an existing corporation into a PBC requires amending the company's governing documents, a move that requires support from shareholders in most jurisdictions. The process involves filing documents or amendments with the state, declaring the company as a benefit corporation, and outlining its specific public benefit objectives.

While PBCs and nonprofits share a commitment to social or environmental objectives, they differ in structure and purpose. PBCs, unlike nonprofits, have shareholders who own the company and expect a return on their investment. PBCs are allowed to pursue profit-making activities, while nonprofits are dedicated to serving a public benefit without making a profit. PBCs demonstrate that a commitment to public benefit can be harmoniously combined with a successful business.

PBCs offer several advantages to companies. By committing to social and environmental objectives, companies can distinguish themselves as businesses with a conscience. They also provide directors with more freedom to adopt a long-term vision for the company. PBCs can also attract impact investors interested in sustainable and socially responsible investments. Several well-known companies have adopted the PBC model, including Warby Parker, Allbirds, and Kickstarter.

---

[8] *Maryland First State in Union to Pass Benefit Corporation Legislation*, CSRWire (April 4, 2010) (www.csrwire.com/press_releases/29332-maryland-first-state-in-union-to-pass-benefit-corporation-legislation).

PBCs align well with the principles of ESG as they offer a tangible way for investors to support companies committed to sustainable and responsible practices. The emergence of PBCs in the IPO market has opened new opportunities for investors, albeit the process and initial foray of PBCs to IPO was new ground for regulators. These IPOs often provide greater transparency on ESG goals, offering investors a clearer picture of a company's commitment to sustainability and social responsibility. The PBC model is still evolving, and as businesses and investors increasingly prioritize sustainability and social responsibility, the PBC model has become a mainstay in the corporate landscape.

Distinguishing PBCs from B Corps

Another important distinction to make is the one between PBCs and Certified B Corporations (B Corps). A PBC is a legal corporate structure, while a B Corps has obtained a private certification provided by the nonprofit B Lab, indicating that the certified company meets high standards of social and environmental performance.

B Lab is a global nonprofit organization founded in 2006 that champions business as a force for good.[9] The organization is known for its rather rigorous certification process, whereby companies become certified B Corporations, or "B Corps." A B Corp is a for-profit company that has been certified by B Lab for its social and environmental impact.[10] This certification is a stamp of approval of the company's commitment to balance profit with purpose. It signifies that a company has met the B Corp standards of social and environmental performance, accountability, and transparency.

---

[9] *About B Lab*, B Corporation (www.bcorporation.net/en-us/movement/about-b-lab/).

[10] *About B Corp Certification*, B Corporation www.bcorporation.net/en-us/certification/).

Companies apply for B Corp Certification for various reasons, including building trust with stakeholders, and attracting talent and mission-aligned investors. A common thread is the desire to demonstrate commitment to social and environmental responsibility. B Corp Certification offers a standardized and recognized way to signal this commitment to stakeholders, including employees, investors, customers, and the public. The certification process encourages companies to adopt a whole-systems approach, considering their social and environmental impact in all aspects of their operations.[11]

The process of obtaining B Corp Certification involves an in-depth evaluation of a company's social and environmental performance. To be certified, a company must achieve a high score on B Lab's Impact Assessment, a comprehensive evaluation of the company's social and environmental performance; pass a due diligence process to verify the information submitted in the Impact Assessment; make a commitment to change corporate governance structure to be accountable to all stakeholders; and exhibit transparency.[12] B Corps must apply for recertification every three years.[13]

As of January 1, 2024, there are 7,996 certified B Corporations across 162 industries and 93 countries.[14] B Lab also creates standards, policies, tools, and programs that support the environmental, social, and governance principles.[15] One example is the B Corp SDG Action Manager, which is a self-assessment tool developed by B Corps and the

---

[11] Id.
[12] Id.
[13] Id.
[14] *Make Business a Force for Good*, B Corporation (www.bcorporation.net/en-us/).
[15] *Programs & Tools Overview*, B Corporation (www.bcorporation.net/en-us/programs-and-tools/).

United Nations Global Compact available to all businesses regardless of B Corp status, for goal setting and progress tracking in support of the UN Sustainable Development Goals.[16]

The B Corp movement is a transformative shift toward a more sustainable and socially responsible economic model. By reviewing the standards, governance, and operations requirements to become a B Corp, businesses can gain an understanding of what it means to be a socially and environmentally responsible business. By harnessing the power of business, the B Corp movement is actively reshaping the global economy to create a system with multi-stakeholder benefits.

CASE STUDY: PROTA FIORI

Prota Fiori is a luxury footwear company that utilizes traditional Italian craftsmanship methods and supports sustainable practices.[17] The company focuses on responsible sourcing of raw materials and embeds responsible design and distribution principles into its business model and brand ethos. "Proto Fiori" translates from the Italian "protect the flowers" and the brand continues to innovate with its use of circular materials, including leather alternatives derived from plant-based biomass such as upcycled apple skins. Prota Fiori is a Public Benefit Corporation and the first woman-owned luxury footwear brand to become a certified B Corp.[18]

---

[16] *SDG Action Manager*, B Corporation (www.bcorporation.net/en-us/programs-and-tools/sdg-action-manager/).
[17] *Our Commitments*, Prota Fiori (https://protafiori.com/pages/our-commitment).
[18] Id.

10.5 Stakeholder Engagement

We've discussed the importance of stakeholders to the success of ESG throughout this book. We'll talk in more detail about the importance of stakeholder relationships in the business world in this section.

A stakeholder is an individual, group, or organization that has a vested interest in the activities of a business. As we've discussed, stakeholders include employees, customers, suppliers, investors, shareholders, governmental and regulatory bodies, members of the community, and society. Stakeholder relations—developing and maintaining strong relationships with stakeholders—is a business imperative, and it is part and parcel of an effective ESG strategy.

Engaging with stakeholders in meaningful ways opens channels for communication and collaboration. It helps businesses understand stakeholder needs, concerns, and expectations, aligning everyone toward shared goals and objectives. It can also make or break trust and credibility, and sink planned projects, as we saw in the failed community relations outreach discussed in Chapter 5. Stakeholder relations can be an asset to a business, or it can be an area of risk, depending on how proactively and effectively outreach is conducted.

We've highlighted the importance of stakeholder mapping in the ESG context. The process can be accomplished through the use of a simple spreadsheet at a smaller scale or through the use of stakeholder relationship management platforms (SRM), customer relationship management programs (CRM), and other specifically designed technology applications. The first step in effective stakeholder management is stakeholder identification. This involves brainstorming to identify all the individuals or groups likely to be affected by the organization's actions. The next step is to assign relationship managers to be responsible for each of the stakeholders. Once that is accomplished, current activities and planned projects should be mapped and regularly updated in advance of announcements for intersections with stakeholders.

Effective stakeholder engagement then requires strategic planning and execution. This involves keeping stakeholders informed about the organization's actions, decisions, and progress; being open and truthful about goals, commitments, and challenges; and involving stakeholders in relevant decision-making processes. Cultivating positive stakeholder relations is fundamental to a business maintaining its social license to operate.

10.6 ESG Strategy, Program Development, and Reporting

As we move from voluntary to mandatory ESG reporting and disclosure with the new regulatory frontier, there is heightened scrutiny of how a company develops their ESG strategy, ESG program, and ESG reporting processes. As ESG is dynamic, all strategy, program, and reporting aspects should flex to incorporate changing circumstances, stakeholder expectations, and regulatory requirements.

To overcome the challenges and risk associated with regulatory disclosure and reporting, companies can adopt best practices that enhance the effectiveness and credibility of their ESG program and related reporting efforts. Clear and measurable goals that align with business objectives and operations, as well as stakeholder expectations, are essential for any ESG strategy.

Transparent reporting of ESG commitments and performance against targets is essential. According to the Morgan Stanley Institute for Sustainable Investing, around 80% of global investors consider a company's sustainability reporting and its commitment to reduce greenhouse gas emissions when making a new investment.[19] We discussed the four pillars of the TCFD standard, which have been incorporated into TNFD, in Chapter 8. Those pillars are instructive as elemental to sound reporting practices: (1) governance, (2) strategy, (3) risk management, and (4) metrics and targets.

We've provided an ESG Program Building process diagram with Ten Key Steps to Advancing the ESG Program (Figure 10-1) as a foundation for businesses developing their programs that can be expanded and built upon as those programs and reporting processes mature. Principles of continuous improvement should be applied to ESG initiatives and operations, particularly as ESG regulatory and disclosure requirements continue to progress.

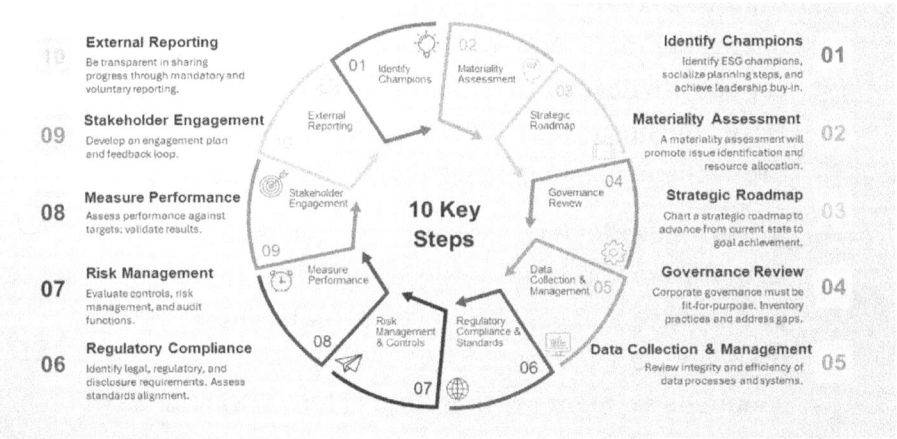

Figure 10-1. ESG Program Building: Ten Key Steps to Advancing the Company ESG Program

---

[19] *Individual Investors' Interest in Sustainability is on the Rise*, Morgan Stanley Institute for Sustainable Investing (January 26, 2024) (www.morganstanley.com/ideas/sustainable-investing-on-the-rise?linkId=356500813).

A best practice is to outline and follow a process that has been socialized and vetted when designing and building an internal program. We'll walk through each of the Ten Key Steps to Advancing the Company ESG Program.

1. **Identify Champions:** Identify **ESG champions, socialize planning steps, and achieve buy-in from leadership** within the company.

2. **Materiality Assessment:** Conduct an **ESG materiality assessment** to ensure proper issue prioritization and resource allocation.

3. **Strategic Roadmap:** Develop a **strategic ESG roadmap** charted to goals, resource allocation, and talent alignment. Assess **internal readiness and capacity** to develop and maintain ESG initiatives across the business. This is an opportunity to lay out the five- and ten-year plans to **scale the strategy**.

4. **Governance Review:** Assess the **governance structure** so it is fit-for-purpose, including an inventory of core policies and board committee oversight, and address ESG coverage gaps. There should be a **cogent presentation** of how ESG issues are covered by corporate governance materials, managed within the company, and overseen by the board of directors.

5. **Data Collection and Management:** Review **ESG data collection, reporting, and disclosure processes**. Review the current protocol for assessing carbon footprint and other environmental and social impacts for integrity and efficiency.

CHAPTER 10   THE ROLE OF THE CORPORATION

6. **Regulatory Compliance and Standards:** Evaluate **ESG legal and regulatory requirements and standards frameworks** for program conformity. **Monitor the evolving landscape** of regulations including reporting and disclosure requirements across the company's jurisdictional map and adjust practices for compliance as needed.

7. **Risk Management and Controls:** Build and test **ESG controls** within risk management, compliance, and audit functions. Ensure those functions incorporate the **strategic roadmap** and **reporting requirements**.

8. **Measure Performance:** Establish **KPIs** and a **timeline for implementation** and **review of ESG commitments.** Track progress and set benchmarks. Establish a continuous improvement plan.

9. **Stakeholder Engagement:** Develop a **stakeholder engagement plan** and feedback loop, including investors/shareholders, regulators, employees, customers, and the community. Build an outreach plan with a RACI (identifying who is to be responsible, accountable, consulted, and informed) model and map partners.

10. **External Reporting:** Differentiate **mandatory and voluntary reporting requirements** and implement review process to ensure consistency across all public statements, filings, and reports. Determine **external transparency and communication plan** to share metrics and socialize progress.

CHAPTER 10   THE ROLE OF THE CORPORATION

Many market-leading companies have been publicly and voluntarily reporting on their environmental, social, and governance results for years. That is particularly true for companies with a manufacturing footprint, companies in the extractives industries, companies in the energy sector, and companies who produce consumer products. The operations of companies with those profiles directly impact resources and communities, and often *rely* on the use of natural resources and the support of local communities to do so.

In recent years, many of those companies have drilled down on climate transition planning and outlined the process and key performance indicators by which they will achieve their stated net-zero and related commitments. The linkage between climate and natural resource scarcity has also become crystalline and an area of investment.

CASE STUDY: UNILEVER

Unilever is a British multinational company in the fast-moving consumer product goods (FMCG) sector. The company was formed in 1929 and has become a market leader as well as a well-known proponent of corporate sustainability practices.[20] Unilever has been reporting on environmental aspects of its operational footprint—over half of which is in developing and emerging markets[21]—since the late 1990s and set initial value chain emissions reduction goals in 2010. Unilever launched its Climate & Nature Fund in 2020, with a commitment to invest EUR€1 billion in climate, nature, and resource projects by 2030.

---

[20] *Welcome to Unilever*, Unilever (www.unilever.com/).
[21] *Partners*, Mastercard Center for Inclusive Growth (www.mastercardcenter.org/about-us/featured-partners/unilever).

CHAPTER 10   THE ROLE OF THE CORPORATION

The company first Issued its Climate Transition Action Plan (CTAP) in 2021 and a subsequent update to the CTAP in 2024, in which Unilever reported that it had accelerated its net-zero goals (Scopes 1, 2, and 3) after achieving a 68% reduction in emissions across operations in 2022 and an overall reduction in GHG impacts across the product life cycle since 2010.[22]

10.7 ESG in Mergers and Acquisitions

ESG factors are increasingly influencing strategic choices for corporations, including mergers and acquisitions (M&A) strategies and decisions. ESG considerations raise risks to navigate as a merger or acquisition target's ESG performance may create legal, financial, or reputational risk to the acquiring company. They also provide opportunities for value creation. Two key issues that have raised the profile of ESG in M&A activity are (1) ESG has proven its ability to influence market dynamics, with companies demonstrating strong ESG performance often commanding premium valuations and stronger financial returns; and (2) lenders have also recognized the potential influence of significant ESG risks on the creditworthiness of businesses and industries.

Traditionally, assessing the full spectrum of ESG risk and translating that into financial terms was challenging. This also corresponded to the often-longer time horizons associated with ESG risk. With a greater understanding of how to measure ESG risk, there is generally now a more accurate quantification of ESG impacts. The incorporation of representations, warranties, material adverse effect clauses, and operating covenants on ESG issues into agreements has become standard in M&A practice.

---

[22] *Climate action: Strategy and goals*, Unilever (www.unilever.com/planet-and-society/climate-action/strategy-and-goals/#:~:text=Our%20Climate%20Transition%20Action%20Plan,emissions%20in%20our%20value%20chain).

CHAPTER 10   THE ROLE OF THE CORPORATION

According to a US study of over 200 corporate investors, financial investors, and M&A debt providers conducted by KPMG in 2023, 74% are already integrating ESG considerations into M&A, and more than half reported the cancellation of deals as a result of poor ESG due diligence findings.[23] Those findings illustrate the risk level assigned to ESG issues in mergers and acquisitions activity. Companies are also becoming more adept at post-M&A integration on ESG issues, particularly in the areas of corporate governance, external communications, and investor relations, which shows the importance of transparency around ESG considerations.

10.8 Creating Green Jobs and Reskilling the Workforce

The transition to a green economy presents a unique opportunity to create sustainable jobs and boost economic growth while working toward sustainability solutions. The shift toward a green economy indicates enormous potential for growth. It is creating new job opportunities and necessitating a change in the workforce. The transition also demands a significant reskilling and upskilling of the workforce.

Green jobs, as defined by the International Labour Organization (ILO), are "decent jobs in any economic sector (e.g., agriculture, industry, services, administration) which contribute to persevering, restoring and enhancing environmental quality."[24] Green jobs can be found in traditional sectors and in emerging sectors. Green jobs are becoming a

---

[23] *2023 US ESG Due Diligence Study*, KPMG (https://kpmg.com/us/en/articles/2023/esg-due-diligence/us-esg-due-diligence-study.html#:~:text=ESG%20M%26A%20due%20diligence%20findings,to%20legal%20and%20reputational%20consequences).

[24] *Frequently Asked Questions on green jobs*, International Labour Organization (www.ilo.org/global/topics/green-jobs/WCMS_214247_EN/lang--en/index.htm).

critical component of economic recovery strategies, with governments and businesses worldwide seizing the opportunity to *build back better* as we've seen in the United States with the Infrastructure Investment and Jobs Act and the Inflation Reduction Act.

The ILO estimates that the transition to a green economy can create approximately 100 million new jobs by 2030.[25] The renewable energy, energy efficiency, and waste management sectors are leading the way in green job creation. The employment shift in the energy sector alone is predicted to add 10.3 million net new green jobs globally by 2030, which will offset the 2.7 million jobs predicted to be lost in the fossil fuel sector.[26]

The UNEP predicts that where just transition policies are put in place, nature-based solutions in areas such as climate change, disaster risk, and food and water security could generate 20 million new jobs, with a significant uptick in rural areas.[27] The focus on nature-based solutions promises to bring new economic opportunity. The ILO, the UNEP, the International Union for Conservation of Nature (IUCN), and other partners are preparing a new report on decent jobs and work in nature-based solutions that is intended to be "the flagship report for monitoring job creation through the conservation and restoration economy."[28]

---

[25] *The Just Ecological Transition: An ILO solution for creating 100 million jobs by 2030*, International Labour Organization (May 24, 2022) (www.ilo.org/global/topics/green-jobs/news/WCMS_846279/lang--en/index.htm).

[26] *How many jobs could the clean energy transition create?* World Economic Forum (May 25, 2022) (www.weforum.org/agenda/2022/03/the-clean-energy-employment-shift-by-2030/).

[27] *Nature-based Solutions can generate 20 million new jobs, but "just transition" policies needed*, United Nations Environment Programme (December 8, 2022) (www.unep.org/news-and-stories/press-release/nature-based-solutions-can-generate-20-million-new-jobs-just).

[28] *The Just Ecological Transition: An ILO solution for creating 100 million jobs by 2030*, International Labour Organization (May 24, 2022) (www.ilo.org/global/topics/green-jobs/news/WCMS_846279/lang--en/index.htm).

The World Economic Forum estimates that four in ten workers will need to be reskilled.[29] The primary skill category gaps identified by the WEF include (1) science skills, (2) architectural and planning skills, (3) green engineering and tech skills, (4) agriculture skills, (5) environmental justice skills, and (6) systems skills.[30] Given the rapid technological advancements and evolving industry requirements, the current workforce lacks the necessary technical skills and often the training opportunities to fill the new positions. To get there, we'll need to methodically close the green skills gap.

There is a pressing need for upskilling and reskilling initiatives from governments, educational institutions, and businesses. Education and training programs, including vocational and technical programs, will have to expand to bridge the skills gap. By upskilling their employees, businesses can future-proof their operations and manage a successful transition to the green economy.

Leading Sustainability Within the Corporation

Corporate sustainability functions have taken on new dimensions over the decades as CSR and ESG have expanded. Many companies with sustainability programs tap their sustainability leaders to flex and assume broader social responsibility and environmental, social, and governance responsibilities. That function is typically led by a Chief Sustainability Officer (CSO).

The role of the CSO for a company has never been more critical or more complex as the scale and influence of sustainability and ESG continues to grow. Many public and private companies have long-standing sustainability programs led by a CSO. In recent years, we've

---

[29] *These are the skills young people will need for the green jobs of the future*, World Economic Forum (August 23, 2021) (www.weforum.org/agenda/2021/08/these-are-the-skills-young-people-will-need-for-the-green-jobs-of-the-future/).
[30] Id.

CHAPTER 10   THE ROLE OF THE CORPORATION

seen more early stage and emerging companies building with purpose and incorporating a sustainability function from the outset, often led by a CSO or a fractional CSO depending on the growth stage of the company. A fractional CSO position shares similar characteristics to those of other fractional C-suite roles such as a Chief Marketing Officer (CMO) and a Chief Legal Officer (CLO).

Three factors that are currently shaping CSO priorities include the (1) expectation to align the sustainability strategy to the broader corporate strategy, (2) increased pressure and interest from investors, and (3) new regulatory requirements.[31] The CSO is more often than not a member of the C-suite and the corporate officer responsible for sustainability. That includes all internal and external activities that touch on sustainability and that usually include a myriad of issues, from facility planning to product design to supply chain to marketing strategy to community, government, and investor relations. The sustainability program can serve as a coordinating function and as a centralized clearinghouse within the company. The CSO maintains a two-way information flow on sustainability and ESG issues with leadership and teams across the organization.

The typical model for managing and leading sustainability and ESG efforts across a matrixed organization has been the hub and spoke. The CSO assumes the hub and coordinates with the spokes of all the relevant functions across the company. In that capacity, the CSO is responsible for the vision and strategy and for coordinating implementation across the organization. The hub and spoke model often takes on added dimensions depending on the nature of a company's operations and global footprint.

---

[31] *2023 Weinreb Group Chief Sustainability Officer Report* (Weinreb Group) (https://weinrebgroup.us2.list-manage.com/track/click?u=4963d34c40b0c288833bf994d&id=abd12bddbb&e=c72bd2673c).

Figure 10-2 provides a visual representation of the hub and spoke model for leading the sustainability function within an organization. It highlights the primary business partners and functions with which the CSO dynamically interacts on a regular basis, and it flexes as needed to engage with other functions.

Figure 10-2. Hub and Spoke Model for Leading Sustainability

10.9 Continuous Innovation

> *Competing in overcrowded industries is no way to sustain high performance. The real opportunity is to create blue oceans of uncontested market space.*[32]
>
> —W. Chan Kim and Renée Mauborgne

We highlight the importance of continuous innovation across all sectors to solve environmental and social challenges throughout this book. In this chapter, we'll cover two influential and distinct strategies: the Blue Ocean Strategy and the Blue Ocean Economy.

---

[32] *Blue Ocean Strategy*, Harvard Business Review (October 2004).

CHAPTER 10   THE ROLE OF THE CORPORATION

Blue Ocean Strategy

Blue Ocean Strategy (BOS), a revolutionary concept in business management, is transforming the way companies operate, prompting them to think beyond conventional boundaries. Simultaneously, it contributes to combating climate change, promoting equity in the economy, and enabling unprecedented market access. This section delves into the intricacies of this strategy, its challenges and opportunities, and its profound implications for a sustainable future.

Blue Ocean Strategy, conceptualized by W. Chan Kim and Renée Mauborgne, professors at INSEAD, advocates the creation of new, uncontested market spaces, metaphorically termed as "blue oceans."[33] Unlike "red oceans," which represent existing, highly competitive markets, "blue oceans" symbolize untapped, competition-free markets. The strategy encourages companies to innovate and shift from competing in overcrowded markets to creating new markets, thereby making competition irrelevant.

The strategy emphasizes the creation of uncontested market spaces—the blue oceans. At the heart of Blue Ocean Strategy is the concept of *value innovation*. This concept strives for simultaneous differentiation and cost reduction, creating unique value for both the company and its customers.[34] By redefining product features or services, companies can offer superior value at a lower cost, thereby unlocking new demand. This involves shifting the focus from existing competition to potential markets, opening up opportunities for rapid and profitable growth.

---

[33] *Blue Ocean Strategy*, INSEAD (www.insead.edu/executive-education/strategy/blue-ocean-strategy).
[34] Id.

CHAPTER 10   THE ROLE OF THE CORPORATION

Blue Ocean Strategy holds immense potential for shaping a sustainable future. By fostering innovation and sustainability, it can drive the transition toward a green economy, create jobs, and promote social equity. By fostering innovation, it encourages companies to develop sustainable products and services that not only meet consumer needs but also address environmental challenges. For example, the emergence of renewable energy technologies represents a blue ocean market that offers a solution to the global energy crisis while mitigating climate change. Similarly, green buildings, electric vehicles, and circular economy models are examples of blue ocean innovations addressing environmental concerns.

Innovation can contribute significantly to promoting equity in the economy. It encourages companies to create products and services that cater to a broader demographic, including traditionally underserved markets. This not only opens up new revenue streams for businesses but also creates opportunities for social inclusion and economic empowerment. It can also facilitate market access by creating new market spaces and offers a pathway for companies to break free from the constraints of competition, drive innovation, and achieve sustainable growth. By aligning with societal values and environmental sustainability, companies can enhance their reputation, customer loyalty, and long-term profitability.

Blue Ocean Economy

The global economy is constantly evolving, and in recent years, one area that has gained significant attention is the Blue Ocean Economy. It's also been termed the Sustainable Blue Economy (SBE) and the Blue Economy. For our purposes, we'll refer to it as the Blue Ocean Economy or the blue economy. A focus on innovation in the blue economy alongside the much-discussed transition to a green economy is warranted.

CHAPTER 10 THE ROLE OF THE CORPORATION

The blue economy refers to the sustainable use of ocean resources for economic growth, improved livelihoods, and a healthier ocean ecosystem for current and future generations.[35] This includes various activities like shipping, fishing, aquaculture, coastal tourism, offshore renewable energy, and marine biotechnology. Overfishing, pollution, and climate change have led to a significant decline in coral reefs, threatening livelihoods and marine biodiversity. The blue economy centers around the sustainable use of ocean resources and driving related economic growth.

The blue economy is estimated to be worth more than USD$1.5 trillion per year globally.[36] It provides around 31 million jobs and supports the livelihoods of more than three billion people around the world.[37] Two-thirds of the global economy is moderately or highly dependent on ocean resources.[38] Its growth potential is impressive. In fact, the Organisation for Economic Co-operation and Development (OECD) predicts that the ocean-based industry could double in size to USD$3 trillion by 2030.[39]

This growth is expected to be driven by an increasing investment by the public and private sectors in nature-based solutions to climate change that are provided by the oceans, including carbon sequestration, coastal

---

[35] *Action Brief an ocean of opportunities: How the Blue Economy can Transform Sustainable Development in Small Island Developing States,* United Nations Development Programme (February 2023) (www.undp.org/sites/g/files/zskgke326/files/2023-02/UNDP-RBAP-Blue-Economy-Action-Brief-2023.pdf).

[36] *High Level Panel for A Sustainable Ocean Economy,* United Nations (https://oceanpanel.org/). See also *In the same boat: ocean finance, inclusivity and social equity,* United Nations Environment Programme Finance Initiative (April 7, 2022) (www.unepfi.org/themes/ecosystems/in-the-same-boat-ocean-finance-inclusivity-and-social-equity/).

[37] Id.

[38] Id.

[39] *Ocean economy and innovation,* Organisation for Economic Co-operation and Development (www.oecd.org/ocean/topics/ocean-economy/#:~:text=Previous%20OECD%20analysis%20projected%20a,USD%203%20trillion%20in%202030).

CHAPTER 10 THE ROLE OF THE CORPORATION

protection, and biodiversity conservation.[40] The UN High Level Panel for a Sustainable Ocean Economy predicts that the oceans economy can deliver 21% of the greenhouse gas emission reductions needed to meet the Paris Agreement target of limiting the global temperature rise to 1.5°C by 2050.[41]

The United Nations promotes the sustainable use of the world's oceans. Sustainable Development Goal (SDG) 14 is *Life Below Water* and it is focused on conserving and sustainably using the oceans, seas, and marine resources. We discussed the historic agreement for a High Seas Treaty, reached by the United Nations in 2023, in Chapter 3. The goal of the treaty is to protect wildlife, ensure equal access to marine genetic resources, allocate more funds to marine conservation, and establish new rules for deep sea mining. The United Nations Environment Programme (UNEP) has launched the Sustainable Blue Economy Initiative to facilitate sustainable, climate-resilient, and inclusive blue economy policies and strategies.[42] The initiative aims to reduce human impacts on marine and coastal ecosystems and promote clean technologies, renewable energy, and circular material flows.[43]

There has been a series of blue bonds issued for ocean resource management projects, including offshore renewable energy, issued in recent years by sovereign governments, banks, and companies.[44] The United Nations Environment Programme Finance Initiative (UNEPFI),

---

[40] Id.

[41] *The Ocean as a Solution to Climate Change*, High Level Panel for a Sustainable Economy, United Nations (https://oceanpanel.org/wp-content/uploads/2023/09/Full-Report_Ocean-Climate-Solutions-Update-1.pdf).

[42] *Sustainable Blue Economy*, United Nations Environment Programme (www.unep.org/topics/ocean-seas-and-coasts/ecosystem-based-approaches/sustainable-blue-economy#:~:text=UNEP%20Finance%20Initiative%20(UNEP%20FI,finance%20a%20sustainable%20blue%20economy).

[43] Id.

[44] *Blue bonds: Accelerating Sustainable Ocean Business*, UN Global Compact (https://unglobalcompact.org/take-action/ocean/communication/blue-bonds-accelerating-sustainable-ocean-business).

CHAPTER 10 THE ROLE OF THE CORPORATION

the Global Compact, and international finance organizations collaborated to develop a set of principles and a practioner's guide for blue ocean finance.[45] The guide builds upon existing global market standards and materials previously issued by banks and international organizations and is intended to help to

- "Define blue economy typology and eligibility criteria;
- Suggest key performance indicators;
- Showcase latest case studies from the field; and
- Highlight the critical need for increased financing to achieve Sustainable Development Goal 14, and other global sustainability targets."[46]

The blue economy holds tremendous potential on environmental and social fronts. It has the potential to contribute significantly to Sustainable Development Goals, climate change mitigation, and economic growth. It will require concerted efforts from governments, international organizations, and dedicated private sector actors.

CHAPTER TAKEAWAYS

- Businesses should review the List of Key Emerging ESG Trends and Risks for Businesses
- Corporate directors should be mindful of their fiduciary duties and ensure their boards have the requisite expertise and composition for ESG oversight

---

[45] *New guidance on blue bonds to help unlock finance for a sustainable ocean economy*, United Nations Environment Program Finance Initiative (September 6, 2023) (www.unepfi.org/themes/ecosystems/new-guidance-on-blue-bonds-to-help-unlock-finance-for-a-sustainable-ocean-economy/).
[46] Id.

CHAPTER 10   THE ROLE OF THE CORPORATION

- Businesses should review and implement best practices as they launch or update their ESG programs
- Companies should plan for the necessary shifts required to adapt to the green economy and energy transition
- Companies should adopt and implement ESG policies, practices, and requirements across their value chain
- Regularly reviewing corporate purpose, stakeholder engagement plans, progress against ESG commitments, and adjusting as necessary is a business requisite
- Continuous innovation is necessary for competitiveness and long-term strategy, and it can unlock vast market opportunity in the ESG arena

CHAPTER 11

The Role of the Financial Sector

The momentum for ESG-aligned investments is increasing, and ESG has unlocked new opportunities, including in emerging markets. Responsible and ethical investing in these markets can promote sustainable economic growth. More asset managers and investors are recognizing the potential of ESG investing to deliver superior returns alongside a profound societal impact.

11.1 Rise in ESG-Aligned Investing

According to PwC's *Asset and Wealth Management Revolution 2022* report, which tapped 250 institutional investors and asset managers worldwide, representing nearly half of global AUM, ESG-related assets under management are expected to climb to USD$33.9 trillion by 2026 and are on pace to constitute 21.5% of total global AUM before 2027.[1] The report also

---

[1] *ESG-focused institutional investment seen soaring 84% to US$33.9 trillion in 2026, making up 21.5% of assets under management: PWC report*, PwC (October 10, 2022) (www.pwc.com/gx/en/industries/financial-services/asset-management/publications/asset-and-wealth-management-revolution-2022.html?WT.mc_id=CT11-PL1000-DM2-TR2-LS4-ND1-TTA9-CN_gx-fy22-xlos-esg-awm-esg-revolution-pressrelease).

© Kristyn Noeth 2024
K. Noeth, *The ESG and Sustainability Deskbook for Business*,
https://doi.org/10.1007/979-8-8688-0261-4_11

CHAPTER 11 THE ROLE OF THE FINANCIAL SECTOR

notes that nearly 81% of institutional investors in the United States and 83.6% in Europe plan to increase their allocations to ESG products in the short term.[2]

The perception of ESG as part of fiduciary duties is increasingly gaining acceptance among investors. Three-quarters of investors now consider ESG to be part of their fiduciary duties, indicating that ESG factors are becoming an integral part of investment decision-making.[3] That fundamentally changes how investors view ESG and how it is prioritized in the investment community.

11.2 Growth in Sustainable Finance

Sustainable finance is an evolving discipline that combines principles of financial management with sustainable development and environmental considerations. Its primary goal is to encourage businesses, investors, and financial institutions to include environmental, social, and governance considerations in their investment decisions. This practice is increasingly being recognized as a component in the global effort to achieve sustainable development and combat climate change.

Sustainable finance has the dual goal of increasing value in the long term and supporting sustainable activities. The approach was originally based on the recognition that economic activities have profound impacts and that the financial sector has a role to play in shaping these impacts. Sustainable finance has evolved significantly over the past few decades. In the early stages of sustainable finance, the approach was known as negative screening. It was a filtering process, as it was mainly about avoiding investments in companies or sectors with poor performance on specific environmental and social issues.

---

[2] Id.
[3] Id.

CHAPTER 11   THE ROLE OF THE FINANCIAL SECTOR

It has expanded with the recognition that sustainable finance also unlocks tremendous economic opportunity. The field is rapidly expanding. In addition to impact investing and ESG funds, a variety of financial instruments are used to channel funds toward sustainable projects and initiatives. These instruments have historically included the following:

- **Green bonds:** These are debt instruments where the proceeds are used to finance or refinance environmentally friendly projects. The World Bank issued the first labeled green bond in 2008, and the market has since grown exponentially, with various types of issuers tapping into this market.[4]

- **Social bonds:** These bonds are used to finance projects that have positive social outcomes, such as improving education or health care.

- **Sustainability bonds:** These bonds combine elements of both green and social bonds, with proceeds used to finance a combination of green and social projects.

- **Sustainability-linked loans:** These are loans that incentivize borrowers to achieve predetermined sustainability performance targets.

- **Debt-for-nature swaps:** These are deals in which a developing nation's debt is cut in return for ecosystem protection. There are other developing ESG debt mechanisms.

---

[4] *What You Need to Know about IFS's Green Bonds*, The World Bank (December 8, 2021) (www.worldbank.org/en/news/feature/2021/12/08/what-you-need-to-know-about-ifc-s-green-bonds).

CHAPTER 11   THE ROLE OF THE FINANCIAL SECTOR

As highlighted in Chapter 7, issuance of green bonds surpassed the USD$1 trillion mark in 2020. That level has ebbed and flowed over the years and S&P Global anticipates the volume for issuance of green, social, sustainability, and sustainability-linked bonds will clock in at around USD$1 trillion in 2024.[5] The forecast for a growth in transition bonds to finance activities to reach climate goals and in blue bonds to support sustainable ocean practices is notable and reflects the expansion of the market. The family of bonds has also been referred to collectively as outcome bonds.

Sustainable finance is playing an increasingly important role in the global economy. It is helping to channel capital toward sustainable businesses and projects, helping to drive the transition to a more sustainable global economy. Public entities including sovereigns and cities, as well as development banks, have long been engaged in the sustainable finance market as an aspect of economic policy. As we covered in Chapter 7, regulators are developing taxonomies for sustainable finance to provide a classification system of activities and other oversight mechanisms.

This form of finance aims at generating long-term value through the recognition and management of risks associated with ESG issues and the identification of opportunities through better alignment of financial services with societal goals and sustainable economic development. It is poised to figure even more prominently in the global economy. Sustainable finance can be leveraged to close the estimated USD$5–7 trillion per year investment gap needed across all sectors to meet the Sustainable Development Goals (SDGs).[6]

---

[5] *Sustainability Insights Research: Sustainable Bond Issuance to Approach $1 Trillion in 2024*, S&P Global Ratings (February 2024) (www.spglobal.com/ratings/en/research/pdf-articles/240213-sustainability-insights-research-sustainable-bond-issuance-to-approach-1-trillion-in-2024-101593071).

[6] *The Sustainable Development Agenda*, United Nations (www.un.org/sustainabledevelopment/development-agenda/).

CHAPTER 11   THE ROLE OF THE FINANCIAL SECTOR

Barclays estimates that the ESG debt market has the potential to grow to USD$800 billion.[7] The former Credit Suisse was the early mover and for a time the only commercial bank engaging in debt-for-nature swaps. A cohort of multinational development banks and climate funds announced the inception of a global task force to scale debt-for-nature swaps at COP28.[8] Given the increasing recognition of the financial risks associated with climate change and other sustainability challenges, sustainable finance is likely to become an essential tool for managing these risks and ensuring the resilience of the financial system. The challenge will be on the due diligence front to evaluate the integrity and stated benefits of financed projects.

CASE STUDY: CANADA GREEN BOND ISSUANCE

The Government of Canada completed the issuance of a C$4 billion green bond in March 2024. It marks the nation's second green bond and the first sovereign issuer to include nuclear energy as an eligible use for proceeds.[9] Canada issued its first green bond in 2022 and raised C$5 billion to finance green infrastructure and other climate-related projects.[10]

---

[7] Natasha White, *Interest in Credit-Suisse Pioneered Debt Swaps Is Soaring*, Bloomberg (January 22, 2024) (www.bloomberg.com/news/articles/2024-01-22/debt-swaps-created-by-credit-suisse-generate-huge-banker-interest?embedded-checkout=true).

[8] Marc Jones, *Top development banks, funds set up 'debt-for-nature' task force*, Reuters (December 4, 2023) (www.reuters.com/sustainability/sustainable-finance-reporting/top-development-banks-funds-set-up-debt-for-nature-task-force-2023-12-04/).

[9] *Government of Canada - Green Bond Framework*, Government of Canada (www.canada.ca/en/department-finance/programs/financial-sector-policy/securities/debt-program/canadas-green-bond-framework.html).

[10] *Canada issues second green bond*, Department of Finance Canada, Government of Canada (February 28, 2024) (www.canada.ca/en/department-finance/news/2024/02/canada-issues-second-green-bond.html).

CHAPTER 11   THE ROLE OF THE FINANCIAL SECTOR

Nuclear energy has been a controversial topic over the years. As the world grapples with climate change, some jurisdictions have classified nuclear as a sustainable energy. Canada updated its Green Bond Framework in 2023 and, in part, included nuclear expenditures as an eligible investment. [639] Other jurisdictions, like the EU and the UK have included nuclear in their taxonomies as well.

We'll discuss the philanthropic plays in innovative financing toward environmental and social benefit in Chapter 15.

11.3 Managing Climate Risk as a Strategic Priority

The financial sector, particularly investment, banking, and insurance interests, is increasingly recognizing the importance of managing climate risk. This involves addressing the challenges and opportunities presented by climate change, with a focus on risk management, strategic planning, and the role of governance. Climate risks can be categorized into two types:

- **Physical risks** refer to direct threats to property and businesses arising from climate-related events, such as hurricanes, floods, and heatwaves.

- **Transition risks** are related to the changes in policy, consumer behavior, and technologies associated with transitioning to a lower-carbon economy.

National governments and the supranational European Union are also instituting principles and regulations for large financial institutions, given the magnitude of financial impacts associated with climate risk. The varying nature of climate risks necessitates coordination among various regulatory bodies. The European Central Bank (ECB) has undertaken significant efforts around managing climate-related risk in the green transition and extrapolated that work into a broader climate and nature plan, as noted in Table 7-1. An other example of this is the US interagency guidance,

Principles for Climate-Related Financial Risk Management of Large Financial Institutions, issued by the Comptroller of the Currency, the Federal Reserve System, and the Federal Deposit Insurance Corporation on October 30, 2023.[11] This is a high-level framework to provide consistency in oversight of climate-related financial risks for financial institutions with more than USD$100 billion in total consolidated assets.

The *Principles for Climate-Related Financial Risk Management of Large Financial Institutions* provide that financial institutions should employ comprehensive processes for identifying these risks, consistent with methods used to identify other types of emerging and material risks.[12] Given the magnitude and complexity of these risks, it's imperative for financial institutions to implement tailored risk management practices and integrate climate risk into existing risk assessment frameworks to safeguard their operations and ensure long-term viability. As climate risk modeling and data analysis methodologies evolve, institutions should update their risk management frameworks accordingly. Forward-looking strategic planning should also consider the potential impact of climate-related risks on the institution's financial condition, operations, and business objectives over various time horizons.

11.4 Advancing the New Economy and the Clean Energy Transition

The global transition to a clean economy will require a massive shift across sectors. The financial sector can unlock vast amounts of capital and apply innovative, sustainable financial mechanisms to achieve the transition

---

[11] *Principles for Climate-Related Financial Risk Management for Large Financial Institutions*, Federal Register, 88 FR 74183 (October 30, 2023) (www.federalregister.gov/documents/2023/10/30/2023-23844/principles-for-climate-related-financial-risk-management-for-large-financial-institutions).
[12] Id.

CHAPTER 11   THE ROLE OF THE FINANCIAL SECTOR

goals. The International Energy Agency (IEA) estimates that approximately USD$2.8 trillion will be invested in energy in 2023, with about USD$1.7 million of that total amount going to clean energy (e.g., renewable power, nuclear, grid improvements, storage, low-emission fuels, and electrification).[13] The International Renewable Energy Agency (IRENA) estimates that the world needs USD$35 trillion in investment by 2035 for a successful energy transition.[14] That level of investment would create millions of new jobs and bolster economic growth.

Banks are major stakeholders in the net-zero transition. Moving away from financing conventional and GHG-intensive energy projects and providing the necessary financial backing for clean technologies, renewable energy, and carbon capture and storage will quicken the transition to a sustainable economy. In the face of rising global temperatures and the escalating climate crisis, the financial sector has emerged as a pivotal player in the global transition to a new, sustainable economy. This transition, marked by a shift from fossil fuels to clean energy sources, requires vast amounts of capital and innovative financial mechanisms.

The 60 largest banks in the world provided USD$669 billion in financing to fossil fuel companies in 2022, and that number is part of a seven-year total of USD$5.5 trillion since the Paris Agreement was adopted.[15]

---

[13] *World Energy Investment 2023: Overview and key findings*, International Energy Agency (www.iea.org/reports/world-energy-investment-2023/overview-and-key-findings).

[14] *Investment Needs of USD 35 trillion by 2030 for Successful Energy Transition*, International Renewable Energy Agency (March 28, 2023) (www.irena.org/News/pressreleases/2023/Mar/Investment-Needs-of-USD-35-trillion-by-2030-for-Successful-Energy-Transition).

[15] *New Report: Canadian Bank RBC the #1 Financier of fossil fuels, World's Biggest Banks Continue to Pour Billions into Fossil Fuel Expansion,* Rainforest Action Network (April 12, 2023) (www.ran.org/press-releases/new-report-canadian-bank-rbc-the-1-financier-of-fossil-fuels-worlds-biggest-banks-continued-to-pour-billions-into-fossil-fuel-expansion/#:~:text=In%20the%20seven%20years%20since,provided%20%24669%20billion%20in%20financing).

Financial institutions will also have to address how they handle financed emissions, which refers to greenhouse gas emissions not coming directly from the bank's operations but from the projects and companies in which the bank invests or to which it lends capital. Banks will have to provide more transparency into what actually makes up their carbon footprint.

The transition to a low-carbon economy has been heralded as a "mega force" in shaping investment returns, according to a recent report released by the BlackRock Investment Institute.[16] That follows on the previously released 2024 Private Markets Outlook, in which the firm predicted a "massive relocation of capital" and identified specific opportunities presented by the transition.[17] The firm's scenario modeling forecasts that global capital investment in the energy sector to advance the energy transition could double to USD$4 trillion per year through 2050.[18]

There is tremendous global opportunity in transition finance. Barclays, the British multinational banking and financial services company, announced that it would no longer directly finance non-diversified oil and gas projects and set expectations for energy client transition strategies. As part of that announcement, the bank reiterated its commitment to finance USD$1 trillion of sustainable and transition finance by 2030.[19] We've seen some multinational banks teetering on climate commitments, but it is expected that others will move forward to capitalize on the energy

---

[16] *Weekly commentary*, BlackRock Investment Institute (March 11, 2024) (www.blackrock.com/corporate/literature/market-commentary/weekly-investment-commentary-en-us-20240311-low-carbon-transition-themes-in-2024.pdf).

[17] *2024 Private Market Outlook*, BlackRock (www.blackrock.com/institutions/en-us/insights/private-markets-outlook).

[18] Id.

[19] *Barclays focuses capital resources on supporting energy companies to decarbonize*, Barclays (February 9, 2024) (https://home.barclays/news/press-releases/2024/01/barclays-focuses-capital-and-resources-on-supporting-energy-comp/).

CHAPTER 11   THE ROLE OF THE FINANCIAL SECTOR

transition opportunities ahead. And while public markets and banking institutions remain the primary source of capital in the energy transition, private capital investment is expanding particularly in electrification and renewable power deployment.[20]

CASE STUDY: BROOKFIELD GLOBAL TRANSITION FUNDS

Brookfield Asset Management is a publicly traded global alternative asset manager with over USD$850 billion AUM. Brookfield is the largest transition investor among private fund managers. It announced a new USD$10 billion fund, the second Brookfield Global Transition Fund (BGTF II), to support the global energy transition in 2024.[21] The first such fund, BGTF I, closed at USD$15 billion in 2022. In the press release covering the new fund launch, Connor Teskey, CEO of Brookfield Renewable Power & Transition, said, "Corporate demand for decarbonization technologies is now the primary driver of transition investment, delivering significant economic value as well as meaningful environmental benefits."[22]

---

[20] *Financing the energy transition*, S&P Global (January 8, 2024) (www.spglobal.com/commodityinsights/en/market-insights/blogs/energy-transition/010824-financing-the-energy-transition).

[21] *Brookfield Announces $10 Billion First Closing for Second Brookfield Global Transition Fund*, Brookfield Asset Management (February 5, 2024) (https://bam.brookfield.com/press-releases/brookfield-announces-10-billion-first-closing-second-brookfield-global-transition).

[22] Id.

CHAPTER 11   THE ROLE OF THE FINANCIAL SECTOR

CHAPTER TAKEAWAYS

- ESG-related assets under management are expected to climb to USD$33.9 trillion by 2026 and are on pace to constitute 21.5% of total global AUM before 2027[23]

- Strong ESG performance can best position a company for investment

- Businesses should evaluate their climate risk with respect to both physical risks and transition risks, and develop an informed climate transition strategy for operational, regulatory, and investor purposes

- Large financial institutions have a particularly important role to play in the clean energy transition

- Sustainable finance is positioned to expand exponentially and businesses should be mindful of how best to access opportunity

---

[23] *ESG-focused institutional investment seen soaring 84% to US$33.9 trillion in 2026, making up 21.5% of assets under management: PWC report,* PwC (October 10, 2022) (www.pwc.com/gx/en/industries/financial-services/asset-management/publications/asset-and-wealth-management-revolution-2022.html?WT.mc_id=CT11-PL1000-DM2-TR2-LS4-ND1-TTA9-CN_gx-fy22-xlos-esg-awm-esg-revolution-pressrelease).

CHAPTER 12

The Role of International Bodies and UN Programs

> *Climate change presents a major threat to long-term growth and prosperity, and it has a direct impact on the economic wellbeing of all countries.[1]*
>
> —International Monetary Fund

International bodies and forums serve an important function in convening stakeholders and achieving consensus on global sustainability and ESG issues. In this chapter, we'll discuss a selection of the primary United Nations (UN) initiatives and ad hoc governmental groups interacting with the business and investment sectors to drive sustainable change.

---

[1] *The IMF and Climate Change*, International Monetary Fund (www.imf.org/en/Topics/climate-change#highlights).

CHAPTER 12   THE ROLE OF INTERNATIONAL BODIES AND UN PROGRAMS

12.1 Driving International Agreement

The UN is an international organization founded in 1945, currently made up of 193 member states.[2] It was established after World War II with the aim of preventing future conflicts and fostering a culture of peace and cooperation among nations.[3] The core mission of the UN revolves around international peace and security, promoting social and economic development, protecting human rights, and upholding international law.[4] The UN accomplishes these goals through various programs, funds, and specialized agencies, working on a wide array of issues, from health and education to climate change and refugee protection.

The UN addresses environmental, social, and governance challenges through several frameworks and specialized agencies. The United Nations Environment Programme (UNEP) leads environmental efforts, promoting sustainable practices and policies globally. The UN also convenes world leaders at major summits, such as the Conference of the Parties (COP) to the UN Framework Convention on Climate Change (UNFCCC), to negotiate international agreements like the Paris Agreement, aiming to mitigate climate change by reducing greenhouse gas emissions.

To tackle social issues, the UN relies on various entities, like the United Nations Development Programme (UNDP), which works to reduce poverty and inequality. The UN also advocates for human rights through the Office of the High Commissioner for Human Rights (OHCHR) and supports nations in crisis through humanitarian aid provided by the United Nations Children's Fund (UNICEF), the World Food Programme (WFP), and the United Nations Refugee Agency (UNHCR). One of the most significant contributions by the United Nations to the global sustainability agenda is

---

[2] *About Us*, United Nations (www.un.org/en/about-us).
[3] *History of the United Nations*, United Nations (www.un.org/en/about-us/history-of-the-un).
[4] Id.

the development of the Sustainable Development Goals (SDGs), the set of 17 goals adopted by all UN member states in 2015, as part of the 2030 Agenda for Sustainable Development. The 17 SDGs are identified and illustrated in Chapter 3.

We've discussed many of those UN initiatives and their relevance to the private sector throughout this book. There are also ad hoc governmental groups and civil society organizations that provide forums for world leaders to convene and take collective action. The agendas for groups such as the Group of 20 (G20), the Group of 7 (G7), the Organisation for Economic Co-operation and Development (OECD), and the International Organization of Securities Commissions Oversight (IOSCO) have increasingly focused on ESG topics, such as climate risk and investment, the energy transition, and curtailing human rights abuses.

12.2 Integration of the UN Principles for Responsible Investment (PRI)

The United Nations Principles for Responsible Investment (PRI) is a globally recognized initiative that aims to integrate environmental, social, and corporate governance issues into investment decision-making practices. We introduced PRI as part of the discussion on ESG integration with international principles and the UN Sustainable Development Goals in Chapter 3. We'll discuss PRI and how investors follow The Six Principles and the five-part framework in more detail in this chapter.

The PRI initiative was born out of a realization that traditional investment practices often did not take ESG factors into account. In 2005, then-United Nations Secretary-General Kofi Annan convened a group of the world's largest institutional investors as part of a process

CHAPTER 12   THE ROLE OF INTERNATIONAL BODIES AND UN PROGRAMS

to develop principles for responsible investment.[5] The development of these principles was a collective effort involving multiple stakeholders. That collective proposed a set of principles that were then incubated by the United Nations Environment Programme (UNEP) Finance Initiative and the UN Global Compact.[6] The outcome was the establishment of the UN Principles for Responsible Investment, which was publicly launched in 2006.

PRI has significantly influenced the investment industry landscape in its goal to promote a more sustainable global financial system. As discussed in Chapter 3, PRI comprises six fundamental principles and provides a global framework for investors to consider ESG issues and incorporate them into their investment decisions. We provided the text of the principles in Chapter 3, and here is a shorthand of the principles as a reference:

1. Incorporation of ESG issues into investment analysis
2. Active ownership of ESG issues
3. Disclosure of ESG issues by entities scoped for investment
4. Promotion of acceptance and implementation of the principles
5. Collaboration to enhance implantation of the principles
6. Reporting on implementation activities and progress[7]

---

[5] *About the PRI*, UN Principles for Responsible Investment (www.unpri.org/about-us/about-the-pri).

[6] Id.

[7] *What are the Principles for Responsible Investment?* UN Principles for Responsible Investment (www.unpri.org/about-us/what-are-the-principles-for-responsible-investment).

CHAPTER 12   THE ROLE OF INTERNATIONAL BODIES AND UN PROGRAMS

By becoming a signatory to PRI, an organization makes a public commitment to invest responsibly and to follow The Six Principles. These signatories range from large investment firms to smaller financial institutions, all of whom support responsible investment. The PRI community has grown significantly since its inception. As of 2022, PRI has 4,902 signatories with an estimated USD$121.3 trillion AUM.[8]

When the UN Sustainable Development Goals were launched in 2015, PRI set out to support the investment community's contribution in meeting the goals pursuant to SDG 17, which calls for Partnerships for the Goals. [9] PRI published *The SDG Investment Case* in 2017,[10] and later produced the five-part framework for investing with SDG outcomes, as described in Chapter 3.[11] That blueprint offers tools and materials to investors in the public and private markets to align assessment and decision-making processes with the SDGs.[12]

PRI is a leading proponent of responsible investing and is committed to fostering a sustainable global financial system. It continues to work toward achieving this goal by encouraging the adoption of its principles

---

[8] *Annual Report 2022*, United Nations Principles for Responsible Investment (www.unpri.org/annual-report-2022).

[9] *The SDG Investment Case*, UN Principles for Responsible Investment (October 12, 2017) (www.unpri.org/sustainable-development-goals/the-sdg-investment-case/303.article).

[10] *The SDG Investment Case*, UN Principles for Responsible Investment (October 12, 2017) (www.unpri.org/sustainable-development-goals/the-sdg-investment-case/303.article).

[11] *Investing with SDG outcomes: a five-part framework*, UN Principles for Responsible Investment (June 14, 2020) (www.unpri.org/sustainable-development-goals/investing-with-sdg-outcomes-a-five-part-framework/5895.article).

[12] *Investment tools*, UN Principles for Responsible Investment (www.unpri.org/investment-tools).

and fostering good governance, integrity, and accountability.[13] The organization is also focused on "addressing obstacles to a sustainable financial system that lie within market practices, structures, and regulation."[14] This includes working on initiatives that promote transparency and accountability in the financial industry. That work will come into greater focus given the financial reporting and transparency aspects of the new ESG regulatory regimes.

12.3 Business Engagement with the UN Global Compact (Global Compact)

The United Nations Global Compact is closely linked to ESG and sustainability in the corporate sector. In Chapter 1, we discussed the first usage of the term *ESG* in the *Who Cares Wins* report and explored the origins of the Global Compact as well as its Ten Principles that cover human rights, labor, the environment, and anti-corruption. In this section, we'll reference those guiding principles and highlight the bridge provided by the Global Compact between the UN and the business world.

The Global Compact fosters public–private collaborations aimed at tackling global challenges. While it is a voluntary initiative, companies can become signatories to the Global Compact by publicly committing to implement the Ten Principles, supporting the SDGs, and agreeing to produce an annual Communication on Progress, essentially a progress report, to the Global Compact.[15] More than 17,000 companies from over 160 countries are signatories to the Global Compact.[16]

---

[13] *A blueprint for responsible investment*, UN Principles for Responsible Investment (www.unpri.org/about-us/a-blueprint-for-responsible-investment).
[14] Id.
[15] *Participation*, United Nations Global Compact (https://unglobalcompact.org/participation).
[16] Id.

CHAPTER 12   THE ROLE OF INTERNATIONAL BODIES AND UN PROGRAMS

By joining the Global Compact, companies commit to integrating its principles into their corporate strategies, their corporate culture, and their daily operations. The Global Compact directly advocates for responsible business practices and the implementation of sustainability strategies. In its *Guide to Corporate Sustainability*, the Global Compact identifies "five things sustainable companies do" as centered around being a principled business, contributing to a strengthened society, characterized by a leadership commitment, publicly reporting progress, and taking local action.[17]

In addition to guides, the Global Compact offers a global multi-stakeholder network, local networks, and targeted technical tools to companies to further their sustainability programs.[18] These platforms support businesses to take concrete actions, develop more robust reporting programs, and show leadership in sustainability.

12.4 United Nations Environment Assembly (UNEA) and Multilateralism

The United Nations Environment Assembly (UNEA), established by the United Nations Environment Programme (UNEP), is a high-level decision-making body on environmental matters.[19] It brings together ministers of environment, stakeholders, and member states to deliberate on urgent

---

[17] *Guide to Corporate Sustainability*, United Nations Global Compact (www.unglobalcompact.org/docs/publications/U.N._Global_Compact_Guide_to_Corporate_Sustainability.pdf).

[18] *U.N. Global Compact Strategy, 2021-2023*, United Nations Global Compact (https://ungc-communications-assets.s3.amazonaws.com/docs/about_the_gc/U.N.-GLOBAL-COMPACT-STRATEGY-2021-2023.pdf).

[19] *The United Nations Environment Assembly*, United Nations Environment Programme (www.unep.org/environmentassembly/).

CHAPTER 12   THE ROLE OF INTERNATIONAL BODIES AND UN PROGRAMS

environmental issues. The assembly serves as an opportunity for world governments, civil society groups, the scientific community, and the private sector to shape global environmental policy.

The UNEA was formed to provide leadership and catalyze intergovernmental actions on the environment. Its mission is to provide a platform for the international community to address the environmental challenges that the world faces. UNEA's responsibilities include the provision of strategic guidance, policy direction, and the development of international environmental law.

The assembly convenes every two years at the UNEP headquarters in Nairobi, Kenya. The goal is to foster harmony between humanity and nature, improving the lives of the world's most vulnerable people. The sixth session of the United Nations Environment Assembly (UNEA-6) took place in 2024. The session focused on how multilateralism can help tackle the "triple planetary crisis of climate change, nature and biodiversity loss, and pollution and waste."[20]

12.5 Group of 7 (G7) and Group of 20 (G20) Weathervanes

The Group of 7 (G7) and Group of 20 (G20) have emerged as pivotal forums for national government delegations to meet on topics of international consequence. The G7 and the G20 are informal convenings of the world's major economies. The gatherings allow heads of state to discuss and deliberate on global issues, from economic policies to climate change to conflict resolution.

---

[20] *Outcomes of UNEA-6*, United Nations Environment Programme (www.unep.org/environmentassembly/unea6?%2Funea-6=).

CHAPTER 12   THE ROLE OF INTERNATIONAL BODIES AND UN PROGRAMS

G7 and Global Sustainability

The G7 forum for intergovernmental collaboration emerged 50 years ago and is organized around shared values of pluralism, liberal democracy, and representative government.[21] While the group is convened as an informal gathering without the weight of a formal treaty or secretariat, it has significant international influence. The G7 consists of Canada, France, Germany, Italy, Japan, the UK and the United States.[22] The European Union also participates in the G7 as a non-enumerated member. They are all major advanced economies and collectively account for more than half of global wealth as well as 30–44% of GDP, per widely reported statistics.[23]

The G7 holds an annual summit, the purpose of which is to discuss coordinating solutions to critical global issues. Trade, economics, security, non-proliferation, health, climate change, and energy policy have been on the list in recent years. The 2023 summit was held in Hiroshima. The agreed-upon document, the *G7 Hiroshima Leaders' Communiqué*, outlines top-line commitments made at the summit. The environmental and social priority policy issues include

- Staying **steadfast in commitment to the Paris agreement**, including keeping a limit of **1.5°C** global temperature rise within reach through **scaled-up action** in this critical decade;

- Halting and reversing **biodiversity loss** by 2030;

---

[21] *G-7 and G-20*, United States Department of the Treasury (https://home.treasury.gov/policy-issues/international/g-7-and-g-20).
[22] Id.
[23] *Major advanced economies (G7)*, International Monetary Fund (October 2023) (www.imf.org/external/datamapper/profile/MAE).

CHAPTER 12 THE ROLE OF INTERNATIONAL BODIES AND UN PROGRAMS

- Accelerating achievement of the **U.N. Sustainable Development Goals (SDGs)** by 2030, recognizing that **reducing poverty and tackling the climate and nature crisis** go hand in hand;

- Development of **Global Sustainable Disclosure Standards;**

- Preserving the planet by accelerating the **decarbonization of the energy sector** and the **deployment of renewables, ending plastic pollution** and **protecting the oceans**; and

- Deepening cooperation through **Just Energy Transition Partnerships (JETPs)**, the **Climate Club** and new **Country Packages for Forest, Nature and Climate.**[24]

The G7 support for the development of Global Sustainable Disclosure Standards is critically important and received the lion's share of the slim press coverage on the most recent summit. The traction on ESG issues and the agreement reached by these seven powerhouse nations have the potential to move the needle. The *Communiqué* provides the specific commitments made on climate, SDGs, forests, and oceans.[25]

---

[24] *G7 Hiroshima Leaders' Communiqué*, The White House (May 20, 2023) (www.whitehouse.gov/briefing-room/statements-releases/2023/05/20/g7-hiroshima-leaders-communique/#:~:text=We%2C%20the%20Leaders%20of%20the,course%20for%20a%20better%20future).
[25] Id.

CHAPTER 12   THE ROLE OF INTERNATIONAL BODIES AND UN PROGRAMS

G20 and the Green Growth Agenda

The G20's principles are rooted in promoting strong, sustainable, and balanced growth across its diverse member states.[26] It primarily focuses on international economic cooperation while also addressing other pressing issues like trade, sustainable development, climate change, energy and the environment, and anti-corruption.[27] Its member states include the G7 countries, plus Argentina, Australia, Brazil, China, India, Indonesia, Mexico, Russia, Saudi Arabia, South Africa, South Korea, Turkey, and the European Union.[28] The G20 member states account for approximately 85% of the global economy and 75% of the world population.[29]

The G20 maintains an emphasis on global economic policy and has increasingly prioritized green growth over the past decade. This includes a focus on international environmental, energy, and climate policies. The G20 has recognized "the importance of collective action in tackling climate change while promoting transitions towards more flexible, transparent, and cleaner energy systems."[30] The group has acknowledged that climate change drives desertification, biodiversity loss, food insecurity, poor human health, water scarcity, and overall societal insecurity, thus making climate action a great unifier for action.

---

[26] *G-7 and G-20*, United States Department of the Treasury (https://home.treasury.gov/policy-issues/international/g-7-and-g-20).

[27] *About the G20: Group is the main forum for international economic cooperation*, Group of 20 (www.g20.org/en/about-the-g20).

[28] *G-7 and G-20*, United States Department of the Treasury (https://home.treasury.gov/policy-issues/international/g-7-and-g-20).

[29] *About the G20: Group is the main forum for international economic cooperation*, Group of 20 (www.g20.org/en/about-the-g20).

[30] *Sustainability – Climate Sustainability and Energy*, Organisation for Economic Co-operation and Development (www.oecd.org/g20/topics/climate-sustainability-and-energy/).

401

To address these challenges, G20 countries have pledged to make zero-carbon choices, invest in green and efficient energy and transport, and promote nature-based solutions.[31] The group has had a significant role in promoting green finance and investment. It has promoted the development of green finance and investment, with a view to mobilizing private capital for green and climate-resilient infrastructure. And it has specifically pushed for the integration of ESG considerations into financial decision-making processes.

The G20 has launched multiple initiatives to promote green growth, including the Global Biofuel Alliance to foster collaboration and avenues for the use of biofuels. At the 2023 summit in New Delhi, the G20 member countries committed to taking collective action for effective and timely implementation of the G20 2023 Action Plan to Accelerate Progress on the SDGs.[32] The Organisation for Economic Co-operation and Development (OECD) has been instrumental in supporting the work of the G20 on green growth.

12.6 Global Reach of the Organisation for Economic Co-operation and Development (OECD)

The Organisation for Economic Co-operation and Development (OECD) is an international body established to foster economic growth and progress globally. The OECD was established in 1960, succeeding the Organisation for European Economic Co-operation (OEEC), which was formed to

---

[31] Inger Andersen, *The G20 is a global force for sustainability*, Speech Delivered at the G20 Environment and Climate Sustainability Ministerial Meeting (www.unep.org/news-and-stories/speech/g20-global-force-sustainability).

[32] *Previous Summits: 2023 – New Delhi Summit*, Group of 20 (www.g20.org/en/about-the-g20/previous-summit).

CHAPTER 12   THE ROLE OF INTERNATIONAL BODIES AND UN PROGRAMS

administer Marshall Plan aid for the reconstruction of Europe after World War II.[33] There are 38 member countries of the OECD across diverse regions, including North and South America, Europe, Asia, and the Pacific.[34]

A central aspect of the OECD's mission, as stated in the Convention on the OECD, is to achieve sustainable economic growth and employment and a higher standard of living.[35] This is to be achieved by maintaining financial stability, promoting economic growth and employment, and contributing to the expansion of world trade.[36] The organization is instrumental in setting international non-binding standards for economic policies, particularly in taxation and the international taxation rules for multinational companies. The OECD has built on its extensive expertise and produces similar principles and research on green growth, clean and climate-resilient infrastructure, energy regulation, green finance and investment, environmental taxation, and ESG.[37]

The OECD has produced papers to support the G20 reports on fossil fuel subsidies, options for scaling up climate finance, and carbon markets. It also publishes the OECD Environmental Outlook, which provides analyses and projections for environmental change and its economic impacts, and the SDG Pathfinder, which is an open-access tool for content and data on the Sustainable Development Goals.[38] The OECD was an early

---

[33] *About the OECD 60th anniversary*, Organisation for Economic Co-operation and Development (www.oecd.org/60-years/).

[34] *Our global reach*, Organisation for Economic Co-operation and Development (www.oecd.org/about/).

[35] *Convention on the OECD*, Organisation for Economic Co-operation and Development (www.oecd.org/about/document/oecd-convention.htm).

[36] Id.

[37] *Sustainability – Climate Sustainability and Energy*, Organisation for Economic Co-operation and Development (www.oecd.org/g20/topics/climate-sustainability-and-energy/).

[38] *OECD and the Sustainable Development Goals: Delivering on universal goals and targets*, Organisation for Economic Co-operation and Development (www.oecd.org/dac/sustainable-development-goals.htm).

CHAPTER 12   THE ROLE OF INTERNATIONAL BODIES AND UN PROGRAMS

partner to the UN Principles for Responsible Investment and published a joint working document as part of the 2007 Annual OECD Roundtable on Corporate Responsibility that spelled out 35 actions that investors could take to implement The Six Principles.[39] Those actions (Table 12-1) served as a basis for PRI's subsequent development of sector-specific implementation plans and remain applicable today.

Table 12-1. PRI 35 Actions to Implement The Six Principles[40]

1. **Incorporate ESG issues into investment analysis and decision-making processes**
   - Address ESG issues in investment policy statements
   - Support development of ESG-related tools, metrics, and analyses
   - Assess the capabilities of internal investment managers to incorporate ESG issues
   - Assess the capabilities of external investment managers to incorporate ESG issues
   - Ask investment service providers (such as financial analysts, consultants, brokers, research firms or rating companies) to integrate ESG factors into evolving research and analysis
   - Encourage academic and other research on this theme

(continued)

---

[39] *The U.N. Principles for Responsible Investment and the OECD Guidelines for Multinational Enterprises: Complementarities and Distinctive Contributions,* Organisation for Economic Co-operation and Development (2007) (www.oecd.org/investment/mne/38783873.pdf).

[40] *The U.N. Principles for Responsible Investment and the OECD Guidelines for Multinational Enterprises: Complementarities and Distinctive Contributions,* Organisation for Economic Co-operation and Development (2007) (www.oecd.org/investment/mne/38783873.pdf).

Table 12-1. (*continued*)

2. **Be active owners and incorporate ESG issues into ownership policies and practices**
   - Develop and disclose an active ownership policy consistent with the Principles
   - Exercise voting rights or monitor compliance with voting policy (if outsourced)
   - Develop an engagement capability (either directly or through outsourcing)
   - Participate in the development of policy, regulation, and standard setting (such as promoting and protecting shareholder rights)
   - File shareholder resolutions consistent with long-term ESG considerations
   - Engage with companies on ESG issues
   - Participate in collaborative engagement initiatives
   - Ask investment managers to undertake and report on ESG-related engagement

3. **Seek appropriate disclosure on ESG issues by the entities in which they invest**
   - Ask for standardized reporting on ESG issues (using tools such as the Global Reporting Initiative)
   - Ask for ESG issues to be integrated within annual financial reports
   - Ask for information from companies regarding adoption of/adherence to relevant norms, standards, codes of conduct, or international initiatives (such as the UN Global Compact)
   - Support shareholder initiatives and resolutions promoting ESG disclosure

4. **Promote acceptance and implementation of the Principles within the investment industry**
   - Include Principles-related requirements in requests for proposals (RFPs)
   - Align investment mandates, monitoring procedures, performance indicators, and incentive structures accordingly (e.g., ensure investment management processes reflect long-term time horizons when appropriate)
   - Communicate ESG expectations to investment services providers
   - Revisit relationships with service providers that fail to meet ESG expectations
   - Support the development of tools for ESG integration
   - Support regulatory or policy developments that enable implementation of the Principles

(*continued*)

Table 12-1. (*continued*)

5. **Work together to enhance their effectiveness in implementing the Principles**
   - Support/participate in networks and information platforms to share tools, poll resources, and make use of investor reporting as a source of learning
   - Collectively address relevant emerging issues
   - Develop or support appropriate collaborative initiatives
6. **Report on their activities and progress towards implementing the Principles**
   - Disclose how ESG issues are integrated within investment practices
   - Disclose active ownership activities (voting, engagement, and/or policy dialogue)
   - Disclose what is required from service providers in relation to the Principles
   - Communicate with beneficiaries about ESG issues and the Principles
   - Report on progress and/or achievements relating to the Principles using a "Comply or Explain" approach
   - Seek to determine the impact of the Principles
   - Make use of reporting to raise awareness among a broader group of stakeholders

12.7 International Union for Conservation of Nature (IUCN): A Convening of Governments and Civil Society

The International Union for Conservation of Nature (IUCN) is a global organization that has been at the forefront of nature conservation and sustainable use of natural resources for more than seven decades. The IUCN was officially founded in Fontainebleau, France, in 1948, with the support of the United Nations Educational, Scientific and Cultural

CHAPTER 12   THE ROLE OF INTERNATIONAL BODIES AND UN PROGRAMS

Organization (UNESCO).[41] Its purpose is singular: "to advance sustainable development and create a just world that values and conserves nature."[42]

The structure of the IUCN is as an environmental network. It brings together 1,400 member organizations of nations and government agencies at all levels, NGOs, indigenous peoples organizations, scientific organizations, academic institutions, and business associations, along with a network of 15,000 scientists and experts.[43] In addition to the stakeholder forum, the IUCN gathers and analyzes data, conducts research and field projects, and advocates for conservation and education.[44]

The organization maintains the IUCN Green List of Protected and Conserved Areas. The Green List provides local expertise to the conversation of protected and conserved areas worldwide.[45] It also utilizes the IUCN Green List Standard as an international benchmark for performance of conservation objectives with baseline components of "good governance, sound design and planning, and effective management."[46] A protected or conserved area that meets the IUCN Green List Standard is certified, and as of June 2022, there were 77 certified sites around the world.[47]

---

[41] *Our Union*, International Union for the Conservation of Nature (www.iucn.org/our-union).

[42] Id.

[43] Id.

[44] *Our Work*, International Union for the Conservation of Nature (www.iucn.org/our-work).

[45] *IUCN Green List of Protected and Conserved Areas*, International Union for the Conservation of Nature (www.iucn.org/resources/conservation-tool/iucn-green-list-protected-and-conserved-areas).

[46] *Global Standards*, IUCN Green List (https://iucngreenlist.org/standard/global-standard/).

[47] *IUCN Green List of Protected and Conserved Areas*, International Union for the Conservation of Nature (www.iucn.org/resources/conservation-tool/iucn-green-list-protected-and-conserved-areas).

CHAPTER 12   THE ROLE OF INTERNATIONAL BODIES AND UN PROGRAMS

The IUCN is also known for maintaining the IUCN Red List of Threatened Species, which is a comprehensive inventory assessing the conservation status of species worldwide.[48] The IUCN Red List is utilized by governments, NGOs, and others in policymaking and biodiversity impact assessments. There are currently 157,1000 species on the Red List, with extinction threatening more than 44,000 species.[49] The work of the ICUN evidences the global decline in biodiversity and informs conservation efforts. The Red List Partnerships was formed in 2000, with conversation-focused nonprofits like Conservation International and Re:wild, and research and academic institutions partnering to support the Red List and the IUCN's biodiversity assessments initiative.[50] The multi-stakeholder model across governmental and non-governmental organizations is constructive and necessary in confronting biodiversity loss and championing nature-based solutions.

12.8   International Organization of Securities Commissions Oversight (IOSCO)

The International Organization of Securities Commissions Oversight (IOSCO) is a global association of securities regulatory agencies. It is a nongovernmental body established in 1983, and it has 130 member jurisdictions which regulate more than 95% of the world's financial markets, including the major emerging markets.[51] IOSCO sets standards

---

[48] *Red List*, IUCN Red List (www.iucnredlist.org/about/background-history).
[49] Id.
[50] *Red List Partnership*, IUCN Red List (www.iucnredlist.org/about/partners).
[51] *Fact Sheet*, International Organization of Securities Commissions (November 2022) (www.iosco.org/about/pdf/IOSCO-Fact-Sheet.pdf).

CHAPTER 12 THE ROLE OF INTERNATIONAL BODIES AND UN PROGRAMS

for the securities sector and works with the G20 and the Financial Stability Board (FSB) on global regulatory reform.[52] The FSB is the organization that created the now-disbanded Task Force on Climate-related Financial Disclosures (TCFD) in 2015.[53]

Given the application of ESG in the financial markets and IOSCO's promotion of stability and transparency in the markets, the organization has been active on sustainability and ESG disclosures. It has adopted a set of policies and guidelines designed to enhance environmental, social, and governance disclosure by securities issuers and to integrate sustainability considerations into investment decision-making processes. IOSCO announced its official endorsement of the new IFRS sustainability and climate-related disclosure standards, which it called a "major step towards consistent, comparable and reliable sustainability information."[54]

IOSCO has also called for oversight of ESG ratings providers and ESG data product providers[55] and issued a report with detailed recommendations for oversight in 2021.[56] The IOSCO recommendations were based on information gleaned in a fact-finding consultation process that identified areas for improvement, including the lack of reliability of raw ESG data, lack of transparency around ESG ratings methodologies

---

[52] *About IOSCO*, International Organization of Securities Commissions (www.iosco.org/about/?subsection=about_iosco).

[53] *Task Force on Climate-related Financial Disclosures*, Financial Stability Board (www.fsb-tcfd.org/).

[54] *IOSCO endorses the ISSB's Sustainability-related Financial Disclosure Standards*, International Organization of Securities Commissions (July 25, 2023) (www.iosco.org/news/pdf/IOSCONEWS703.pdf).

[55] *IOSCO calls for oversight of ESG Ratings and Data Product Providers*, International Organization of Securities Commissions (www.iosco.org/news/pdf/IOSCONEWS627.pdf).

[56] *Environmental, Social and Governance (ESG) Ratings and Data Products Providers Final Report*, The Board of the International Organization of Securities Commissions (November 2021) (www.iosco.org/library/pubdocs/pdf/IOSCOPD690.pdf).

409

CHAPTER 12   THE ROLE OF INTERNATIONAL BODIES AND UN PROGRAMS

and ESG data products, lack of reliability of ESG ratings and data products and potential conflicts of interest, and lack of communication between ESG ratings and data products providers.[57] The ISOCO report and recommendations carry significant weight, and as we discussed in Chapter 9, regulators are moving ahead to enact oversight requirements for ESG ratings and data product providers.

CHAPTER TAKEAWAYS

- Businesses should assess how best to engage with international and UN organizations to support their sustainability efforts and the Sustainable Development Goals

- The Global Compact and PRI offer valuable resources for companies in developing and implementing meaningful ESG strategies

- The G7, G20, OECD, IUCN, IOSCO, and other international intergovernmental groups set indicators that influence global public policy and regulatory action

- Companies should stay attuned to the priorities set, actions taken, and assessments conducted on climate, nature conservation, human rights, and other ESG issues by international bodies and UN programs

---

[57] Id.

CHAPTER 13

The Role of Business Interest Groups

The global business community builds consensus in support of ESG and often navigates the direction of travel on global topics. There are a number of high-level groups, nonprofits, trade associations and other industry-specific groups that convene and influence corporate action. We'll discuss the role of those organizations and their engagement on ESG topics.

13.1 Convening the World Economic Forum (WEF)

We discussed the World Economic Forum's (WEF) Global Risks Report in detail in Chapter 4, and now we'll provide additional background on the organization. The WEF is recognized as a public–private corporation organized under Swiss law.[1] The WEF was formed as a nonprofit organization in 1971, and it evolved over time with an expanded reach and global agenda.[2]

---

[1] *Leadership and Governance, Related documents*, World Economic Forum (www.weforum.org/about/world-economic-forum-related-documents/).

[2] *Our Mission*, World Economic Forum (www.weforum.org/about/world-economic-forum/).

CHAPTER 13   THE ROLE OF BUSINESS INTEREST GROUPS

The organization brings together leaders from business, politics, academia, and civil society. WEF is committed to improving the state of the world through collective action and dialogue. The WEF's primary source of funding comes from its 1,000 member companies, which tend to be "global enterprise[s] with a minimum turnover exceeding US $5 billion."[3] The level of a company's engagement with WEF activities determines its membership fees, which increase as participation in meetings, projects, and initiatives rises.

The WEF's annual meeting, held in Davos, Switzerland, is the organization's flagship event. It provides a platform for leaders to discuss solutions for pressing global challenges. The annual meeting typically draws around 3,000 invited attendees and the topics are primarily centered around globalization, world markets, conflict resolution, and climate.[4] These include around 600 CEOs, other business leaders, numerous heads of state and other political leaders, investors, economists, celebrities, and journalists.[5]

Annual meeting participants engage in discussions about global issues. WEF has focused on sustainability in the corporate sector through a range of initiatives. A long-term strategy for climate and nature was one of four core themes at the most recent annual meeting in January 2024.[6] The organization also puts forth public materials and articles on sustainability, as well as other key business topics.

---

[3] *Charter for Foundation Members*, World Economic Forum (www3.weforum.org/docs/WEF_FM_Charter.pdf).

[4] *What is Davos?* McKinsey & Company (December 11, 2023) (www.mckinsey.com/featured-insights/mckinsey-explainers/what-is-davos).

[5] Siddharth K, *Davos 2023: The World Economic Forum explained*, Reuters (January 16, 2023) (www.reuters.com/business/davos-2023-world-economic-forum-explained-2023-01-16/).

[6] *World Economic Forum Annual Meeting*, World Economic Forum (www.weforum.org/events/world-economic-forum-annual-meeting-2024/).

CHAPTER 13   THE ROLE OF BUSINESS INTEREST GROUPS

13.2 World Business Council for Sustainable Development (WBCSD) and the SDGs

The World Business Council for Sustainable Development (WBCSD) is a global, CEO-led organization with more than 225 businesses collaborating to accelerate implementation of the UN Sustainable Development Goals and transition to a sustainable world.[7] The council membership represents all business sectors, all major economies, a combined revenue of more than USD$5 trillion, and approximately 19 million employees.[8] The WBCSD aims to make a positive impact on the world's sustainability challenges, including climate change, ecosystems, and the future of work.[9]

The organization is committed to advancing the United Nations Sustainable Development Goals (SDGs) by harnessing the power of the business community to create systemic change and deliver meaningful results. In 2019, the WEF and the WBCSD signed a Memorandum of Understanding (MOU) to work together to "mobilize action on urgent issues related to climate change, biodiversity, food systems, circular economy (including plastics), mobility and other environment- and social-related Sustainable Development Goals (SDGs)."[10]

The WBCSD's strategic objectives revolve around engaging with high-impact sectors, driving business action, and influencing policy.

---

[7] *About us*, World Business Council for Sustainable Development (www.wbcsd.org/Overview/About-us).

[8] *About us*, World Business Council for Sustainable Development (www.wbcsd.org/Overview/About-us).

[9] Id.

[10] *WBCSD and WEF to scale up impact on pressing sustainable development issues*, World Business Council for Sustainable Development (August 20, 2019) (www.wbcsd.org/Overview/News-Insights/General/News/WBCSD-and-WEF-to-scale-up-impact-on-pressing-sustainable-development-issues#:~:text=Geneva%2C%2020%20August%202019%3A%20The,circular%20economy%20(including%20plastics)%2C).

CHAPTER 13   THE ROLE OF BUSINESS INTEREST GROUPS

The WBCSD is another collective action model for sustainable growth and facilitates the sharing of best practices.[11] WBCSD also co-created the GHG Protocol with WRI, as mentioned in Chapter 8.

13.3 Business for Social Responsibility (BSR) as an Early Mover

Business for Social Responsibility (BSR) was one of the early movers in recognizing the opportunity and responsibility for businesses to drive sustainable practices. BSR is a global network and consultancy of about 300 member companies and is focused on "working with business to create a just and sustainable world."[12] BSR was officially established in 1992, and it has origins in the Social Venture Network (SVN) that had been organized in the early days of sustainability.[13] Global office locations include New York, San Francisco, Washington, DC, Hong Kong, Shanghai, Singapore, Tokyo, Copenhagen, London, and Paris.[14]

The organization views sustainability as a strategic advantage for companies and focuses on sustainable business practices. BSR has six focus areas on which it provides materials and guidance to businesses:

1. Climate change
2. Equity, inclusion, and justice
3. Human rights

---

[11] *Our approach*, World Business Council for Sustainable Development (www.wbcsd.org/Overview/Our-approach).
[12] *About Us: Business Transformation for a Just and Sustainable World,* Business for Social Responsibility (www.bsr.org/en/about).
[13] *Our Story,* Business for Social responsibility (www.bsr.org/en/about/story).
[14] *Contact Us,* Business for Social Responsibility (www.bsr.org/en/about/contact).

CHAPTER 13   THE ROLE OF BUSINESS INTEREST GROUPS

4. Nature
5. Supply chain sustainability
6. Sustainability management[15]

BSR has ongoing collaborative initiatives with more than 400 companies across the social responsibility spectrum.[16] These practical initiatives include supporting companies to implement the UN Guiding Principles on Human Rights, and to scale and improve carbon credit programs and other climate solutions.[17] BSR also publishes useful insights on topics such as ESG and sustainability reporting and standards[18] and continuously works with businesses to develop sustainable strategies and solutions.

13.4 Business Roundtable (BRT) and the "Statement on the Purpose of a Corporation"

We discussed Business Roundtable's (BRT) impactful Statement on the Purpose of a Corporation in Chapter 1. We'll provide more background on the organization in this section. BRT is a business association that was formed in 1972 and is based in Washington, DC.[19] Its membership is limited to Corporate Executive Officers (CEOs) and its focus is primarily

---

[15] *Focus Areas*, Business for Social Responsibility (www.bsr.org/en/focus).
[16] *Collaborative Initiatives*, Business for Social Responsibility (www.bsr.org/en/collaboration/groups).
[17] Id.
[18] *Sustainability Insights*, Business for Social Responsibility (www.bsr.org/en/sustainability-insights).
[19] *About Us*, Business Roundtable (www.businessroundtable.org/about-us).

CHAPTER 13   THE ROLE OF BUSINESS INTEREST GROUPS

US policy and advocacy on a broad range of business sector topics, including tax and fiscal policy, technology, workforce and DEI, climate, energy, and corporate governance.[20]

BRT's "Statement on the Purpose of a Corporation," first issued in 1997 and subsequently updated in 2019, was signed by nearly 200 CEOs of some of the largest companies. [21] Its message that "shareholder value is no longer everything" continues to transcend as the organization focuses on how best to provide "good jobs, a strong and sustainable economy, innovation, a healthy environment and economic opportunity for all."[22]

Sustainability is fundamentally about balancing short-term needs with long-term impacts for all stakeholders. The Statement on the Purpose of a Corporation illustrates that balance and the importance of corporate purpose beyond shareholder primacy. There is, of course, ongoing skepticism about whether companies are truly prepared for the level of change required to fulfill their commitment to all stakeholders and about how that calculus impacts a corporation's duties to its shareholders.

13.5   The Ceres Influence

Ceres is an influential nonprofit organization that was formed in 1990. The organization works with "capital market leaders to solve the world's greatest sustainability challenges."[23] Ceres collaborates with investors, companies, and other interest groups on policy solutions and often engages directly on regulatory and legislative proposals.

---

[20] Id.

[21] David Gelles and David Yaffe-Bellany, *Shareholder Value Is No Longer Everything, Top C.E.O.s Say*, The New York Times (August 19, 2019) (www.nytimes.com/2019/08/19/business/business-roundtable-ceos-corporations.html).

[22] Business Roundtable, *Statement on the Purpose of a Corporation* (2022) (https://opportunity.businessroundtable.org/ourcommitment/).

[23] *About Us*, Ceres (www.ceres.org/about-us).

CHAPTER 13   THE ROLE OF BUSINESS INTEREST GROUPS

That engagement and the organization's platform are centered around six core areas:

1. Advancing Climate Solutions
2. Protecting Global Water Resources
3. Building a Just and Inclusive Economy
4. Accelerating Sustainable Capitals Markets
5. Protecting and Restoring Life on Land
6. Advocating for Smart Public Policy[24]

Ceres has been instrumental in several key initiatives. As mentioned previously, Ceres co-launched the Global Reporting Initiative (GRI) in 1997.[25] Ceres is also one of the five investor networks that formed ClimateAction100+ in 2017, which is a group of more than 700 investors with USD$68 trillion in assets that delivers a framework for corporate climate action and assesses and benchmarks corporate climate performance.[26]

The Ceres Roadmap 2030 was published in October 2020, to provide businesses with a ten-year plan for the transition to a just and sustainable economy.[27] The roadmap provides tools and resources for companies at various growth stages, and it is aligned with the UN Sustainable Development Goals.[28] Through concerted action, Ceres is accelerating the adoption of sustainable business practices and mobilizing large-scale economic transformation.

---

[24] *What We Do*, Ceres (www.ceres.org/homepage).
[25] *Our History*, Ceres (www.ceres.org/about-us).
[26] *Global Investors Driving Business Transition*, Climate Action 100+ (www.climateaction100.org/).
[27] *Ceres Roadmap 2030*, Ceres (October 6, 2020) (www.ceres.org/resources/reports/ceres-roadmap-2030).
[28] *How to use the Roadmap*, Ceres (https://roadmap2030.ceres.org/how-use-roadmap).

CHAPTER 13   THE ROLE OF BUSINESS INTEREST GROUPS

13.6 We Mean Business Coalition and Climate Action

The We Mean Business Coalition is a collaboration of seven business-oriented nonprofit organizations, including three organizations that we've discussed in this chapter: Business for Social Responsibility (BSR); Ceres; and the World Business Council for Sustainable Development (WBCSD).[29] The organization was formed in 2014, as a concerted effort to address climate change and accelerate the transition to a sustainable, low-carbon economy.[30]

The coalition's mission is to catalyze business action and accelerate climate action to "create an inclusive net-zero economy, build resilient communities and limit global heating to 1.5°C."[31] In support of that mission, the coalition is mobilizing companies to commit to setting science-based targets and to transition to renewable electricity through the RE100 global corporate renewable energy initiative.[32]

The collaboration of business groups is powerful and lends itself to the collective action model like that utilized by the We Mean Business Coalition. Those efforts are expected to continue to support business efforts to accelerate climate action.

---

[29] *Coalition Partners*, We Mean Business Coalition (www.wemeanbusinesscoalition.org/about/).

[30] Id.

[31] *Homepage*, We Mean Business Coalition (www.wemeanbusinesscoalition.org/).

[32] *Progress*, We Mean Business Coalition (www.wemeanbusinesscoalition.org/progress/). See also Climate Group RE100 (www.there100.org/).

CHAPTER 13   THE ROLE OF BUSINESS INTEREST GROUPS

13.7 State of Play for Trade Associations and Business Groups

Trade associations and business groups serve various convening purposes. Those typically include advocating for public policy, setting industry standards, and establishing best practices. They often provide a forum for regulators, industry, and other stakeholders to come together and discuss industry practices, standards, and oversight. We've included some of the larger and well-known business interest groups in this chapter. They each offer information and resources to companies to advance their climate, socially responsible, and sustainability focused efforts.

There are also myriad specific-industry-focused alliances and coalitions that we have covered throughout this book. Many of these organizations have taken on the mantle of leading industry sectors to become more sustainable and to adopt more responsible business practices. For example, we discuss the Sustainable Apparel Coalition and the controversy surrounding its Higg Index in Chapter 16.

The Roundtable on Sustainable Palm Oil (RSPO), formed in 2004, is another example of a global association focused on developing industry-specific sustainability standards.[33] Palm oil is a globally traded agricultural commodity and it is used in 50% of all consumer goods, such as food, cosmetics, and cleaning products.[34] Approximately 85% of palm oil is grown in Indonesia, Malaysia, and Papua New Guinea, where severe impacts have been documented on deforestation of tropical rainforests and the local communities and workforce.[35] RSPO and its standards have been criticized by stakeholders, including NGOs and environmental

---

[33] *Roundtable on Sustainable Palm Oil* (https://rspo.org/).
[34] *Palm Oil Fact Sheet*, Rainforest Action Network (www.ran.org/palm_oil_fact_sheet/#:~:text=Palm%20oil%20is%20a%20globally,to%20body%20lotion%20and%20biofuels).
[35] Id.

419

CHAPTER 13   THE ROLE OF BUSINESS INTEREST GROUPS

groups, particularly regarding how deforestation, forced labor, and other major issues figure into the standards. While many groups continue to challenge the sustainable palm oil initiative and standards, after nearly a decade of multi-stakeholder meetings, the World Wildlife Fund (WWF) revised its position to support the RSPO. The WWF produces a *Palm Oil Buyers Scorecard* and provides guidance for both companies and customers on its website.[36]

As regulatory, legislative, and executive actions bring enhanced oversight of ESG and sustainability issues, it is expected that trade associations and business groups will take on even more significant duties in advocacy, policymaking and regulatory activity, standards setting, and establishing best practices across industries. The sustainability efforts taken up and endorsed by industry groups will continue to be under scrutiny by stakeholders. The focus is likely to be on developing meaningful, science-based, transparent, and independently reviewed standards and practices.

We've also witnessed a new era of public–private partnerships developed to solve climate and other environmental and social problems. We've discussed a few of those throughout the book. It's important to recognize the sharpened focus of the technology sector on developing climate tech and green tech solutions. Climate tech companies raised approximately USD$51 billion in venture capital and private equity funding in 2023, according to BloombergNEF.[37] The Bay Area climate tech sector is driving innovation in the United States, and recent support by the New York City government has infused approximately USD$3.5 billion

---

[36] *Responsible Purchasing*, World Wildlife Fund (https://wwf.panda.org/discover/our_focus/food_practice/sustainable_production/palm_oil/responsible_purchasing/).

[37] Musfika Mishi, *Over $50 Billion Flow to Climate-Tech Startups in a Stormy Year*, BloombergNEF (February 13, 2024) (https://about.bnef.com/blog/over-50-billion-flow-to-climate-tech-startups-in-a-stormy-year/).

CHAPTER 13   THE ROLE OF BUSINESS INTEREST GROUPS

in venture capital funding to the NYC climate tech sector since 2021.[38] That represents a 250% funding increase over the level of investment just five years prior.[39] Recent investment and incentive activity by local, state, and federal governments are unlocking new potential in climate and green tech.

CHAPTER TAKEAWAYS

- Business interest groups have global influence on corporate ESG practices

- Corporate leaders have long engaged with national and international business groups like WEF and Ceres on sustainability topics

- BRT's Statement on the Purpose of a Corporation remains foundational to responsible business principles

- Trade associations and business groups often develop industry-specific environmental, social, and governance standards and best practices, and that function is set to expand given the global regulatory advancements

- Companies should assess their membership in and how they engage with trade associations and business groups

---

[38] Stephen Lee, *NYC Thrives as an Unlikely Silicon Valley of Climate Technology*, Bloomberg Law (December 19, 2023) (https://news.bloomberglaw.com/environment-and-energy/nyc-thrives-as-an-unlikely-silicon-valley-of-climate-tech).

[39] Id.

CHAPTER 14

The Role of NGOs and Third-Party Mechanisms

Nongovernmental organizations (NGOs) and other certifying and verifying organizations participate in and deliver services in a number of environmental, social, and governance areas. NGOs often administer humanitarian and related programs at a significant scale around the world. Companies often partner with NGOs to deliver on their ESG commitments, particularly with respect to alignment with the Sustainable Development Goals. Similarly, businesses engage with recognized certifying and verifying organizations to assess corporate practices and operations, such as water stewardship and agricultural practices.

14.1 Multipurpose Role of the Nongovernmental Organization (NGO)

NGOs are typically organized as nonprofit entities, have a broad mission, and have an international footprint. The scale of their operations distinguishes them from the rest of operating nonprofits. While they operate independently from the government, these organizations further environmental, social, sustainable development, economic empowerment, and humanitarian missions through a variety of activities, including direct action, advocacy, collaboration, and oversight.

These global organizations are often change agents. Their work has helped shift the focus of corporations, governments, and consumers toward more sustainable and responsible practices. They focus on driving policy changes, fostering collaborations, ***providing technical expertise***, and raising public awareness about sustainability and climate issues.

NGOs are influential on social and environmental issues and ***shape public policies*** related to ESG and sustainability. They advocate before legislators, engage in legal battles, and lead campaigns to drive policy changes. They can be instrumental in pushing for the adoption of stricter environmental regulations, better labor standards, and more ethical business practices.

These organizations often ***collaborate*** with corporations, governments, and other stakeholders to drive sustainable practices. They also leverage their networks to mobilize resources and support for programmatic initiatives.

NGOs ***raise public awareness*** about ESG and sustainability issues. They educate the public about the environmental, social, and economic impacts of unsustainable practices. They also provide information and resources to help individuals and organizations make more sustainable choices.

Many corporations partner with global NGOs. Companies regularly support the operational work of NGOs across the globe. These partnerships typically align with a company's social responsibility strategy and operations in that NGOs share technical and regional expertise on key social metrics that can inform better planning and outcomes. Given their missions and global platforms, several NGOs link their work to the UN Sustainability Goals (SDGs) and track progress against SDG targets. By working with NGOs, companies can support their strategy alignment to

CHAPTER 14   THE ROLE OF NGOS AND THIRD-PARTY MECHANISMS

the SDGs. Some of the most well-known and long-serving NGOs include the following:

- **CARE International** is an anti-poverty NGO that focuses on alleviating poverty and injustice, with a specific focus on women and girls.[1] In 2021, CARE and partners supported 100.2 million people with 1,495 projects in 102 countries.[2] CARE aligns its work with the UN Sustainable Development Goals and has been tracking progress against SDG targets since 2015.[3] The organization was established in 1945.

- **International Rescue Committee (IRC)** assists refugees affected by humanitarian crises like disasters and conflict, including due to climate change.[4] Areas of focus include safety, health, economic well-being, empowerment, and education. In 2021, the NGO reached 31.5 million people worldwide.[5] The IRC was established in 1933.

---

[1] *Who we are*, CARE International (www.care-international.org/who-we-are).
[2] *CARE International Impact, Accountability & Learning Report 2021*, CARE International (December 20, 2022) (www.care-international.org/resources/care-international-impact-accountability-learning-report-2021).
[3] Id.
[4] *What we do*, International Rescue Committee (www.rescue.org/what-we-do?gad_source=1&gclid=CjOKCQiAtOmsBhCnARIsAGPa5yYIZo2yxUVcpSuRk35sgOhBTB6urH3CGz7TeDSEHriu62SB39AEFdQaArVpEALw_wcB&gclsrc=aw.ds).
[5] *2021 Annual Report Executive Summary*, International Rescue Committee (October 12, 2022) (www.rescue.org/report/2021-annual-report-executive-summary).

CHAPTER 14   THE ROLE OF NGOS AND THIRD-PARTY MECHANISMS

- **Oxfam** is an anti-poverty organization that focuses on water and sanitation, gender equality, climate justice, accountable governance, and humanitarian action.[6] In 2021–2022, Oxfam reached approximately 15.6 million people across 90 countries.[7] The organization was established in 1942.

- **Save the Children** works to improve the lives of children through health care, education, emergency aid, and economic development.[8] In 2022, Save the Children and its member organizations reached almost 48.8 million children across 116 countries, responded to 107 humanitarian emergencies, and worked on 129 policy and legislative measures.[9] Save the Children was established in 1919.

14.2 Leading the Way on Science-Based Targets

There has been recent activity to apply the global NGO model to address climate change. The Science Based Targets initiative (SBTi) is a collaborative project that guides corporations in setting scientifically

---

[6] Id.

[7] *The Future is Equal, Oxfam International Annual Report 2021/2022*, Oxfam International (www.oxfam.org/en/what-we-do/about/our-finances-and-accountability/annual-reports-and-financial-statements).

[8] *What We Do*, Save the Children (www.savethechildren.net/).

[9] *Accountability*, Save the Children (www.savethechildren.net/about-us/accountability).

grounded greenhouse gas reduction targets.[10] The initiative was formed as a partnership between the Carbon Disclosure Project (CDP), the United Nations Global Compact (UNGC), the World Wide Fund for Nature (WWF), the World Resources Institute (WRI), and the We Mean Business Coalition.[11] Launched in 2015, the initiative has gained widespread recognition and adoption among companies globally. As of January 2024, SBTi is an independent organization with charitable status granted by the Charity Commission for England and Wales.[12]

SBTi is described as "a corporate climate action NGO" and it can be thought of as a third-party mechanism that enables businesses to set science-based targets for reducing their greenhouse gas emissions, helping to steer the global economy toward a zero-carbon future.[13] As of the beginning of 2024, 4,264 companies had set science-based targets and 2,981 made net-zero commitments through the SBTi.[14] The primary objective of the SBTi is to support corporations in establishing a clear course for sustainable growth by defining a cogent plan to curtail their greenhouse gas emissions.

By setting science-based targets, companies can establish ambitious emission reduction targets that are grounded in climate science. The initiative provides a structured process and sector-specific methodologies for corporations to align their carbon reduction targets with the goals of the Paris Agreement, which aims to limit global warming to well below 2°C

---

[10] *About Us*, Science Based Targets initiative (https://sciencebasedtargets.org/about-us).

[11] Id.

[12] *SBTi doubles corporate climate validations in one year as scale up gathers pace*, Science Based Targets (January 30, 2024) (https://sciencebasedtargets.org/news/sbti-scale-up-gathers-pace?utm_source=social&utm_medium=organic&utm_campaign=linkedinbulletinFeb24).

[13] *About Us*, Science Based Targets initiative (https://sciencebasedtargets.org/about-us).

[14] *Companies Taking Action*, Science Based Targets initiative (https://sciencebasedtargets.org/companies-taking-action).

CHAPTER 14   THE ROLE OF NGOS AND THIRD-PARTY MECHANISMS

above pre-industrial levels and to pursue efforts to limit the temperature increase to 1.5°C. The latest SBTi progress report chronicles that 68% of all companies participating in SBTI with science-aligned targets were aligned with the 1.5°C goal of the Paris Agreement.[15]

SBTi has been scrutinized for its stance on the use of carbon credits for abatement of Scope 3 emissions from supply chains, which was informed by a consultative effort. As discussed in Chapter 4, the underlying complexities of the carbon markets continue to be a topic of discussion. The SBTi model is being applied by the Science Based Targets Network (SBTN), which is a group of experts from more than 60 NGOs, as well as business and consulting groups, working to establish science-based targets in the protection of natural systems.[16] Setting science-based targets is becoming a standard business practice, driving corporate decarbonization, biodiversity protection, and natural capital conservation on a global scale.

CASE STUDY: SALESFORCE

Salesforce is a US-headquartered cloud software company.[17] As part of its ongoing efforts to decarbonize its supply chain and reduce Scope 3 emissions, Salesforce has incorporated a requirement to set science-based targets into its supplier contracts.[18] Salesforce committed that 60% of its suppliers

---

[15] *SBTi Progress Report 2021*, Science Based Targets initiative (https://sciencebasedtargets.org/reports/sbti-progress-report-2021).

[16] *Our mission*, Science Based Targets Network (https://sciencebasedtargetsnetwork.org/our-mission/).

[17] *Our Story*, Salesforce (www.salesforce.com/company/our-story/).

[18] *Salesforce Urges Suppliers to Reduce Carbon Emissions, Adds Climate to Contracts*, Salesforce Sustainability News & Stories (April 29, 2021) (www.salesforce.com/news/stories/salesforce-urges-suppliers-to-reduce-carbon-emissions-adds-climate-to-contracts/).

CHAPTER 14   THE ROLE OF NGOS AND THIRD-PARTY MECHANISMS

(by emissions) that provide "purchased goods and services, capital goods, upstream transportation and distribution, waste generated in operations, and upstream leased assets will set science-based targets by 2024."[19]

In collaboration with CDP, the UN Global Compact, WRI, and WWF, SBTi released the SBTi Glossary on February 28, 2024. It is one of the most comprehensive glossaries of sustainability terms available.[20]

14.3 Organizational Advocacy Platforms

In addition to conducting operations, many NGOs also aim to influence public policy. There are legal and regulatory restrictions in many jurisdictions on the type and amount of campaigning activity that nonprofit entities can undertake, and nonprofit entities are keen to abide by those rules. Generally, NGOs seek to make policy changes, and their opinion and voice in advocating on an issue can be powerful.

These service-based organizations often have direct experience with and exposure to the issues on which they engage, and that often informs public and political opinion. NGOs can be catalysts for change. They often lead civil society participation in joint initiatives with the public sector.

One example is the role Global Witness, a UK-based NGO working to investigate and expose the link between natural resource exploitation, conflict, corruption, and human rights abuses,[21] played in establishing

---

[19] *Supplier Engagement Case Study—Salesforce*, Science Based Targets (https://sciencebasedtargets.org/companies-taking-action/case-studies/supplier-engagement-case-study-salesforce).
[20] *Glossary*, Science Based Targets (https://sciencebasedtargets.org/glossary).
[21] *About us*, Global Witness (www.globalwitness.org/en/about-us/).

CHAPTER 14   THE ROLE OF NGOS AND THIRD-PARTY MECHANISMS

the Kimberely Process in the diamond industry. Global Witness was instrumental in bringing forth evidence that the diamond trade had been used to fund conflicts in Africa, by terrorist groups to finance their activities, and for other money laundering purposes.[22] Global Witness and other NGOs met with diamond trading and producing countries and representatives from the diamond industry in Kimberely, South Africa, in May 2000.[23] The outcome of that convening was an international diamond conflict-free certification scheme, launched in 2003 and known as the Kimberely Process.[24] The Kimberely Process currently has 59 participants, representing 85 nations, which in total represent 99% of global diamond production and trade.[25]

14.4 Independent Role of Certifying Organizations

Certain NGOs also serve the function of verifying or certifying corporate programs. Key areas include, but are not limited to, human rights, fair labor, and water conservation. These NGOs facilitate partnerships, provide technical expertise, and help implement sustainability projects. They also act as so-called watchdogs, holding corporations and governments accountable for their actions. NGOs often advocate for stricter environmental regulations, better labor practices, and more transparent corporate governance.

---

[22] *Human Rights Activism and the Role of NGOs*, Council of Europe (www.coe.int/en/web/compass/human-rights-activism-and-the-role-of-ngos).

[23] Id.

[24] Id.

[25] *The Kimberley Process, the fight against 'conflict diamonds,'* European Commission (https://fpi.ec.europa.eu/what-we-do/kimberley-process-fight-against-conflict-diamonds_en).

CHAPTER 14   THE ROLE OF NGOS AND THIRD-PARTY MECHANISMS

Companies routinely partner with NGOs to certify or verify operational practices. The partnership provides an independent assessment of performance on environmental, social, and governance issues and also lends credibility to representations made about the company's ESG record. We'll explore three examples of NGOs working as certifying and verifying organizations for the corporate sector and take a look at the impact of those engagements:

1. **Rainforest Alliance**

    The Rainforest Alliance (RFA) is an NGO that operates in 190 countries to advance biodiversity conservation and sustainable livelihoods.[26] The Rainforest Alliance (RFA) certification program is a sustainable agriculture initiative focused on farming and supply chain standards for coffee, cocoa, tea, bananas, and other food commodities in more than 70 nations.[27] It was designed to promote environmentally responsible, socially equitable, and economically viable agricultural production.[28] The RFA certification program sets forth comprehensive standards for farmers, encouraging them to adopt better farming practices that reduce environmental impact, such as conserving biodiversity and minimizing chemical use. Farmers

---

[26] *Our Impacts*, Rainforest Alliance (www.rainforest-alliance.org/impact/?_ga=2.267958670.1835243492.1704500666-1573444450.1704141423&_gl=1*14msvkr*_gcl_au*MTk4Mzg2NjY2Mi4xNzA0MTQxNDI2*_ga*MTU3MzQ0NDQ1MC4xNzA0MTQxNDIz*_ga_NFQ21FT91S*MTcwNDUwMDczMi4xLjEuMTcwNDUwMTEzMy4wLjAuMA).

[27] *2020 Certification Program*, Rainforest Alliance (www.rainforest-alliance.org/for-business/2020-certification-program/).

[28] Id.

431

who participate in this program receive training and support to improve their farming methods.[29] The certification focuses on environmental stewardship and the livelihoods of farmers and their communities by promoting fair wages and safe working conditions.

The Rainforest Alliance seal, which appears on certified products and promotional materials, shows consumers that a product has been sourced from farms certified by RFA.[30] Many of the major consumer packaged goods (CPG) companies pledge to buy from RFA-certified farms. In fact, more than 6,000 companies work with RFA to source certified ingredients and improve their sustainability practices.[31]

2. **Verité**

Verité is an NGO focused on labor and work to eliminate supply chain abuses so that people around the world can work under safe, fair, and legal conditions.[32] The organization was established in 1995. Verité's mission is to promote workers'

---

[29] Id.
[30] *New Rain Forest Alliance Certification Seal*, Rainforest Alliance (www.rainforest-alliance.org/business/marketing-sustainability/new-seal/).
[31] *Our Impacts*, Rainforest Alliance (www.rainforest-alliance.org/impact/?_ga=2.267958670.1835243492.1704500666-1573444450.1704141423&_gl=1*14msvkr*_gcl_au*MTk4Mzg2NjY2Mi4xNzAOMTQxNDI2*_ga*MTU3MzQONDQ1MC4xNzAOMTQxNDIz*_ga_NFQ21FT91S*MTcwNDUwMDczMi4xLjEuMTcwNDUwMTEzMy4wLjAuMA).
[32] *About Verité*, Verité (https://verite.org/about/).

CHAPTER 14 THE ROLE OF NGOS AND THIRD-PARTY MECHANISMS

rights through advocacy, corporate engagement, and partnerships with various stakeholders, including businesses, governments, and civil society organizations.[33] It aims to empower workers, increase transparency, and drive changes in labor practices to fight against labor injustices such as child labor, forced labor, and human trafficking for labor exploitation and to create a more equitable global workforce.[34]

The organization functions through a comprehensive approach that includes (i) researching and gathering data on labor conditions globally; (ii) providing training and resources to enhance labor standards; (iii) providing consulting services to companies to improve their supply chain labor practices; and (iv) "conduct[ing] comprehensive audits, investigations, and screenings to provide insight into working conditions in global supply chains."[35] Major companies have partnered with Verité to conduct assessments such as Forced Labor & Human Trafficking Risk Assessments and Supply Chain Risk Screening.[36]

---

[33] Id.
[34] Id.
[35] *Servies*, Verité (https://verite.org/services/).
[36] *Assessments*, Verité (https://verite.org/services/assessments/).

433

3. **Alliance for Water Stewardship**

The Alliance for Water Stewardship (AWS) is a global membership organization based in Scotland focused on global and local leadership on water stewardship.[37] Its members include businesses, NGOs, and the public sector.[38] AWS was founded in 2008, by World Wildlife Fund (WWF), The Nature Conservancy, Water Stewardship Australia, Pacific Institute, Water Environment Federation, UNEP, UN Global Compact's CEO Water Mandate, European Water Partnership, and Water Witness International.[39]

Its mission is to promote water stewardship and responsible water use.[40] AWS functions primarily by certifying organizations through the AWS Standard, which is an International Water Stewardship Standard, guiding the implementation of responsible and transparent water management practices.[41] AWS offers training, supports members in achieving their water stewardship goals, and fosters a community of practice to share knowledge and experiences in water stewardship. AWS also engages in advocacy and policy dialogues to influence water governance frameworks

---

[37] *About the Alliance for Water Stewardship*, Alliance for Water Stewardship (https://a4ws.org/about/).
[38] Id.
[39] *Alliance for Water Stewardship Projects*, World Wildlife Fund (www.worldwildlife.org/projects/alliance-for-water-stewardship).
[40] *About the Alliance for Water Stewardship*, Alliance for Water Stewardship (https://a4ws.org/about/).
[41] *The AWS International Water Stewardship Standard*, Alliance for Water Stewardship (https://a4ws.org/the-aws-standard-2-0/).

at various levels. Operations for a variety of companies, including major multinationals, across sectors such as food and beverage, chemicals, agriculture, mining, textiles, and manufacturing, are certified by AWS.[42]

CHAPTER TAKEAWAYS

- NGOs are agents for social change, and that civil society model has been adapted successfully across the nonprofit and corporate sectors

- Companies are setting science-based targets as a sound basis for ESG decision-making, commitments, and strategy

- Engaging with certifying and verifying organizations is an important aspect of responsible business practices

- NGOs and other nonprofits can lend expertise and experience in working with local communities around the world, and collaboration can improve stakeholder relations

---

[42] *Certification*, Alliance for Water Stewardship (https://a4ws.org/certification/).

CHAPTER 15

The Role of the Philanthropic Sector

We covered corporate philanthropy in Chapter 5. Now, we'll turn first to institutional philanthropy, which is primarily the large private grant-making foundations, and then to a discussion of public charitable organizations that receive corporate and foundation grants to fund their operations. Specifically, we'll discuss the transformative role of philanthropy in advancing climate, sustainability, and social impact efforts, and how that influences corporate social programs and philanthropy.

15.1 Philanthropic Impact and Influence

Philanthropic organizations, particularly large private foundations, have always been at the forefront of social change. Many have a long-established history and have played a significant role in raising awareness, driving policy changes, and implementing various programs aimed at improving society's quality of life. Philanthropic organizations often fund educational programs, awareness campaigns, and research initiatives aimed at increasing public understanding and engagement with these issues.

CHAPTER 15   THE ROLE OF THE PHILANTHROPIC SECTOR

Large philanthropic foundations have made significant contributions to environmental conservation and sustainability. Through their funding and advocacy efforts, they have supported projects aimed at protecting biodiversity, promoting sustainable agriculture, mitigating climate change, and raising awareness about environmental issues. Philanthropic organizations have also played a vital role in addressing social issues such as poverty, inequality, and discrimination. Through their funding and collective action efforts, they have supported initiatives aimed at empowering marginalized communities, promoting democracy and social justice, and advancing human rights.

15.2 Mission Shift to Achieve Impact

As the world grapples with complex challenges, the philanthropic sector has undergone a recent shift.[1] Civil rights and the global pandemic have caused many philanthropic organizations to recalibrate[2] and to reshape giving and investment priorities. Private foundations have begun addressing global challenges on a much more impactful scale. This trend, coupled with the sector's inherent focus on social impact and sustainability, is revolutionizing the way nonprofits approach their missions. Philanthropic organizations are focusing on "how to create transformational, systems-level impact."[3]

---

[1] Lysa Ratliff, *The Changing Landscape of Philanthropy – Bolder Moves for Greater Impact*, The Center for Effective Philanthropy (https://cep.org/the-changing-landscape-of-philanthropy-bolder-moves-for-greater-impact/).

[2] Gabriel Kasper, Justin Marcoux, and Jennifer Holk, *The T-Rex and the Snowshoe Hare: What's Next for Philanthropy in the 2020s*, Stanford Social Innovation Review (January 24, 2023) (https://ssir.org/articles/entry/the_t_rex_and_the_snowshoe_hare_whats_next_for_philanthropy_in_the_2020s#).

[3] *Shifting Systems Initiative*, Rockefeller Philanthropy Advisers (www.rockpa.org/project/shifting-systems/).

Increasingly, those efforts are intersecting with ESG principles. The urgency of climate action has also become an essential catalyst for philanthropy's evolution. Organizations are reconfiguring their mission to directly confront this crisis. Its impacts are far-reaching and disproportionately affect the most vulnerable communities, exacerbating societal challenges such as declining health, education access, equality, and food security. As the negative impacts of climate change become increasingly apparent, philanthropies have begun to prioritize climate action in their funding strategies, as we'll discuss throughout this chapter.

CASE STUDY: THE EARTHSHOT PRIZE

The Earthshot Prize is a company registered in England and Wales and a registered charity.[4] It was created by Prince William and The Royal Foundation. Modeled on President Kennedy's "Moonshot" challenge in 1962, The Earthshot Prize gives a global platform to the most promising innovative solutions to our greatest environmental challenges. Recognizing the climate crisis, The Earthshot Prize recently announced its new "Launchpad" platform which aims to catalyze more investment in climate innovation. In its pilot phase, Launchpad spotlights 25 Earthshot solutions across six continents with more than GBP£400 million in associated funding.[5]

---

[4] The Earthshot Prize (https://earthshotprize.org/).
[5] *The Earthshot Prize 'Launchpad' Platform to Catalyze More Investment In Climate Innovation*, The Earthshot Prize (March 11, 2024) (https://earthshotprize.org/news/the-earthshot-prize-launchpad-platform-to-catalyze-more-investment-in-climate-innovation/).

15.3 Increased Level of Funding and Grants Activity

The amount of philanthropic capital available for deployment has the potential to make positive change. For example, the level of total private foundation assets in the United States was USD$1.057 trillion in 2022,[6] and that asset level rose to USD$1.25 trillion in the first half of 2023.[7] When coupled with strategic initiatives and collaboration, particularly among private foundations in support of public action, that capital can have a profound impact.

As the world grapples with the escalating crisis of climate change, philanthropic organizations have been increasingly shifting their missions and funding strategies. The objective is to catalyze climate action initiatives that seek to mitigate the impacts of climate change and foster a sustainable and resilient future. The growth in philanthropic support for climate action mirrors the rising global concern over climate change. In the last few years, philanthropic funding for climate change mitigation has increased at a significant rate, outpacing the growth of overall philanthropic giving. In 2021, philanthropic giving to climate change mitigation increased to 25%, while overall philanthropic giving only rose 8%.[8] However, total giving to climate change mitigation still represents a very small percentage of total philanthropic giving—less than 2% of the USD$750 billion given by foundations.[9]

---

[6] *U.S foundation assets fell $246 billion in value in 2022, study finds*, Philanthropy News Digest, Candid (https://philanthropynewsdigest.org/news/u.s.-foundation-assets-fell-246-billion-in-value-in-2022-study-finds).

[7] *Foundation assets rose $100 billion in first half of 2023, study finds*, Philanthropy News Digest, Candid (https://philanthropynewsdigest.org/news/foundation-assets-rose-100-billion-in-first-half-of-2023-study-finds).

[8] *Funding trends 2022: Climate change mitigation philanthropy*, ClimateWorks Foundation (www.climateworks.org/report/funding-trends-2022/).

[9] Id.

CHAPTER 15   THE ROLE OF THE PHILANTHROPIC SECTOR

Historically, philanthropic organizations have demonstrated an ability to respond to the emergence of global crises. The foundation response to the global pandemic exemplifies this capacity. Philanthropic organizations repositioned their programs relatively quickly given the nature and size of the organizations, reconceptualized their partnerships, and rethought how they support their grantees. This ability to pivot and adapt is now being harnessed to address the climate crisis.

Climate change disproportionately affects marginalized communities and those least responsible for contributing to it.[10] As philanthropic organizations shift more of their focus toward climate action, there is a growing recognition that climate change is not just an environmental issue but also a matter of justice.[11] This has led to an increased emphasis on climate justice in philanthropic funding strategies. The inclusion of climate justice in philanthropic practices is a significant step toward developing climate solutions that are equitable and inclusive. It involves redirecting funds toward grassroots organizations and initiatives that are closely connected to frontline communities. This helps to incorporate a wider range of voices and perspectives in climate change mitigation efforts.

To achieve greater impact, philanthropic organizations are beginning to increase their funding for climate action, mobilize additional capital from private and public sources, and expand joint efforts. As the urgency of climate action escalates, there has been a resounding call for philanthropy

---

[10] *EPA Report Shows Disproportionate Impact of Climate Change on Socially Vulnerable Populations in the United States*, US Environmental Protection Agency (September 2, 2021) (www.epa.gov/newsreleases/epa-report-shows-disproportionate-impacts-climate-change-socially-vulnerable#:~:text=WASHINGTON%20(Sept.,%2C%20flooding%2C%20and%20other%20impacts).

[11] Aaron Dorfman, *Listening to the Experts: A Campaign to Redirect Climate Justice and Just Transition*, National Committee for Responsive Philanthropy (July 12, 2023) (www.ncrp.org/publication/listening-to-the-experts-a-campaign-to-redirect-climate-justice-and-just-transition-summer-2023).

CHAPTER 15   THE ROLE OF THE PHILANTHROPIC SECTOR

to lead and take bolder action. The scale and urgency of the climate crisis necessitates an unprecedented level of philanthropic commitment and collaboration.

> **CASE STUDY: PROTECTING OUR PLANET CHALLENGE**
>
> The **Protecting Our Planet Challenge** is a USD$5 billion funding commitment by philanthropic organizations to protect at least 30% of the world's land and ocean areas by 2030—the UN's 30x30 target. This includes USD$1 billion of dedicated funding toward ocean and marine protection. Partners include Bezos Earth Fund, Arcadia, Bloomberg Philanthropies, the Gordon and Betty Moore Foundation, Nia Tero, Rainforest Trust, Re:wild, Wyss Foundation, and the Rob and Melani Walton Foundation.[12]

15.4 Innovative Financing Vehicles

Philanthropic and charitable giving have long been the traditional routes for addressing social challenges. However, in the face of growing global issues such as health disparities, climate change, and economic inequality, there is a pressing need for innovative financing mechanisms. These

---

[12] *World leaders pledge US$ 5 billion to protect nature*, United Nations Development Programme (September 22, 2021) (www.undp.org/press-releases/world-leaders-pledge-us-5-billion-protect-nature); *Philanthropies pledge $5 billion to 'Protecting Our Planet Challenge,'* Philanthropy News Digest (September 22, 2021) (https://philanthropynewsdigest.org/news/philanthropies-pledge-5-billion-to-protecting-our-planet-challenge); *WCS Welcomes the Launch of the 'Protecting Our Planet Challenge,'* Wildlife Conservation Society (September 22, 2021) (https://newsroom.wcs.org/News-Releases/articleType/ArticleView/articleId/16688/WCS-Welcomes-the-Launch-of-the-Protecting-Our-Planet-Challenge.aspx).

CHAPTER 15   THE ROLE OF THE PHILANTHROPIC SECTOR

mechanisms are designed to channel funds from private sources, financial markets, and other stakeholders to create scalable and effective solutions for global development and humanitarian challenges.

Traditional philanthropy often involves grant-making and donations directed toward nonprofit operating organizations that work directly with communities to address societal issues. While grants continue to be the primary source of giving for the major private foundations, many foundations and philanthropists have found the model constricting in that it often lacks the scalability and sustainability needed to tackle complex, systemic challenges.

Philanthropy has the potential to bring about cross-sectoral collaboration and innovation. By bridging gaps in funding and supporting innovative solutions, philanthropies can play a pivotal role in accelerating the global transition to a sustainable and carbon-neutral future. Innovative finance goes beyond grant-making and donations. It involves developing new financial instruments and mechanisms that can unlock larger pools of capital from diverse sources, including private sector investors, financial markets, and philanthropists.[13]

Innovative financing has emerged or, some might say, re-emerged, as a promising trend in philanthropy, and a select group of philanthropic funding organizations are implementing innovative finance approaches.[14] It offers a new approach to funding by blending public, private, and philanthropic resources in strategic ways to solve complex societal

---

[13] *Innovative Financing for Development: Scalable Business Models that Produce Economic, Social, and Environmental Outcomes*, Innovative Financing Initiative, Global Development Incubator (September 2014) (www.citigroup.com/rcs/citigpa/akpublic/storage/public/innovative_financing_for_development.pdf).

[14] Gabriel Kasper & Justin Marcoux, *The Re-Emerging Art of Funding Innovation*, Stanford Social Innovation Review (Spring 2014) (https://ssir.org/articles/entry/the_re_emerging_art_of_funding_innovation#).

CHAPTER 15   THE ROLE OF THE PHILANTHROPIC SECTOR

problems.[15] Unlike traditional philanthropy, which is often subject to the specific intent of donors, innovative financing creates more sustainable and scalable solutions by leveraging market-based strategies.

The need for innovative finance has never been greater. The United Nations estimates a funding gap between USD$5.4 and $6.3 trillion to achieve the Sustainable Development Goals (SDGs).[16] Given this enormous gap, it's clear that traditional philanthropy alone won't be enough to meet these ambitious targets. New approaches can help to mobilize additional resources, encourage cross-sector collaboration, and drive greater social impact.

Select philanthropic organizations have explored how best to utilize innovative financing mechanisms. These mechanisms are designed to leverage private investment, promote efficiency, and drive measurable social impact. The following are examples of those mechanisms, some of which we covered in previous sections on sustainable finance initiatives:

- **Impact investing:** Impact investing refers to making investments for the dual purpose of generating a positive financial return and a measurable social or environmental impact. It's a form of double or even triple bottom-line investing, where the investor seeks not only financial return but also social and/or environmental benefits.[17]

---

[15] *Innovative finance*, International Labour Organization (www.ilo.org/empent/areas/social-finance/WCMS_747999/lang--en/index.htm).

[16] *Annual cost for reaching the SDGs? More than $5 trillion*, United Nations (September 2023) (https://news.un.org/en/story/2023/09/1140997#:~:text=The%20cost%20of%20achieving%20ambitious,%241%2C383%20per%20person%2C%20per%20year).

[17] *What You Need to Know About Impact Investing*, Global Impact Investing Network (https://thegiin.org/impact-investing/need-to-know/).

CHAPTER 15   THE ROLE OF THE PHILANTHROPIC SECTOR

- **Blended finance:** Blended finance is another innovative financing mechanism that combines grant funds with private capital to achieve specific impact investment opportunities.[18] By blending different types of capital, it can help to de-risk investments and attract more private sector participation.[19]

- **Social, environmental, and development impact bonds:** Bonds are pay-for-success financing mechanisms that tie financial returns to the achievement of specific social outcomes. In these models, private investors provide upfront capital for social programs, and if the programs achieve their outcome targets, the investors receive their principal plus a return from an outcome funder, usually a government or philanthropic organization.[20]

- **Debt swaps:** Debt swaps are a type of innovative finance where a creditor forgives a portion of a country's debt in exchange for investments in social or environmental programs. This mechanism, also known

---

[18] *Blended Finance*, Organization for Economic Development (www.oecd.org/dac/financing-sustainable-development/blended-finance-principles/).

[19] *How can 'blended finance' help fund climate action and development goals?* The London School of Economics and Political Science (November 30, 2022) (www.lse.ac.uk/granthaminstitute/explainers/how-can-blended-finance-help-fund-climate-action-and-development-goals/).

[20] *Innovative Finance: Mobilizing Capital for Maximum Impact*, Wharton Social Impact Initiative, University of Pennsylvania (July 2016) (https://esg.wharton.upenn.edu/wp-content/uploads/2022/09/Innovative-Finance-Report_July-2016.pdf).

as debt-for-development swaps, provides a win-win solution by reducing the debt burden on developing countries while mobilizing additional resources for development programs.[21] These mechanisms are primarily applied to climate and nature, and the UN Development Programme believes they could potentially be expanded to other Sustainable Development Goal areas.[22]

These mechanisms require a high level of collaboration and coordination among diverse stakeholders, including governments, private sector investors, philanthropists, and nonprofit organizations. To truly maximize the potential of innovative finance, the philanthropic sector is adopting a more proactive and strategic approach to build capacity in terms of expertise and infrastructure, promote collaboration among partners, and measure impact.

A leading example of a private foundation employing innovative financing models is the venerable Ford Foundation. The Ford Foundation refocused its mission to address global inequality in 2015. In 2017, it announced plans to commit USD$1 billion of its USD$12 billion endowment to mission-related investments (MRIs) to catalyze social impact, which was the largest commitment of its kind by a private foundation.[23]

---

[21] Kristalina Georgieva, Marcos Chamon, Vimal Thakoor, *Swapping Debt for Climate or Nature Pledges Can Help Fund Resilience*, International Monetary Fund (December 14, 2022) (www.imf.org/en/Blogs/Articles/2022/12/14/swapping-debt-for-climate-or-nature-pledges-can-help-fund-resilience).

[22] *A new Wave of Debt Swaps for Climate or Nature*, United Nations Development Programme (www.undp.org/future-development/signals-spotlight/new-wave-debt-swaps-climate-or-nature).

[23] Ford Foundation commits $1 billion from endowment to mission-related investments, Ford Foundation (April 5, 2017) (www.fordfoundation.org/news-and-stories/news-and-press/news/ford-foundation-commits-1-billion-from-endowment-to-mission-related-investments/).

Darren Walker, the widely respected President of the Ford Foundation, wrote an important piece in *The New York Times*, entitled "Why Giving Back Isn't Enough," in 2015. In that opinion, he included words spoken by the Reverend Dr. Martin Luther King Jr. nearly 50 years earlier: "Philanthropy is commendable, but it must not cause the philanthropist to overlook the circumstances of economic injustice which make philanthropy necessary."[24] Walker's later work, *From Generosity to Justice: A New Gospel of Wealth*, published in 2023, suggests that now is the time for a systemic change in philanthropy and places a focus on inclusive capitalism.[25]

CASE STUDY: BREAKTHROUGH ENERGY

Breakthrough Energy is an umbrella of several organizations with a focus on philanthropy, private investment, and public policy aligned to accelerate breakthroughs to mitigate climate change and reduce global GHG emissions. Helmed by Bill Gates and announced at COP21 held in Paris in 2015, the public–private partnership model is being deployed to advance innovation in sustainable technologies particularly where the risk of failure is high and the return on investment (ROI) time frame is 20 years. Breakthrough Energy was seeded with a USD$2 billion investment by Bill Gates and launched with the participation of a group of well-known individual investors as well as the University of California system.[26]

---

[24] Darren Walker, *Why Giving Back Isn't Enough*, The New York Times (December 27, 2015) (www.nytimes.com/2015/12/18/opinion/why-giving-back-isnt-enough.html). See also Dr. Martin Luther King, Jr., *Strength to Love* (1963).

[25] *From Generosity to Justice: A New Gospel of Wealth*, Ford Foundation (www.fordfoundation.org/news-and-stories/big-ideas/the-future-of-philanthropy/from-generosity-to-justice/).

[26] *Our Work*, Breakthrough Energy (https://breakthroughenergy.org/our-work/).

CHAPTER 15   THE ROLE OF THE PHILANTHROPIC SECTOR

15.5 Private Foundations Making their Mark in Climate Philanthropy

The landscape of climate philanthropy is changing. Many foundations have recognized that the scale of the climate crisis requires a commensurate response. Private foundations generally take the form of an independent foundation, a family foundation, or a corporate foundation. They are stepping up their commitments and pledging significant funds to climate change mitigation and resilience efforts. The growing involvement of private foundations in climate change initiatives is a promising development in the fight against the global climate crisis. By providing vital funding, fostering innovation, and promoting collaboration, foundations are helping to accelerate the transition to a sustainable, low-carbon economy.

Charitable foundations can convene a multisector network to foster collaboration and innovation on climate change. By pooling resources and knowledge, they can help to scale up proven solutions, drive policy change, and spur technological innovation. A recent public–philanthropic partnership provides an example of how organizations can collaborate to advance climate resilience.

The US National Climate Resilience Framework is a White House initiative aimed at climate adaptation and resilience.[27] It acknowledges the growing impacts of climate change on citizens' lives, livelihoods, industries, and natural resources. The framework emphasizes the need to considerably diminish emissions while equipping communities with the necessary resources to brace for the escalating effects of climate change.[28]

---

[27] *Fact Sheet: Biden-Harris Administration Hosts First-Ever White House Climate Resilience Summit and Releases National Climate Resilience Framework,* The White House (September 18, 2023) (www.whitehouse.gov/briefing-room/statements-releases/2023/09/28/fact-sheet-biden-harris-administration-hosts-first-ever-white-house-climate-resilience-summit-and-releases-national-climate-resilience-framework/).
[28] Id.

Contemporaneous with the announcement by the White House, several prominent charitable foundations, such as the Kresge Foundation, Gordon and Betty Moore Foundation, David and Lucile Packard Foundation, and Walton Family Foundation, pledged their shared commitment to climate resilience.[29] These foundations are leading a philanthropic initiative focused on four key actions:

1. **Developing a Shared Framework:** Collaborating across the philanthropic sector and with community organizations and other stakeholders to establish a framework for climate resilience philanthropy alongside public and private sector action

2. **Mobilizing New Commitments:** Rallying additional philanthropic commitments to close current funding gaps and expand the climate philanthropy network, emphasizing organizations that represent those most vulnerable to climate change

3. **Enhancing Knowledge and Learning:** Strengthening the knowledge base and accelerating learning across the climate resilience field to drive strategy, action, and philanthropic investment

4. **Improving Communications:** Making climate resilience more comprehensible and achievable by promoting simpler language about resilience requirements and benefits[30]

---

[29] *Philanthropic leaders commit to building more climate resilient communities,* The Kresge Foundation (September 28, 2023) (https://kresge.org/news-views/philanthropic-leaders-commit-to-building-more-climate-resilient-communities-as-the-biden-harris-administration-releases-first-ever-national-framework/).
[30] Id.

CHAPTER 15   THE ROLE OF THE PHILANTHROPIC SECTOR

The commitment from these philanthropic foundations builds upon a range of climate resilience and adaptation strategies. It's a proactive model for collaboration between the public and philanthropic sectors. It may also serve as a blueprint for multisector and cross-sector climate action, which is necessary to address the pressing challenges of climate change.

CASE STUDY: METHANESAT

MethaneSAT is a satellite launched into the Earth's atmosphere in March 2024.[31] Methane is the second-largest contributor to global warming behind carbon dioxide and accounts for 30% of global warming since pre-industrial times, as we noted in Chapter 4. The nonprofit Environmental Defense Fund (EDF) partnered with business interests and the philanthropy Bezos Earth Fund to launch MethaneSAT. The satellite will track global methane emissions, and research estimates that curbing methane emissions could slow the rate of warming from GHG emissions by 30%.[32]

---

[31] *Four Things to Know About MethaneSAT—The Satellite Taking Climate Surveillance to New Heights,* Bezos Earth Fund (March 4, 2024) (www.bezosearthfund.org/news-and-insights/what-to-know-about-methanesat-taking-climate-surveillance-new-heights)

[32] *Study: Cutting Methane Emissions Quickly Could Slow Climate Warming Rate by 30%,* Environmental Defense Fund (April 27, 2021) (www.edf.org/media/study-cutting-methane-emissions-quickly-could-slow-climate-warming-rate-30); *Four Things to Know About MethaneSAT—The Satellite Taking Climate Surveillance to New Heights,* Bezos Earth Fund (March 4, 2024) (www.bezosearthfund.org/news-and-insights/what-to-know-about-methanesat-taking-climate-surveillance-new-heights).

15.6 Public Charities Programming for Impact and Alignment with the SDGs

As we discussed in Chapter 1, businesses partner with nonprofit organizations as part of their social responsibility strategy. The first steps to doing so are identifying nonprofits with a charitable mission that line up with the company's social responsibility goals, and to conduct due diligence on those organizations, such as review of the nonprofit's performance and compliance with applicable laws.

Companies can enhance their ESG practices by collaborating with nonprofits that support the specific UN Sustainable Development Goals mapped by the company as discussed in Chapter 3. The SDGs offer a framework and actionable targets toward the achievement of the 17 interconnected goals. For-profit and nonprofit partnerships are part of the multisector effort required to achieve the SDGs in a timely manner.

Nonprofits, by their very nature, are intertwined with many of the SDGs. In fact, most nonprofits are already contributing toward achieving these goals through their daily operations and missions, even if they don't explicitly discuss the linkage. Nonprofits can use the SDGs as a guide for setting specific goals and measuring their impact, thereby increasing their efficiency and effectiveness. This alignment not only aids nonprofits in their quest for a sustainable future but also enables them to be part of a global movement, thereby multiplying their reach and influence.

There are many nonprofit organizations committed to making an impact on local, regional, and global scales. They are part of a tremendous network of organizations working to leverage resources toward a more sustainable and equitable world. Companies should review their giving practices and undertake a similar mapping exercise to Table 3-3 from Chapter 3, as well as explore the use of innovative funding mechanisms and opportunities to collaborate with nonprofit partners for greater impact.

CHAPTER 15   THE ROLE OF THE PHILANTHROPIC SECTOR

CHAPTER TAKEAWAYS

- The mission and focus of institutional philanthropy are adjusting to address current global environmental and social challenges

- Philanthropy giving levels to mitigate climate change impacts and develop climate resilience have increased, but more is needed

- Philanthropic organizations are innovating and adjusting their funding and financial models

- Partnerships with philanthropic organizations on key ESG issues and in collaboration to meet the Sustainable Development Goals can amplify impact

CHAPTER 16

The Pathways to Decarbonizing Key Sectors

In this chapter, we'll explore current thinking, collaboration, and innovation to decarbonize key industrial sectors. Just as scientists track the data on the key emitting sectors, we also have considerable knowledge and know-how to decarbonize those sectors. Historical impediments to lower GHG emissions are being removed from the decarbonization path, including

1. Lack of investment and financing
2. Slow regulatory approvals and permitting processes
3. Competition from existing and conventional sources
4. Difficult to abate sectors and inefficient alternatives
5. Challenges to adapting infrastructure
6. Supply chain disruptions
7. Hurdles to get to market and scale

CHAPTER 16   THE PATHWAYS TO DECARBONIZING KEY SECTORS

We have just lived through the hottest year on record. The case for climate action is beyond doubt and requires planning now. Going forward, we will have to stand more firmly on two legs: a safe and healthy climate for all to live in, and a strong, resilient economy, with a bright future for business and a just transition for all.[1]

—Wopke Hoekstra, European Commissioner for Climate Action

As we've discussed, public and private investment, particularly in terms of capital deployment for new technologies and infrastructure shifts, as well as tailored regulations and industry engagement are advancing decarbonization pathways. Collaboration across sectors has also facilitated decarbonization goals and enables those goals to be met at scale. The WEF First Movers Coalition is an illustrative case study on building consensus and capacity toward action.

CASE STUDY: WORLD ECONOMIC FORUM (WEF) FIRST MOVERS COALITION

The **First Movers Coalition** is a global group of companies "leveraging their purchasing power to decarbonize the world's heavy-emitting sectors" and coordinated by the World Economic Forum (WEF).[2] WEF estimates that by 2050, 50% of emissions reductions needed to get to net zero will come from technologies that are not yet available at a necessary scale.[3] There are more

---

[1] *Commission presents recommendation for 2040 emissions reduction target to set the path to climate neutrality in 2050,* European Commission (February 6, 2024) (https://ec.europa.eu/commission/presscorner/detail/en/ip_24_588).

[2] *First Movers Coalition,* World Economic Forum (https://initiatives.weforum.org/first-movers-coalition/home).

[3] Id.

than 90 active members in the coalition and the focus is on seven core sectors and sub-sectors: aluminum; aviation; cement and concrete; shipping; steel; trucking; and carbon dioxide removal. The concept is to build market demand for clean tech to accelerate commercial viability.

16.1 Decarbonizing the Energy Sector

The energy sector is an essential component of our modern world, powering everything from our homes and businesses to transportation and industry. However, it is also the primary contributor to global greenhouse gas (GHG) emissions, a key driver of climate change. While advances are being made at a quicker pace than ever before, challenges to decarbonization remain, including regulatory considerations within the energy sector and the steps necessary for its decarbonization.

Coal, oil, and natural gas have been the primary feedstocks for electricity generation. Infrastructure and grid systems were designed around those sources. Renewable energy sources such as solar, wind, hydropower, geothermal, and biomass are considered the greenest options for energy generation. Those sources have a much lower impact on the environment, producing fewer GHG emissions compared to their fossil fuel counterparts.

The cost of renewables has dropped substantially over the past decade. The percentage of global electricity generated from renewables expanded to 29% in 2020, from 27% in 2019.[4] In the United States, renewable sources generate approximately 20% of all electricity, with the following breakdown: wind (9.2%); hydropower (6.3%); solar (2.8%); biomass (1.3%); and geothermal (0.4%)[5]

---

[4] *Global Energy Review 2021*, International Energy Agency (www.iea.org/reports/global-energy-review-2021/renewables).

[5] *Renewable Energy*, Office of Energy Efficiency & Renewable Energy, US Department of Energy (www.energy.gov/eere/renewable-energy#:~:text=modernize%20the%20grid.-,Renewable%20Energy%20in%20the%20United%20States,that%20percentage%20continues%20to%20grow).

CHAPTER 16   THE PATHWAYS TO DECARBONIZING KEY SECTORS

The world added 50% more renewable capacity from 2022 to 2023.[6] The IEA estimates that we will add more renewable capacity in the five years from 2023 to 2028 than the total installed since the inaugural renewable power plant was constructed over 100 years ago.[7] It is predicted that renewables will surpass coal to become the largest single source of electricity generation by 2025.[8] That growth will continue with the expectation that renewables will account for more than 42% of global electricity generation by 2028.[9] Figure 16-1 is the IEA depiction of the transformation of the global renewable power mix predicted by 2028.

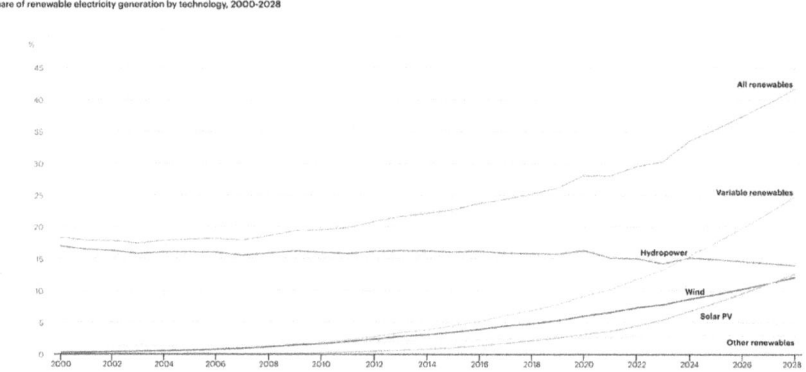

Figure 16-1. *IEA Share of Renewable Electricity Generation by Technology*[10]

---

[6] *Massive expansion of renewable power opens door to achieving global tripling goal set at COP28*, International Energy Agency (January 11, 2024) (www.iea.org/news/massive-expansion-of-renewable-power-opens-door-to-achieving-global-tripling-goal-set-at-cop28?utm_source=linkedin+newsletter&utm_medium=social&utm_campaign=linkedin+newsletter&utm_content=22Jan).

[7] *Renewables 2023: Analysis and forecasts to 2028*, International Energy Agency (January 2024) (www.iea.org/reports/renewables-2023).

[8] Id.

[9] Id.

[10] IEA, Share of renewable electricity generation by technology, 2000-2028, IEA, Paris www.iea.org/data-and-statistics/charts/share-of-renewable-electricity-generation-by-technology-2000-2028, IEA. License: CC BY 4.0.

456

CHAPTER 16   THE PATHWAYS TO DECARBONIZING KEY SECTORS

Decarbonization Challenges and Potential Solutions

While the shift to renewable energy sources is essential to mitigating climate change, the process of decarbonization isn't without its challenges. These include ensuring grid stability, managing the variability of renewable energy, and addressing the cost and accessibility of renewable technologies.

- **Challenge of Grid Stability**

    One of the significant challenges of decarbonization is maintaining grid stability while sufficiently reducing GHG emissions. As reported in a Center for Strategic & International Studies Brief: "Solar and wind power are available zero-carbon emission technologies, but their variability can challenge grid stability if not properly balanced by sufficient storage and power."[11]

- **Price and Accessibility of Renewable Technologies**

    Another significant challenge is ensuring that decarbonization is "equitable and accessible"[12] to low-income countries. While renewable sources have become more cost-competitive, many low-income countries and communities do not have access to them and still rely on high-emitting sources.[13]

---

[11] Sarah Ladislaw and Stephen J. Naimoli, *Climate Solutions Series: Decarbonizing the Electric Power Sector*, Center for Strategic & International Studies (May 12, 2020) (www.csis.org/analysis/climate-solutions-series-decarbonizing-electric-power-sector).
[12] Id.
[13] Id.

- **Regulatory and Public Policy Considerations**

    Decarbonization efforts must also navigate regulatory considerations, which can vary significantly from one jurisdiction to another. These can include environmental regulations, energy market regulations, and policies related to renewable energy generation and transmission. While strides have been made, the ability to bring renewable technologies to market has historically been hampered by the regulatory environment. Policies such as streamlined permitting, renewable standards, and carbon pricing can create a more conducive environment for the deployment of low-carbon technologies and practices.

Despite these obstacles, several potential solutions can facilitate the decarbonization process. These include advancing energy efficiency, improving materials efficiency, expanding the use of carbon capture, utilization, and storage (CCUS) technologies, and increasing the use of low-carbon binding materials in industries such as cement production.

In the short term, actions can be taken to reduce emissions from the energy sector. These include increasing energy efficiency, promoting the use of renewable energy, and implementing policies that incentivize decarbonization. In the long term, solutions for decarbonization include the development of new technologies, the transformation of energy systems, and the implementation of comprehensive climate policies. These solutions require significant investment and collaboration among government, industry, and society.

Key policy measures supporting adoption and expansion of markets for renewable and low-carbon fuels include

1. **Incentives:** Incentives, including tax credits, rebates, or grants for businesses to use or produce renewable and low-carbon fuels, can make them an economically sound option.

2. **Mandates:** Policies that require fuel suppliers to include a certain percentage of renewable and low-carbon fuels in their fuels mix can help drive demand and innovation.

3. **Standards:** Regulations that ensure the sustainability of feedstocks used to produce fuels can mitigate negative environmental and market impacts.

Role of Innovation and Technology

Innovation and technology are elemental to the decarbonization of the energy sector. Advancements in renewable energy technologies, carbon capture and storage, and energy efficiency can significantly reduce GHG emissions.

While the energy sector is a significant contributor to greenhouse gas emissions, it also holds a primary key to mitigating climate change. Decarbonizing the energy sector is a complex process that requires a comprehensive approach. Through a combination of policy, technology, and innovation, we can transition to a cleaner, more sustainable energy future.

CHAPTER 16 THE PATHWAYS TO DECARBONIZING KEY SECTORS

16.2 Decarbonizing the Transportation Sector

Given the GHG emissions profile, the transportation sector is key to transitioning to a cleaner economy. As the world strives to achieve net-zero emissions and transition to a low-carbon future, decarbonizing the transportation industry is a prime objective. It will also require a combination of technological advancements, policy support, and innovative approaches.

The transportation sector heavily relies on liquid fuels made from petroleum sources, which contribute to its high GHG emissions. The ease of transport and storage, along with the energy density of these fuels, has made them the preferred choice. The decarbonization of the transportation sector does not come with a one-size-fits-all solution. Different modes of transportation present varying degrees of difficulty in transitioning to low-carbon alternatives.

Road Transportation: Advancing Technology

Electrification: A Promising Path for Lighter Vehicles

Electrification offers a relatively straightforward solution for smaller vehicles that travel shorter distances with lighter loads.[14] The lightweight nature of these vehicles allows for the inclusion of batteries without significantly impacting their performance. Electric motors and drivetrains bring inherent efficiency advantages, compensating for the added weight of batteries.[15] City buses, urban delivery vehicles, and equipment at ports

---

[14] Nicola De Blasio, Charles Hua, and Alejandro Nunez-Jiminez, *Sustainable Mobility: Renewable Hydrogen in the Transport Sector*, Harvard Kennedy School Belfer Center for Scientific and International Affairs (June 2021) (www.belfercenter.org/sites/default/files/2021-06/HydrogenPB3.pdf).

[15] *All-Electric Vehicles*, US Department of Energy (www.fueleconomy.gov/feg/evtech.shtml).

are leading the way in heavy vehicle decarbonization, as they can be recharged at central locations or wireless pads along their routes. However, the challenge lies in electrifying longer-distance and heavier-load vehicles.[16]

Electric vehicles (EVs) have gained significant traction and adoption in recent years, and usage is predicted to multiply over the next two decades. Passenger EVs on the roads are expected to climb from 27 million at the beginning of 2023 to over 100 million by 2026 and more than 700 million by 2040.[17] This will be aided by government action to facilitate the adoption of new technologies.

For example, California will phase out gasoline-powered cars with a new mandate that all in-state sales of new passenger cars and trucks be zero-emission by 2035.[18] The transportation sector in California produces more than half of the state's CO_2 emissions[19]—largely because of the car-dependent culture due to inadequate public transportation, geographic patterns of development, and aging infrastructure—and the adoption of EVs is seen as a method to reduce the emissions profile.

---

[16] Id.

[17] BloombergNEF, *Electric Vehicle Fleet Set to Hit 100 Million by 2026, but Stronger Push Needed to Stay on Track for Net Zero* (June 8, 2023) (https://about.bnef.com/blog/electric-vehicle-fleet-set-to-hit-100-million-by-2026-but-stronger-push-needed-to-stay-on-track-for-net-zero/).

[18] *Governor Newsom Announces California Will Phase Out Gasoline-Powered Cars & Drastically Reduce Demand for Fossil Fuel in California's Fight Against Climate Change*, Office of Governor Newsom (September 23, 2020) (www.gov.ca.gov/2020/09/23/governor-newsom-announces-california-will-phase-out-gasoline-powered-cars-drastically-reduce-demand-for-fossil-fuel-in-californias-fight-against-climate-change/).

[19] Id.

CHAPTER 16   THE PATHWAYS TO DECARBONIZING KEY SECTORS

The US Environmental Protection Agency announced a final rule with new national pollution standards for passenger cars, light-duty trucks, and medium-duty vehicles for model years 2027–2032 and beyond.[20] The economic review of the standards includes an expected equivalent reduction of seven billion tons of carbon dioxide emissions and a USD$100 billion net societal benefit including USD$13 billion in annual public health benefits.

Medium and Heavy Trucking: The Middle Ground

Decarbonizing medium and heavy trucking presents a unique set of challenges. Vehicles operating on set routes in limited areas are more amenable to electrification. For instance, city buses and urban delivery vehicles can utilize centralized charging infrastructure. However, longer distances and heavier loads require innovative solutions. The weight of the battery and the high power demands for fast charging pose significant hurdles. The development of high-capacity chargers is underway to facilitate the charging of tractor trailers.[21] Grid upgrades will be necessary, particularly in rural areas, to accommodate the heavy loads associated with vehicle charging.

---

[20] *Biden-Harris Administration finalizes strongest-ever pollution standards for cars that position U.S. companies and workers to lead the clean vehicle future, protect public health, address the climate crises, save drivers money,* US Environmental Protection Agency (March 20, 2024) (www.epa.gov/newsreleases/biden-harris-administration-finalizes-strongest-ever-pollution-standards-cars-position); *Final Rule: Multi-Pollutant Emissions Standards for Model Years 2027 and Later Light-Duty and Medium-Duty Vehicles,* US Environmental Protection Agency (www.epa.gov/regulations-emissions-vehicles-and-engines/final-rule-multi-pollutant-emissions-standards-model).

[21] *High-Power Medium- and Heavy-Duty Electric Vehicle Charging,* National Renewable Energy Laboratory (www.nrel.gov/transportation/medium-heavy-duty-vehicle-charging.html).

CHAPTER 16   THE PATHWAYS TO DECARBONIZING KEY SECTORS

Stronger action is needed on heavy transport, such as heavy-duty trucks and fleets. That is the segment of the transportation industry that is most out of sync with any net-zero planning. Overcoming challenges such as upfront costs of batteries, responsible battery sourcing, emerging battery technology, and adequate charging infrastructure will be necessary to accelerate the adoption of EVs. Continuing to build out capacity for renewable fuels such as hydrogen and the related infrastructure, particularly for longer haul road transport, is necessary. In the United States, the IRA has platformed hydrogen for road transport and provided the public incentives to advance renewable fuels as part of the decarbonization solution.

Aviation and Maritime Transport: Energy Density Matters

Aviation and maritime shipping represent two ends of the spectrum in terms of GHG intensity. While aviation is the most emission-intensive mode of transport, maritime shipping is comparatively less GHG-intensive.[22] Both sectors face challenges due to their need for long-haul travel and limited geographical opportunities for frequent refueling. The energy density of conventional fuels used in these sectors is consequential, making low-carbon alternative fuels essential.

In aviation, efficiency gains have already reduced per-mile emissions, with new planes boasting up to 20% greater efficiency than older models.[23] Biomass-derived jet fuel shows promise, but its limited supply hampers widespread adoption.[24] In maritime shipping, liquified natural gas (LNG)

---

[22] *Aviation*, International Energy Agency (www.iea.org/energy-system/transport/aviation).

[23] Id.

[24] *Alternative Aviation Fuels: Overview of Challenges, Opportunities, and Next Steps*, US Department of Energy (www.energy.gov/eere/bioenergy/articles/alternative-aviation-fuels-overview-challenges-opportunities-and-next-steps).

CHAPTER 16   THE PATHWAYS TO DECARBONIZING KEY SECTORS

offers a lower-carbon option that meets the new low-sulfur fuel requirements, but it is less favorable from a footprint standpoint. Additionally, research is being conducted on sustainable marine fuels, hybridization, and energy efficiency and optimization as longer-term alternatives.[25]

In November 2023, Virgin Atlantic made the first transcontinental flight using Sustainable Aviation Fuel (SAF) as it traveled from Heathrow airport in London to JFK airport in New York.[26] SAF is a biofuel that has similar properties to conventional jet fuel but has a substantially lower carbon footprint, and it has been deemed safe and effective for powering aircraft.[27]

Road Ahead

Substituting petroleum-based fuels with low- and zero-carbon alternatives is a vital strategy for achieving decarbonization in transportation. Ethanol, biofuels, hydrogen, and electricity are among the substitutes currently in use. However, ensuring that these alternative fuels are produced without indirectly increasing emissions or causing other trade-offs, such as food for fuel, poses a challenge.

Enhancing energy efficiency can yield significant emission reductions in the transportation sector. Improving the "efficiency of conventional vehicles, aircraft, and ships through lighter materials, more efficient

---

[25] *Maritime Decarbonization*, US Department of Energy (www.energy.gov/eere/maritime-decarbonization).

[26] *Heathrow to New York JFK*, Virginia Atlantic (November 28, 2023) (https://corporate.virginatlantic.com/gb/en/media/press-releases/worlds-first-sustainable-aviation-fuel-flight.html#:~:text=Virgin%20Atlantic's%20historic%20flight%20on,fuel%2C%20compatible%20with%20today's%20engines%2C).

[27] *Sustainable Aviation Fuels*, Bioenergy Technology Office, US Department of Energy (www.energy.gov/eere/bioenergy/sustainable-aviation-fuels#:~:text=SAF%20is%20a%20biofuel%20used,compared%20to%20conventional%20jet%20fuel).

motors, and design changes is an essential step."[28] Electric motors, known for their superior efficiency compared to internal combustion engines, offer potential energy savings.[29]

In addition to zero-carbon fuels and improved efficiency, other strategies can contribute to decarbonization efforts in the transportation sector. Reducing vehicle miles traveled, land use planning that promotes smart development, and congestion pricing can help reduce emissions.[30] The development of new modes of transport and the use of innovative technologies can offer faster and more sustainable alternatives. Electrification holds promise for lighter vehicles, while alternative fuels and efficiency measures target heavier modes of transportation. Synergies between sectors and technological advancements can further bolster decarbonization efforts.

While advanced technologies exist to decarbonize the transportation sector, their scalability and cost-effectiveness need further validation. As policymakers navigate the complexities of the transportation sector, alignment between policies and technological advancements will be required. By embracing sustainable solutions and seizing opportunities for innovation on a large scale, the transportation sector can factor into and advance a net-zero carbon transition.

16.3 Decarbonizing the Agricultural Sector

Agriculture is a mainstay of our civilization. It is a significant contributor to greenhouse gas (GHG) emissions, primarily through crop cultivation and livestock production. The agriculture sector also has the potential

---

[28] *Decarbonizing U.S. Transportation*, Center for Climate and Energy Solutions (www.c2es.org/document/decarbonizing-u-s-transportation/).

[29] *Electric Vehicle Basics*, Natural Resources Defense Council (www.nrdc.org/bio/madhur-boloor/electric-vehicle-basics).

[30] *Decarbonizing U.S. Transportation*, Center for Climate and Energy Solutions (www.c2es.org/document/decarbonizing-u-s-transportation/).

CHAPTER 16   THE PATHWAYS TO DECARBONIZING KEY SECTORS

to transition from a significant emitter to a powerful carbon sink, sequestering up to 1.6 metric tons of global annual CO_2 equivalent emissions.[31] In this section, we aim to provide a picture of the agricultural sector's contribution to climate change, its greenhouse gas emissions profile, and the necessary steps to decarbonize this sector.

GHG Emissions in Agriculture

Agriculture is a major contributor to global GHG emissions, responsible for nearly a third of the total emissions.[32] These emissions primarily originate from two sources: crop cultivation and livestock production. Crop cultivation contributes to CO_2 emissions through the use of synthetic fertilizers and the conversion of forests and grasslands into agricultural land. Livestock produce significant amounts of methane, a potent GHG, through enteric fermentation.[33]

Agricultural emissions vary significantly across the globe, largely due to differences in agricultural practices and environmental conditions. In Africa and Latin America, agriculture contributes around 60% of total GHG emissions.[34] In Asia, the share is around 30%, while in Europe and North America, it's approximately 10%.[35]

---

[31] *Decarbonizing the agriculture sector: are developing countries part of the solution?* European University Institute (www.eui.eu/events?id=549047#:~:text=However%2C%20the%20agriculture%20sector%20holds,carbon%20prices%20through%20sustainable%20options).

[32] M. Crippa, E. Solazzo, D. Guizzardi, et al. *Food systems are responsible for a third of global anthropogenic GHG emissions*, Nature (March 8, 2021) (www.nature.com/articles/s43016-021-00225-9).

[33] *Enteric Fermentation—Greenhouse Gases*, US Environmental Protection Agency (www3.epa.gov/ttnchie1/ap42/ch14/final/c14s04.pdf).

[34] *Emissions due to agriculture*, Food and Agriculture Organization of the United Nations (www.fao.org/3/cb3808en/cb3808en.pdf).

[35] Id.

The sector also holds notable potential for climate change mitigation. Agricultural lands can function as carbon sinks and carbon sequesters.[36] The primary sustainable practices that can help agriculture transition into a carbon sink include enhancing soil carbon sequestration, reducing synthetic fertilizer use, improving livestock feed to reduce methane emissions, and capturing methane emissions.[37]

Pathways to Decarbonization

- **Tackling Food Waste**

    Food waste contributes an estimated 8–10% of global anthropogenic emissions.[38] By reducing food waste, we can significantly cut down on these emissions and also address the issue of food insecurity.

- **Shifting Dietary Patterns**

    Animal-based foods produce approximately twice the emissions of plant-based foods.[39] A shift in dietary patterns can significantly reduce emissions from the agriculture sector.

---

[36] P. Smith, *Soils as carbon sinks: the global context*, Soil Use and Management (January 18, 2006) (https://bsssjournals.onlinelibrary.wiley.com/doi/abs/10.1111/j.1475-2743.2004.tb00361.x).

[37] *NRCS Climate-Smart Mitigation Activities*, US Department of Agriculture Natural Resources Conservation Service (www.nrcs.usda.gov/conservation-basics/natural-resource-concerns/climate/climate-smart-mitigation-activities).

[38] *UNEP Food Waste Index Report 2021*, United Nations Environment Programme (March 4, 2021) (www.unep.org/resources/report/unep-food-waste-index-report-2021).

[39] *Here's How Much Food Contributes to Climate Change*, Scientific American (September 13, 2021) (www.scientificamerican.com/article/heres-how-much-food-contributes-to-climate-change/).

CHAPTER 16   THE PATHWAYS TO DECARBONIZING KEY SECTORS

- **Implementing Nature-Based Solutions**

    The adoption of nature-based solutions, such as improved forestry practices and land-use changes, can abate a significant portion of agriculture's GHG emissions. Regenerative agriculture is an approach to improving soil health which creates ecosystem resilience and sequesters more carbon in soil.[40] Many companies and nonprofits are working together to advance regenerative agricultural practices.

- **Technological Innovations**

    Existing technologies that can reduce emissions and environmental impact. Precision agriculture is one example. Precision agriculture uses advanced technology to optimize the application of farm inputs and it can significantly reduce agricultural emissions.[41] This approach also reduces the need for water and energy, thereby further reducing environmental impact.

Another example is methane capture technology, which can reduce emissions and generate revenue for farms. Anaerobic digesters by which bacteria break down organic matter[42] can significantly reduce emissions from manure management by capturing and converting methane into

---

[40] *Regenerative Agriculture 101*, Natural Resources Defense Council (November 29, 2021) (www.nrdc.org/stories/regenerative-agriculture-101#what-is).

[41] David Cutress, *Can Precision Farming Help Mitigate Climate Change?* Welsh Government (April 29, 2020) (https://businesswales.gov.wales/farmingconnect/news-and-events/technical-articles/can-precision-farming-help-mitigate-climate-change).

[42] *How Does Anaerobic Digestion Work?* US Environmental Protection Agency (www.epa.gov/agstar/how-does-anaerobic-digestion-work).

renewable energy.[43] This technology has been growing in practice, and many of the newer projects in the United States are designed to produce compressed natural gas for pipeline injection to take advantage of carbon credit-trading opportunities such as provided for under California's Low Carbon Fuel Standard program.[44]

With a significant portion of global GHG emissions attributable to the agriculture sector, decarbonization planning is underway. Barriers to be addressed include funding for research and development, capital costs, and needed changes to agricultural regulations and policies to incentivize sustainable practices. Transitioning toward a more sustainable and resilient agricultural system requires a layered approach encompassing technological innovation, policy measures, financial incentives, and changes in consumption patterns.

16.4 Decarbonizing the General Manufacturing Sector

Our modern way of life is heavily dependent on the products of the manufacturing sector. These industries, however, are a major contributor to global carbon dioxide (CO_2) emissions and have extensive environmental implications. Decarbonization of the manufacturing sector is imperative for combating climate change and achieving a sustainable

---

[43] *The Benefits of Anaerobic Digestion*, US Environmental Protection Agency (www.epa.gov/agstar/benefits-anaerobic-digestion).

[44] *Number of on-farm anaerobic digesters systems used to decompose organic waste has increased over time*, US Department of Agriculture Economic Research Service (www.ers.usda.gov/data-products/chart-gallery/gallery/chart-detail/?chartId=106096#:~:text=An%20anaerobic%20digester%20is%20an,that%20started%20operating%20that%20year.).

future. The manufacturing sector, including steel, cement, and chemicals, is a significant contributor to global CO_2 emissions. These heavy industries account for nearly 40% of global carbon dioxide emissions.[45] The magnitude of these emissions is closely linked to the sector's reliance on fossil fuels and natural resources.

The exponential growth of these industries, driven by the increasing global population and the associated demand for infrastructure materials, has led to a surge in emissions. For instance, global demand for steel has tripled since 1971, making steel manufacturers responsible for 6-7% of the world's greenhouse gases.[46] Similarly, cement production has tripled since 1995, accounting for about 7% of the world's total emissions.[47]

Challenges to Decarbonization

Decarbonizing the manufacturing sector is a complex task due to technical and economic factors. The manufacturing sector's reliance on fossil fuels not just as a source of energy but also as feedstocks transformed into final products poses additional challenges for decarbonization. These industries often require high heat and process emissions of carbon dioxide, making the reduction of emissions a significant challenge.

Technically, where possible, eliminating CO_2 emissions from manufacturing processes would require a complete change in process, which presents technical and economic challenges. Economically, the manufacturing sector is characterized by capital intensity and long asset life. These factors make the transition to low-carbon processes financially daunting for many manufacturers.

---

[45] Samantha Gross, *The challenge of decarbonizing heavy industry*, Brookings Institution (June 2021) (www.brookings.edu/articles/the-challenge-of-decarbonizing-heavy-industry/).

[46] *Material efficiency in clean energy transitions*, International Energy Agency (www.iea.org/reports/material-efficiency-in-clean-energy-transitions).

[47] Id.

Regulatory measures and incentives promoting decarbonization in the manufacturing sector have the potential to move the needle. Policies aimed at reducing emissions, promoting the use of clean energy sources, and encouraging sustainable manufacturing practices could greatly facilitate the transition to a low-carbon economy.

Pathways to Decarbonization

Despite the challenges, several pathways can lead to substantial emission reductions in the manufacturing sector. These include improved energy efficiency; fuel switching; carbon capture, utilization, and storage (CCUS); and process improvements.[48]

- **Energy efficiency and fuel selection:** Improving energy efficiency should always be a primary goal. Fuel switching and increasing electrification can significantly reduce emissions from on-site fossil fuel combustion.

- **Carbon capture, utilization, and storage:** CCUS technologies are important to reducing emissions particularly in sub-sectors such as steel, cement, chemicals, hydrogen, and refining. The application of CCUS technologies has been limited by a lack of investment and commercialization.

---

[48] *Decarbonizing U.S. Industry*, Center for Climate and Energy Solutions (www.c2es.org/document/decarbonizing-u-s-industry/).

- **Process changes and improvements:** Redesigning and improving existing processes can lead to a reduction in greenhouse gas emissions. For example, switching inputs or raw materials can lead to significant emission reductions.[49]

By understanding the sector's contribution to climate change, recognizing the challenges to decarbonization, addressing infrastructure and feedstock issues, considering regulatory implications, and implementing key steps toward decarbonization, businesses can innovate and steer the manufacturing sector toward a more sustainable future.

16.5 Decarbonizing the Building Industry

The building industry is an integral part of society, providing us with structures for living, working, and industrial production. In 2020, the output of building and construction accounted for about 13% of the world's gross domestic product (GDP).[50] The building and construction sector accounts for 36% of final energy use and approximately 39% of energy and process-related carbon emissions.[51]

It's important to understand the emission trends and projections in the building sector to develop effective strategies for decarbonization. Residential and commercial buildings incur significant energy use for heating, cooling, lighting, and systems operating. In the United States,

---

[49] Id.

[50] M. Santamouris and K. Vasilakopoulou, *Present and future energy consumption of buildings: Challenge and opportunities towards decarbonization*, Science Direct (2021) (www.sciencedirect.com/science/article/pii/S2772671121000024).

[51] *Global Status Report for Buildings and Construction 2019*, International Energy Agency (www.iea.org/reports/global-status-report-for-buildings-and-construction-2019).

from 1990 to 2015, carbon dioxide (CO_2) emissions from fossil-fuel combustion attributed to commercial buildings and residential buildings increased by 7.8% and 20.4% respectively.[52]

Decarbonizing the building sector requires reducing energy demand and carbon intensity. This includes a focus on energy sources away from fossil fuels to electricity and improving energy efficiency. Improving energy efficiency includes optimizing building design for natural lighting, sourcing construction materials with lower embodied carbon, high-efficiency heating and cooling systems, high-performance appliances, and changing consumer behavior to reduce energy demand. The shift to low-carbon building materials is another basic piece of the decarbonization process. This includes sourcing low-carbon construction materials and increasing the mix of renewable and recycled materials used in refurbishment and new construction.

Green Buildings and Sustainable Urban Planning

Green buildings, also known as sustainable buildings, are a feature of architectural design and urban planning. Civilizations have aimed to capitalize on weather patterns and the use of local, natural resources for centuries. The contemporary green building movement began to take shape in the 1970s in response to the energy crisis and growing environmental concerns.

This approach has been utilized for residential, commercial, and industrial building design. Green buildings are designed and constructed with a commitment to environmental responsibility and resource efficiency. These buildings aim to minimize their environmental footprint

---

[52] *Decarbonizing U.S. Buildings*, Center for Climate and Energy Solutions (www.c2es.org/document/decarbonizing-u-s-buildings/).

CHAPTER 16   THE PATHWAYS TO DECARBONIZING KEY SECTORS

by efficiently utilizing resources such as energy and water and reducing waste and pollution. The goal of green buildings extends beyond the construction phase and includes the entire life cycle of a building, from planning and design to operation, maintenance, renovation, and eventually deconstruction.[53]

Green buildings are one aspect of sustainable design and urban planning. With the United Nations predicting that 68% of the world's population will live in urban areas by 2050,[54] creating sustainable urban spaces is of principal importance. Green buildings are conceptualized, designed, and constructed with several key characteristics in mind. These include

1. **Energy efficiency:** Aiming to reduce energy consumption through the use of energy-efficient materials and technologies. These include passive solar design, high-performance windows, and improved insulation in walls, ceilings, and floors.

2. **Water efficiency:** Striving to minimize water usage through water-efficient fixtures and appliances, rainwater harvesting systems, and greywater recycling. This includes minimizing material waste during construction, reducing waste and water use during operation, and even considering waste during the building's eventual deconstruction.

---

[53] *What is a green building?* Singapore Green Building Council (www.sgbc.sg/resources/live-work-play-green/green-building).

[54] *68% of the world population projected to live in urban areas by 2050, says U.N.*, United Nations Department of Economic and Social Affairs (May 16, 2018) (www.un.org/development/desa/en/news/population/2018-revision-of-world-urbanization-prospects.html#:~:text=News-,68%25%20of%20the%20world%20population%20projected%20to%20live%20in,areas%20by%202050%2C%20says%20UN&text=Today%2C%2055%25%20of%20the%20world's,increase%20to%2068%25%20by%202050).

CHAPTER 16   THE PATHWAYS TO DECARBONIZING KEY SECTORS

3. **Material efficiency:** Prioritizing the use of sustainable, renewable, and recycled materials.

4. **Indoor environmental quality:** Creating a healthy indoor environment by minimizing pollutants, optimizing ventilation, and providing comfortable spaces that consider populations most sensitive to pollutants.

An interesting development in recent years has been the incorporation of green building principles into the design and refurbishment of sports arenas.[55] For example, professional sports teams and leagues across basketball, baseball, soccer, football, and hockey have undertaken holistic systems approaches to new and upgraded stadiums and arenas. That generally results in cost savings from more efficient operations. Importantly, it also results in net-positive environmental and social impact, provides awareness for millions of fans, and engenders goodwill in the community.

There are several certifications available that validate a building's sustainability. Two of the most well-recognized certifications include (1) the Leadership in Energy and Environmental Design (LEED) certification, developed by the US Green Building Council; and (2) the BREEAM (Building Research Establishment Environmental Assessment Method), international standards developed in the United Kingdom.[56]

---

[55] Will Henshall and Joey Lautrup, *How Sports Stadiums Are Going Green*, Time CO2 Futures (February 9, 2024) (https://time.com/collection/time-co2-futures/6692813/sustainable-sports-stadiums/).

[56] *Sustainable Building Certifications*, World Green Building Council (https://worldgbc.org/sustainable-building-certifications/).

CHAPTER 16   THE PATHWAYS TO DECARBONIZING KEY SECTORS

Continuous Innovation in Green Design and ESG Business Strategy

Collaborative research is helping to advance the field of sustainable building design. It involves a myriad of stakeholders, including national laboratories, private companies, and universities. Key areas of research include energy and atmosphere studies, materials and resources, and indoor environmental quality.[57]

Companies can do their part and incorporate sustainable design as a priority in designing and upgrading facilities and other buildings, such as offices. This can have a massive positive impact on the corporate footprint. Numerous examples of green corporate buildings can be found worldwide, each unique in its design and approach to sustainability. Some notable examples include

- **Iberdrola Tower, Spain:** The Iberdrola Tower is a financial and business building in Bilbao and acclaimed for sustainable design and energy efficiency.

- **Bahrain World Trade Center, Bahrain:** The World Trade Center in Bahrain was the first skyscraper to utilize wind turbines.

- **Shanghai Tower, China:** Shanghai Tower is one of the world's tallest buildings and has platinum LEED certification.

- **The Edge, Netherlands:** The Edge is known as the world's most sustainable office building, and it utilizes a range of technologies to optimize its energy use and minimize its environmental impact.

---

[57] *Green Building*, US Environmental Protection Agency (https://archive.epa.gov/greenbuilding/web/html/about.html).

CHAPTER 16   THE PATHWAYS TO DECARBONIZING KEY SECTORS

Green building is constantly evolving, with new technologies and design approaches emerging regularly. Future trends include the increased use of smart technologies for energy management, the incorporation of biophilic design to enhance occupant well-being, and the growing emphasis on creating net-zero energy buildings.

Challenges to and the Future of Decarbonization in the Building Sector

There are significant obstacles to realizing the ambition of decarbonizing the global building sector. These include regulatory and financing barriers, rising materials costs for both commercial and residential construction, and the scale of technological change required in replacing and refurbishing systems. Some jurisdictions, such as New York City, have updated their building regulations with new energy efficiency and phased in greenhouse gas emissions limits.[58] The future of the building sector in a decarbonized world will require a shift in the way buildings are designed, constructed, and operated. This will involve a greater focus on energy efficiency, the use of renewable energy, and the implementation of intelligent efficiency technologies.

16.6 Decarbonizing the Fashion Industry

The fashion industry, with all its glamor, has a darker side with its significant contribution to climate change. On the global stage, fashion's greenhouse gas emissions are alarmingly high. The fashion industry accounts for about 10% of total global carbon emissions, which is "more

---

[58] *Local Law 97*, NYC Sustainable Buildings (www.nyc.gov/site/sustainablebuildings/ll97/local-law-97.page#:~:text=Under%20this%20groundbreaking%20law%2C%20most,coming%20into%20effect%20in%202030).

than all international flights and maritime shipping combined."[59] At the current pace, the fashion industry's greenhouse gas emissions are predicted to surge more than 50% by 2030.[60]

To decarbonize this sector, the industry must undergo a drastic transformation, shifting its focus from the fast fashion market and overproduction to sustainable and ethical practices. This section explores the fashion sector's climate change profile, the challenges faced in its decarbonization, and the necessary steps to make this sector more sustainable.

The fashion industry's contribution to climate change is complex, involving everything in the full life cycle of a garment, from raw material production to the use of fair labor to final disposal. According to a McKinsey study, the sector was responsible for approximately 2.1 billion metric tons of greenhouse gas (GHG) emissions in 2018, equating to around 4% of the global total.[61] To put this into perspective, the fashion industry has the same annual greenhouse gas emissions level as the entire economies of France, Germany, and the UK combined.[62]

Raw Material Production and Processing

The production and processing of raw materials for fashion garments contribute significantly to the sector's carbon footprint. For example, the cultivation of cotton, a commonly used natural fiber in the industry,

---

[59] *How Much Do Our Wardrobes Cost the Environment?* The World Bank (September 23, 2019) (www.worldbank.org/en/news/feature/2019/09/23/costo-moda-medio-ambiente).
[60] Id.
[61] *Fashion on climate*, McKinsey & Co. (August 26, 2020) (www.mckinsey.com/industries/retail/our-insights/fashion-on-climate).
[62] Id.

requires vast amounts of energy, water, and chemical fertilizers, all of which lead to substantial carbon emissions.[63] Similarly, synthetic fibers like polyester, derived from petrochemicals, also contribute to GHG emissions.

Manufacturing and Distribution

The manufacturing process, involving dyeing, finishing, and assembling of apparel, accounts for about two-thirds of the emissions from producing a new garment.[64] Furthermore, the distribution of these products worldwide, often involving long-distance shipping, adds to the industry's carbon footprint.

Consumer Use and Disposal

The consumer's role in the fashion industry's carbon footprint is often overlooked. The washing, drying, and ironing of clothes contribute to energy consumption and, consequently, carbon emissions. Moreover, the disposal of unwanted or worn-out garments, often in landfills or through incineration, leads to further emissions.

Challenges in Decarbonizing the Fashion Sector

While the need for decarbonization in the fashion industry is clear, several challenges hinder this transition.

---

[63] F.A. Esteve-Turrillas and M. de la Guardia, *Environmental impact of Recover cotton in textile industry*, ScienceDirect (January 2017) (www.sciencedirect.com/science/article/abs/pii/S0921344916302828).

[64] *Measuring Fashion: Insights from the Environmental Impact of the Global Apparel and Footwear Industries*, Quantis (https://quantis.com/report/measuring-fashion-report/).

- **Fast Fashion and Overproduction**

  Fast fashion, characterized by rapid and cheap production of high volumes of clothing, is a significant contributor to the industry's carbon emissions. This business model thrives on constant consumption and discarding of trends, leading to overproduction and excessive waste. We discussed the widespread greenwashing challenges faced by companies in Chapter 7.

- **Supply Chain Complexity**

  The fashion industry's supply chain is complex and distributed worldwide, involving multiple stages from raw material cultivation to retail. This complexity makes it difficult to monitor and control carbon emissions throughout the entire process.

- **Cost and Technological Constraints**

  Implementing sustainable practices and technologies often requires significant upfront investment, which can be a deterrent for many fashion companies. Moreover, some of the necessary technologies for sustainable production and recycling of garments are still in their infancy and may not be readily available or economically viable.

- **Responsible Sourcing and Human Rights in Fashion**

  Responsible sourcing involves choosing environmentally friendly materials and ensuring sustainable cultivation and manufacturing practices. The fashion sector must also ensure ethical and responsible practices throughout the supply chain. This includes protecting human rights, paying fair wages, and providing safe working conditions as we'll discuss in the next chapter.

Key Steps to Decarbonize the Fashion Sector

By adopting sustainable practices, improving supply chain transparency, and encouraging responsible consumption, the fashion sector can transform itself into a force for good, contributing to a more sustainable and equitable world. To achieve a sustainable and decarbonized fashion industry, several key steps are necessary.

1. **Adopting Sustainable Business Models**

    The traditional linear business model of take-make-dispose should be replaced with a circular model that values reuse and recycling, and should incorporate environmental justice and responsible sourcing parameters. More fashion brands are adopting practices like clothing rental, resale, repair, and refurbishment to extend the lifespan of their products and reduce waste.

2. **Improving Material Mix**

    Brands can reduce their carbon footprint by choosing sustainable materials for their products. This involves using more recycled fibers, organic cotton, or innovative materials with a lower environmental impact.

3. **Enhancing Energy Efficiency**

   Improving energy efficiency in manufacturing and retail operations can significantly reduce carbon emissions. This involves using renewable energy sources, optimizing manufacturing processes, and implementing energy-saving measures in brick-and-mortar retail stores.

4. **Encouraging Sustainable Consumer Behavior**

   Consumer choice may be determinative in the failure or success of decarbonization in the fashion industry. Brands can encourage sustainable consumer behavior by promoting conscious purchasing, providing information about the environmental impact of products, and offering sustainable alternatives.

5. **Collaborating for Change**

   Collaboration across the industry is vital to drive decarbonization. This includes sharing best practices, pooling resources for research and development, and working together to advocate for supportive policies and regulations.

The Sustainable Apparel Coalition (SAC) is a trade association of fashion-related groups and interests, including approximately 300 apparel, footwear, and textile companies that represent USD$845 billion in annual revenue.[65] The SAC developed the Higg Index as "a suite of five tools that assess and measure the social and environmental performance of the

---

[65] *Our Members*, Sustainable Apparel Coalition (https://apparelcoalition.org/our-members/).

CHAPTER 16   THE PATHWAYS TO DECARBONIZING KEY SECTORS

value chain and the environment impacts of products."[66] The Higg Index was becoming "a de facto global standard"[67] and a benchmark for brands, when it came under scrutiny for greenwashing largely related to the index's treatment of the use of synthetic materials and fast fashion practices. The Norwegian Consumer Authority (NVA) warned H&M Group against using the Higg Index to support its environmental marketing claim.[68] After that regulatory notice, many brands stopped using the Higg Index and SAC suspended its use pending an independent review.[69]

There is a role for trade associations and business groups in decarbonizing key sectors. Trade associations and business groups have long set industry standards as they often play a convening role of industry exports, government agencies, and other stakeholders. As we discussed in Chapter 13, these initiatives can be successful when they are science-based, transparent, and independently reviewed.

---

[66] *Higg Index Tools*, Sustainable Apparel Coalition (https://apparelcoalition.org/tools-programs/higg-index-tools/).

[67] Hiroko Tabuchi, *How Fashion Giants Recast Plastic as Good for the Planet*, The New York Times (June 12, 2022) (www.nytimes.com/2022/06/12/climate/vegan-leather-synthetics-fashion-industry.html?searchResultPosition=1).

[68] Michelle Codiva, *H&M Published and Removed Higg Index Environmental Scorecard After Quartz Called It Out for Misleading Data*, The Science Times (June 30, 2022) (www.sciencetimes.com/articles/38478/20220630/h-m-published-opposite-result-higg-index-environmental-scorecard-quartz.htm). See also Amanda Shendruk, *The controversial way fashion brands gauge sustainability is being suspended*, Quartz (June 29, 2022) (https://qz.com/2180322/the-controversial-higg-sustainability-index-is-being-suspended).

[69] *Fashion brands paused use of sustainability index tool over greenwashing claims*, The Guardian (June 28, 2022) (www.theguardian.com/fashion/2022/jun/28/fashion-brands-pause-use-of-sustainability-index-tool-over-greenwashing-claims).

CHAPTER 16   THE PATHWAYS TO DECARBONIZING KEY SECTORS

CHAPTER TAKEAWAYS

- Climate change is a paramount concern worldwide
- Rising global temperatures are causing irreversible and disparate impacts on human health and natural systems
- The effects of climate change are interconnected
- We have not yet made enough process to halt global temperature change to meet the Paris Agreement goals
- There are viable pathways to decarbonize key sectors to make impactful change
- Businesses must work with investors and regulators to remove the historic challenges to decarbonizing, especially in difficult-to-abate sectors
- Continued innovation, investment, and multistakeholder collaboration is critical to success

Index

A

Accelerated Filers (AFs), 253, 255
Accounting and Corporate Regulatory Authority (ACRA), 268
Ad hoc governmental groups, 391, 393
Administrative Procedure Act (APA), 286
Advertising Standards Authority (ASA), 281
Agricultural sector
  GHG emissions, 465–467, 469
  methane capture technology, 468
  nature-based solutions, 468
  shifting dietary patterns, 467
  tackling food waste, 467
  technological innovations, 468
Aligned corporations, 355, 356
Alliance for Water Stewardship (AWS), 434, 435
Analytics providers, 337, 338
Anti-bribery and corruption (ABAC), 199
Antitrust, 290
  ESG, 290
  UK, 291
  US, 291
Artificial intelligence (AI), 167
Asset and Wealth Management Revolution 2022 report, 379
Assets under management (AUM), 3
Aviation and maritime transport, 463–464

B

Bahrain World Trade Center, 476
Banks, 206, 376, 377, 386, 387
B Corporations (B Corps), 162, 356, 358–360
Beijing Stock Exchange (BSE), 249
Biodiversity
  climate change, 135
  decline, 134
  definition, 131
  funds, 137
  investments, 137
  loss, 135
  marine, 133
  terrestrial, 132

INDEX

Biodiversity Beyond National Jurisdiction (BBNJ), 72
Biological diversity, 134
B Lab, 358, 359
Blended finance, 445
Bloomberg, 18, 337, 339
Blue economy, 377
   activities, 375
   ocean resources, 375
   worth, 375
Blue Ocean Economy, 372, 374–377
Blue Ocean Strategy (BOS), 373–374
Board of Education Retirement System (BERS), 206
Breakthrough Energy, 447
Brookfield asset management, 388
Brookfield Global Transition Fund (BGTF II), 388
Brooks, 10
Brundtland Commission, 58, 60–62, 85
Building industry
   challenges, 477
   GDP, 472
   green buildings, 473–475
   green design and ESG business strategy, 476, 477
   reducing energy demand and carbon intensity, 473
   residential and commercial, 472
Building Research Establishment Environmental Assessment Method (BREEAM), 475

Business ethics, 198–199
Business for Social Responsibility (BSR), 414, 415, 418
Business judgment rule, 359
Business Roundtable (BRT), 4, 415–416
Business values, 167, 198, 199

C

California Air Resources Board (CARB), 259
California State Teachers' Retirement System (CalSTRS), 207
Cap-and-trade systems, 124
Carbon accounting, 327–328
Carbon Border Adjustment Mechanism (CBAM), 125, 265–267
Carbon capture for storage (CCS), 236
Carbon capture for utilization (CCU), 236
Carbon capture, utilization, and storage (CCUS), 458, 471
Carbon credits, 68, 76, 124, 127, 128, 300, 415, 428, 469
Carbon dioxide (CO_2), 75, 113, 118, 124, 267, 473
Carbon dioxide removal (CDR), 263, 264, 455
Carbon Disclosure Project (CDP), 31, 314, 323–324, 427

Carbon markets
   challenges, 128
   corporate participation, 127
   definition, 123
   foundation, 123
   global, 124
   principles, 128
   regional, 124, 126
   VCM, 127
Carbon offsetting, 127, 128
Carbon Removal Certification System, 267–268
CARE International, 425
Ceres, 206, 322, 416–418
Ceres Roadmap 2030, 417
Certification process, 141, 268, 343, 358, 359
Certified Emission Reductions (CERs), 124
Charitable foundations, 448, 449
Chief Executives for Corporate Purpose (CECP), 170, 171
Chief Legal Officer (CLO), 371
Chief Marketing Officer (CMO), 371
Chief Sustainability Officer (CSO), 370, 371
Children's Investment Fund Foundation (CIFF), 325
Circular economy, 144, 145
   aims, 146
   concept, 145
   environmental benefits, 148
   future, 148
   implementation, 149
   principles, 146
   regulation, 276–278
Civil society, 69, 83, 393, 433
Clean Air Act (CAA), 123, 216, 302, 303
Clean Development Mechanism (CDM), 68, 75
Clean Water Act (1972), 217, 302
Climate accountability laws, 258–260, 295
Climate action, 69, 76, 88, 418, 439–441, 454
Climate and ESG Task Force, 297
Climate and Resilience Law, 264, 265, 280
Climate change, 42, 43, 103, 135, 391, 401
Climate change court cases, 301
Climate Disclosure Standards Board (CDSB), 312, 317
Climate litigation, 302
   cases, 303, 304
   courts, 303
   human activity, 302
   judicial challenges, 304
   private plaintiff actions, 302
   public actions, 302
   types, 302
Climate-related financial disclosures, 269
Climate-related reporting rules, 268–269
Climate risks, 31, 39, 100, 112, 207, 213, 252, 307, 384, 385

INDEX

Climate tech companies, 420
Climate Transition Action Plan (CTAP), 367
CO_2 emissions, 115, 118, 461, 466, 469, 470, 473
Common but differentiated responsibilities (CBDR), 64
Community engagement, 173–175
Community relations, 143, 152, 173–175, 202, 361
Companies, 284, 296, 308, 309, 330, 354, 359, 366, 370, 424, 431
Company ESG Program, 363–365
Company programs, 172
Competition and Markets Authority (CMA), 290, 291
Comprehensive Environmental Response, Compensation, and Liability Act (1980), 218
Conference of the Parties (COP), 64, 69, 80, 392
Consumer awareness, 130
Consumer companies, 305
Consumer packaged goods (CPG), 42, 432
Consumer safety, 168
Convention on Biological Diversity (CBD), 64–65
COP28 Dubai Agreement
   challenges, 84
   outcomes, 81, 83
   provisions, 81
Corporate culture, 157, 177, 178, 199, 397
Corporate directors and officers, 351–354
Corporate Executive Officers (CEOs), 26, 307, 415, 434
Corporate governance, 182
   benefits, 180, 181
   principles, 180
Corporate Human Rights Benchmark (CHRB), 164, 165
Corporate leaders, 185, 421
Corporate leadership, 5, 27, 32, 186, 194, 215, 282
Corporate management, 183
Corporate matching, 172
Corporate philanthropy, 171
Corporate planning, 36
Corporate purpose, 354, 356, 416
Corporate reporting, 8, 138
Corporate sector, 16, 27, 31, 34, 80, 162, 412, 431
Corporate Social Responsibility (CSR)
   activities, 11
   definition, 11
   factors, 11
   impact, 13
   modern, 12
   principles, 12
   role, 12
   success, 11
   vs. ESG, 15, 16
Corporate sustainability, 7–10, 31, 44, 92, 193, 321, 370

functions, 370
planning, 31
Corporate Sustainability Due Diligence Directive (CSDDD/CS3D), 56, 240, 241, 274
Corporate Sustainability Reporting Directive (CSRD), 224, 236–238, 256, 294
  companies, 237, 239
  compliance timeline, 239, 240
  reporting obligations, 237, 238
Corporations, 138, 162
Credit ratings agencies (CRAs), 333, 335, 341, 343
Customary international law, 49
Customer relationship management programs (CRM), 361

D

Data analytics, 136
Data collection and management, 330, 364
Data companies, 339
Data product providers, 409
Debt-for-development swaps, 446
Debt-for-nature swaps, 381, 383
Debt swaps, 445
Decarbonization
  agriculture, 465–469
  building industry, 472–477
  energy sector, 455–459
  fashion industry, 477–483
  GHG emissions, 453
  manufacturing sector, 469
  transportation sector, 460–465
Department of Energy (DOE), 263, 264
Disinformation, 104
Diversity, equity, and inclusion (DEI), 10, 151, 155–158, 185, 272
Double materiality, 249, 294
Duty of Vigilance Law, 275

E

Earth Summit, 61
  achievements, 63, 65–67
  criticism, 67
  objective, 62
  sustainable development, 62
Economic Development Board (EDB), 269
Economic uncertainties, 31, 104
Education and training programs, 370
Effective community engagement, 173
Electric motors, 460, 465
Electric vehicles (EVs), 374, 461
Electrification, 386, 388, 460, 462, 465
Elemis, 162
Emerging Growth Companies (EGCs), 255, 256

INDEX

Emissions-trading system (ETS), 125, 266, 267
Endangered Species Act (1973), 217, 302
Energy efficiency, 82, 98, 122, 369, 458, 459, 464, 471, 473, 474, 482
Energy Information Administration (EIA), 80
Energy Policy Act (2005), 218
Energy sector
   CCUS technologies, 458
   challenge of grid stability, 457
   coal, oil, and natural gas, 455
   GHG emissions, 455
   greenhouse gas (GHG) emissions, 455
   IEA estimates, 456
   incentives, 459
   innovation and technology, 459
   mandates, 459
   price and accessibility of renewable technologies, 457
   regulatory and public policy, 458
   renewable, 455
   renewable and low-carbon fuels, 459
   renewable electricity generation, 456
   standards, 459
Energy transition, 250, 260, 385–388

Environmental and social priority policy issues, 399
Environmental Defense Fund (EDF), 450
Environmental impact assessment (EIA), 73
Environmental issues, 57, 60, 66, 97, 98, 184, 424
Environmental justice
   circular economy, 147
   definition, 147
Environmental law, 7, 57, 61, 147, 302
Environmental Protection Agency (EPA), 216, 462
Environmental regulations, 215, 458
Environmental reporting, 98, 238, 323
Environmental, Social, and Governance (ESG), 362
   assets, 25
   benefits, 21–23
   business planning, 28, 29
   business strategy, 21
   board oversight, 187, 188
   climate change, 40–43
   committee structure, 193–195
   company management, 182
   competitive advantage, 27
   compliance, 200
   composition, 189
   corporate leadership, 27

INDEX

corporate sustainability, 7
   Global Compact, 8
   GRI, 8
vs. CSR, 15
decision analysis, 27
definition, 1, 2
diverse boards, 191–193
duties, 36–38
evolution, 3, 4
evolve globally, 38
factors, 5, 6, 215
fiduciary duties, 351–354
framework
   environmental factors, 98
   SDG, 99
governance factors, 178, 179
   SDG, 179
governance roadmap, 195
importance, 187
initiatives, 290
integration, 188
investments, 25
issues, 213, 216, 306, 334, 335
linking pay, 185
market and economic
   impacts, 29–31
metrics, 17
perception, 380
performance evaluation, 197
principles, 28, 439
regulations, 26, 33, 34
regulatory requirements,
   35, 36
regulatory vacuum, 34, 35

reporting, 214, 309
risks, 36, 199, 201, 205, 308–310,
   333, 339, 367
score, 334, 335
SDG, 90
shareholder activism, 32
social factors, 152
   SDG, 153
sources, 18, 20, 21
stakeholders, 5, 31
standards, 56, 136, 271, 311
vs. sustainability, 16
sustainability reporting, 43
sustainability topics, 223
value alignment, 18
ESG Ratings Methodology, 169
Equator Principles, 314
European Commission (EC), 231,
   236, 237, 243, 244, 266, 276,
   279, 344
European Financial Reporting
   Advisory Group (EFRAG),
   315, 329
European Green Deal, 231
   aim, 232
   EU's commitment, 236
   GHG emissions, 236
   goals, 233, 234
   milestones, 232, 233, 235
European Securities Markets
   Authority (ESMA), 345
European Sustainability Reporting
   Standards (ESRS), 237,
   294, 315

INDEX

Exclusive Economic Zones (EEZs), 71
Executive action, 215, 223, 293, 420
Executive compensation, 158, 184–185, 205, 270, 271
Extended Producer Responsibility (EPR), 277
External reporting, 365

F

Fashion industry
  adopting sustainable business models, 481
  to climate change, 478
  collaborating for change, 482
  consumer use and disposal, 479
  cost and technological constraints, 480
  enhancing energy efficiency, 482
  fast fashion and overproduction, 480
  GHG emissions, 478
  greenhouse gas emissions, 477, 478
  improving material mix, 481
  manufacturing and distribution, 479
  production and processing of raw materials, 478, 479
  responsible sourcing and human rights, 480
  SAC, 482, 483
  supply chain complexity, 480
  sustainable consumer behavior, 482
Fast-moving consumer product goods (FMCG), 366
Federal Trade Commission (FTC), 301
Fiduciary duties
  duty of care, 352
  duty of good faith, 353
  duty of loyalty, 352–353
  ESG, 351
  fundamentals, 351–352
  obligations, 352
Financial backing, 386
Financial Conduct Authority (FCA), 245, 246, 283, 346
Financial institutions, 91, 314, 325, 326, 384, 385, 387, 395
Financial markets, 19, 32, 243, 244, 408, 409, 443
Financial materiality, 238, 293, 294
Financial sector, 31, 380, 384–386
Financial Stability Board (FSB), 202, 318, 319, 409
Financial statement disclosures, 254
Flexibility mechanisms, 75
Foreign Corrupt Practices Act (1977), 219
Frameworks, 38, 44, 311

G

Gender equality, 87, 159, 162, 163, 289
Gender inequality, 96, 160
Gender pay gap, 160–163, 288, 289
GHG emissions, 74, 80, 81, 100, 110, 117–120, 122, 124, 219, 236, 238, 251, 253, 256, 259, 266, 302, 328, 469, 478
  in agriculture, 465–467
Global Biodiversity Framework (GBF), 64
Global business community, 411
Global Compact, 8, 377, 396, 397
Global economy, 374, 375
Global Environment Facility (GEF), 66, 325
Global Goal on Adaptation (GGA), 84
Global issues, 27, 399, 412, 442
Global regulations, 19, 193, 308, 309
Global regulatory activity snapshot, 223, 226, 228, 230
Global Reporting Initiative (GRI), 8, 202, 271, 294, 312, 314, 322, 323, 417
Global Risks Report 2023, 104, 105
Global Risks Report 2024, 101–102
Global Sustainability Standards Board (GSSB), 322
Global Sustainable Disclosure Standards, 400
Global transition, 385, 386
Global Witness, 429, 430
Good governance, 181, 184, 407
Google, 264
Governance regulation, 215–220
Governance review, 364
Government and stock exchange board diversity rules
  European Union (EU), 282, 283
  Republic of Korea, 284, 285
  SEC, 287
  United Kingdom, 282–284
  United States (US)
    conservative groups, 286
    NASDAQ Board Diversity Rule, 285, 287
    NASDAQ companies, 285
    prior judicial test, 285, 286
Green bonds, 221, 222, 381, 383, 384
Green buildings, 473–475
  BREEAM, 475
  energy efficiency, 474
  indoor environmental quality, 475
  LEED certification, 475
  material efficiency, 475
  principles, 475
  water efficiency, 474
Green Climate Fund (GCF), 68, 82
Green economy, 231, 232, 368, 369, 374
Green growth, 401–403

INDEX

Greenhouse gases (GHGs), 100
   carbon dioxide, 113, 118
   definition, 112
   emissions by country, 120
   human activities, 115
   impact, 116
   industrial gases, 114
   methane, 113, 118
   nitrous oxide, 114
   per capita emissions, 121–122
   protocol, 328, 329
   regulation, 117
   sector, 118, 119
   types, 113
   water vapor, 114
Green jobs, 368, 369
Greenwashing, 208–210, 246, 248
Greenwashing litigation, 304–305
Greenwashing regulation
   European Union (EU), 279
   France, 280
   green bond standard, 280
   United Kingdom (UK), 281
   United States (US), 281, 282
Grid stability, 457
Gross domestic product (GDP), 472
Group of 7 (G7), 393, 398–400
Group of 20 (G20), 393, 398, 401, 402

H

High Seas Treaty, 72, 73, 376
HSBC Asset Management, 137

Hub and spoke model, 371, 372
Human Capital Disclosure Rule, 272, 273
Human Capital Management (HCM)
   aspects, 154
   dimensions, 154
   metrics, 155
Human Capital Management Disclosures & Practices, 271
Human rights
   businesses, 55, 56
   definitions, 164
   international, 48
   labor practices, 163
Human rights regulations, 274, 275
Hydrofluorocarbons (HFCs), 114

I

Iberdrola Tower, 476
Identify champions, 364
Impact investing, 381, 444
Indices, 339
Industrial gases, 114
Inflation Reduction Act (IRA), 260, 262, 263
Innovation, 374
Innovative financing mechanisms, 442–447
Institutional investors, 5, 22, 32, 137, 287, 380
Institutional Shareholder Services (ISS), 192, 338

Intergovernmental Panel on
    Climate Change (IPCC), 40
  assessment cycles, 110
  climate policy, 111, 112
  recommendations, 108
  structure, 109
Internal Revenue Service (IRS),
    171, 263
International agreements, 57, 61,
    73, 117, 124, 392, 393
International Auditing and
    Assurance Standards Board
    (IAASB), 201
International cooperation, 57, 62,
    67, 77, 107, 168
International Energy Agency
    (IEA), 386
International environmental law
  multinational solution, 57
  origin, 57
International Ethics Standards
    Board for Accountants
    (IESBA), 202
International Finance Corporation
    (IFC), 314, 316
International Financial Reporting
    Standards Foundation
    (IFRS), 312, 313
International Forum of
    Independent Audit
    Regulators (IFIAR), 202
International Integrated Reporting
    Council (IIRC), 39, 312
International Labour Organization
    (ILO), 368, 369
International Organization of
    Securities Commissions
    (IOSCO), 202, 319,
    393, 408–410
International Renewable Energy
    Agency (IRENA), 386
International Rescue Committee
    (IRC), 425
International Seabed Authority
    (ISA), 71
International Standard on
    Sustainability Assurance
    (ISSA), 201
International Sustainability
    Standards Board (ISSB), 39,
    193, 202, 250, 257, 269, 312,
    313, 317
  characteristics, 319, 320
  global sustainability
    standards, 318
  IFRS S1 and IFRS S2, 318, 319
  key objectives, 318
  pragmatic and investor-centric
    approach, 318
  stakeholders, 321
  standards and
    frameworks, 317
  strides, 321
International Union for
    Conservation of Nature
    (IUCN), 369, 406, 408

495

INDEX

Investment service providers, 404
Investor engagement, 205
Investors, 334, 340–342

J

Joint Implementation (JI), 75
Just Energy Transition Partnerships (JETPs), 400

K

Key Performance Indicators (KPIs), 38, 140, 366
Key risk indicators (KRIs), 200
Kimberely Process, 430
Kunming-Montreal Global Biodiversity Framework, 134
Kyoto Protocol
   background, 74
   climate change, 76
   COP3 meeting, 74
   criticism, 76
   extension, 77
   objectives, 75
   US position, 77

L

Labor practices, 163–165
Large Accelerated Filers (LAFs), 253, 255

Leadership in Energy and Environmental Design (LEED) certification, 475, 476
Leading sustainability, 371, 372
Legislation, 292
Liquified natural gas (LNG), 463
L'Oréal, 183
Low-carbon economy, 68, 387, 448, 471
LSEG Data & Analytics, 338

M

Manufacturing sector
   CCUS, 471
   CO_2 emissions, 469
   decarbonization
      challenges, 470, 471
      pathways, 471, 472
   energy efficiency and fuel selection, 471
   exponential growth, 470
   process changes and improvements, 472
Marine biodiversity, 72, 133
Marine conservation, 73, 376
Materiality assessment, 238, 294, 364
Medium and heavy trucking, 462, 463
Memorandum of Understanding (MOU), 413

INDEX

Mergers and acquisitions (M&A), 22, 367–368
Methane, 113, 118
Methane capture technology, 468
MethaneSAT, 450
Millennium Development Goals (MDGs), 85
Misinformation, 104
Mission-related investments (MRIs), 446
Moody's Corporation, 338
MSCI, 284, 335–337, 339, 341

N

National Ambient Air Quality Standards (NAAQS), 216
National Association of Securities Dealers Automated Quotations (NASDAQ), 190, 282, 284–288
National Environmental Policy Act (1970), 218
National governments, 219, 384, 398
Nationally determined contributions (NDCs), 78, 117, 125
Natura & Co, 174
Natural capital
  definition, 132
  depletion, 133
  funds, 137
  GDP, 133
  investments, 136
Natural Resources Conservation Service (NRCS), 262
Nature Restoration Law, 244, 245
New York City Employees' Retirement System (NYCERS), 206
New York Stock Exchange (NYSE), 190, 191
Nitrogen trifluoride (NF_3), 114
Nitrous oxide, 75, 114, 115, 118
Non-Accelerated Filers (NAFs), 255
Non-financial materiality, 294
Non-Financial Reporting Directive (NFRD), 56, 224, 237, 239, 295
Nongovernmental organizations (NGOs), 42, 423
  certifying corporate programs, 430
  collaboration, 424
  companies, 424
  operations scale, 423
  public awareness, 424
  public policies, 424
  SDGs, 424
Nonprofit organizations, 12, 90, 172, 303, 323, 358, 411, 416, 418, 451
Nuclear energy, 383

497

INDEX

O

Office of the High Commissioner for Human Rights (OHCHR), 392
Organisation for Economic Co-operation and Development (OECD), 129, 375, 393, 402
    aspect, 403
    organization, 403
    papers, 403
    PRI, 404
    principles, 404–406
Organisation for European Economic Co-operation (OEEC), 402
Organizations, 214, 243, 264, 271, 280
Oxfam, 426

P, Q

Packaging
    consumer awareness, 130
    supply chain, 129
    sustainability trends, 130
    transformation, 131
Paris Agreement, 69
    climate change, 79
    provisions, 78, 79
Patagonia, 14
Perfluorocarbons (PFCs), 114
Philanthropy
    blended finance, 445
    civil rights, 438
    climate, 448–450
    climate action, 439
    debt swaps, 445
    foundations, 438
    funding, 440–442
    global pandemic, 438
    impact and influence, 437, 438
    impact investing, 444
    innovative financing mechanisms, 442–447
    public charities programming, 451
    social, environmental, and development impact bonds, 445
Political activity laws, 199
Polycrisis, 106, 107
Principles for Responsible Investment (PRI), 91, 393
    development, 394
    organization, 396
    principles, 394
    SDGs, 395
    signatories, 395
    six principles, 91
    sustainable global financial system, 395
    The SDG Investment Case, 395
Private foundations
    climate philanthropy, 448–450
    developing a shared framework, 449

INDEX

enhancing knowledge and learning, 449
improving communications, 449
mobilizing new commitments, 449
Proactive community engagement, 173
Product governance, 169
Program Building Checklist, 363
Prota Fiori, 360
Provisional agreement, 239–241, 267, 274, 279, 344
Proxy voting, 32, 204–207
Public and private investment, 135, 454
Public and private sectors, 42, 89, 98, 136, 375, 449
Public Benefit Corporations (PBCs), 356
   advantages, 357
   vs. B Corps, 358, 360
   creation, 357
   IPO, 358
   nonprofits, 358
   traditional corporations, 356
Public charities programming, 451
Public-private partnerships, 263, 420, 447

R

Rainforest Alliance (RFA), 431, 432
Ratings, 333, 334
   agencies, 340–343
   cost, 342
   definition, 345
   EU, 344
   influence, 340
   Key Issue Framework, 335, 336
   lack of transparency, 341
Ratings providers, 335, 337, 338, 409
   conflicts of interest, 341
   data limitations, 341
   data products, 336
   draft proposal, 346
   inconsistency, 341
   index providers, 339
   jurisdictions, 346
   practices, 346
   proposed regulation, 344, 345
Regulations, 33, 236, 292
Regulators, 293, 297, 335, 340, 343, 382
Regulatory compliance and standards, 365
Regulatory disclosure and reporting models
   jurisdictions, 295
   mandatory reporting, 295
   SEC, 295
   tools, 296
   transparency/accountability, 296, 297
   types, 296
   voluntary reporting, 295

INDEX

Regulatory efforts, 210, 288–289, 308, 343
Regulatory enforcement actions, 301
   Climate and ESG Task Force, 297, 298
   general factors, 299
   management's discussion and analysis (MD&A), 300
   risk factors, 300
   SEC, 298–300
Regulatory environment, 213–215, 330
Renewable energy credits (RECs), 254
Requests for proposals (RFPs), 405
Resource Conservation and Recovery Act (1976), 217
Responsible sourcing
   benefits, 139–140
   definition, 139
   elements, 140–141
   sustainability, 141
Rio Conference, 61
Risk management and controls, 365
Risk management systems, 274, 330
Road transportation
   electrification, 460–462
   medium and heavy trucking, 462, 463
Robust compliance, 330

Roundtable on Sustainable Palm Oil (RSPO), 419, 420
Royal National Institute of Blind People (RNIB), 157

S

Safe Drinking Water Act (1974), 218
Salesforce, 428
S&P Global, 338, 339, 341, 382
Save the Children, 426
Schneider Electric, 138
Science-based targets, 157, 418, 427–429
Science Based Targets initiative (SBTi), 426, 427
Science Based Targets Network (SBTN), 428
SEC climate disclosure rule, 253
   broader market, 257, 258
   companies, 257, 258
   compliance deadlines, 255
   disclosures, 253
   key provisions, 253–255
   vs. other jurisdictions, 256, 257
   phased implementation, 255
Securities and Exchange Commission (SEC), 35, 190, 204, 252, 257, 270, 271, 286, 295, 297
Securities Exchange Act, 190, 286
Shanghai Stock Exchange (SSE), 249

INDEX

Shanghai Tower, 476
Shareholder activists, 204
Shareholder litigation, 306–308
Shareholders, 1, 5, 32, 185, 202, 308
Shenzhen Stock Exchange (SZSE), 249
Singapore Exchange Regulations (SGX RegCo), 268
Smaller Reporting Companies (SRCs), 255, 256, 285
Social bonds, 381
Social cost analysis, 219, 220
Socially responsible investing (SRI), 3
Social regulation, 215
Social Venture Network (SVN), 414
Stakeholder engagement, 29, 95, 187, 202, 203, 361, 362, 365
Stakeholder relationship management platforms (SRM), 361
Stakeholders, 5, 31, 44, 309
Standards, 311, 312, 317, 322
Stock Exchange of Hong Kong (SEHK), 251
Stock exchange rules, 193, 223
Stockholm Conference, 60, 61
Strategic roadmap, 232, 364
Strategies, 16, 21, 122, 143, 200, 304, 349
Substantial funding, 137
Sulphur hexafluoride (SF_6), 114

Summits, 392, 399
Super pollutant, 219
Supply chains, 19, 55, 129, 133, 164, 274, 275
Supply chain transparency
  consumers, 167
  disclosure, 166
  ESG strategy, 167
  visibility, 166
Sustainability, 7, 10, 60, 416
  bonds, 381
  program, 10, 370, 371, 397
  regulation, 223, 231, 265
Sustainability Accounting Standards Board (SASB), 39, 271, 311, 313
Sustainability Disclosure Requirements (SDR), 225, 245
  aim, 249
  entity/product-level, 248
  FCA, 246
  implementation, 246, 247
  investors, 246
  labels, 247, 248
  provisions, 247
  regulatory framework, 246
  rules, 248
Sustainable Aviation Fuel (SAF), 464
Sustainable Blue Economy (SBE), 374, 376
Sustainable buildings, 473, 476

INDEX

Sustainable development
  Brundtland Commission, 58
  Brundtland report, 58
  components, 59
  definition, 59
  sustainability, 60
Sustainable Development Goals
    (SDGs), 10, 98, 153, 214,
    376, 377, 382, 393, 395, 400,
    413, 417, 423, 444
  challenges, 96
  concepts, 87
  ESG, 90
  ESG factors, 93–95
  17 goals, 85–88
  public charities
      programming, 451
  targets, 89
Sustainable finance
  Barclays, 383
  bonds, 382
  debt-for-nature swaps, 383
  economic opportunity, 381
  global economy, 382
  goals, 380
  instruments, 381
  long-term value, 382
  negative screening, 380
  public entities, 382
Sustainable Finance Disclosure
    Regulation (SFDR), 224,
    243, 244, 289
Sustainable Ocean Economy, 376

Sustainable Signals, 33
Sustainalytics, 336, 337, 341

T

Task Force on Climate-related
    Financial Disclosures
    (TCFD), 39, 257, 259, 313,
    317, 319, 329, 409
Taskforce on Nature-related
    Financial Disclosures
    (TNFD), 136, 315,
    317, 324–326
Taxonomy, 220, 222
Teachers' Retirement System
    (TRS), 206
Terrestrial biodiversity, 132
Tesco, 157
The Earthshot Prize, 439
The Edge, 476
The Ford Foundation, 446, 447
The Higg Index, 482, 483
The Norwegian Consumer
    Authority (NVA), 483
The Sustainable Apparel Coalition
    (SAC), 419, 482, 483
The US National Climate Resilience
    Framework, 448
Third-party mechanism, 427
Toxic Substances Control Act
    (1976), 217
Trade associations and business
    groups, 419–421, 483

Traditional philanthropy, 443, 444
Transition finance, 387
Transparent reporting, 362
Transportation sector
    aviation and maritime transport, 463, 464
    electric motors, 465
    electrification, 465
    energy efficiency, 464
    liquid fuels, 460
    net-zero emissions and transition, 460
    road transportation, 460–463
    zero-carbon fuels, 465
Trends and risks, 349–351
Triple bottom line, 4, 444

U

Unilever, 281, 366
United Nations (UN), 8, 391
United Nations Children's Fund (UNICEF), 392
United Nations Conference on Environment and Development (UNCED), 61
United Nations Conference on Trade and Development (UNCTAD), 168
United Nations Convention on the Law of the Sea (UNCLOS)
    achievements, 72
    provisions, 71
United Nations Development Programme (UNDP), 325, 392
United Nations Educational, Scientific and Cultural Organization (UNESCO), 406
United Nations Environment Assembly (UNEA), 397, 398
United Nations Environment Programme (UNEP), 40, 60, 322, 369, 376, 392, 394, 397
United Nations Environment Programme Finance Initiative (UNEPFI), 376
United Nations Framework Convention on Climate Change (UNFCCC), 41, 63, 109, 392
    challenges and criticisms, 70
    civil society, 69
    climate change, 68
    implementation, 69
    Kyoto Protocol, 69
    programs and mechanisms, 68
United Nations Global Compact (UNGC), 2, 8, 360, 396, 427
United Nations Refugee Agency (UNHCR), 392
United States Department of Agriculture's (USDA), 262
Universal Declaration of Human Rights (UDHR), 8, 48, 49, 159

INDEX

Urban planning, 174, 473–475
US interagency guidance, 384
Uyghur Forced Labor Protection Act (UFLPA), 275

V

Value alignment, 18, 205
Value innovation concept, 373
Value Reporting Foundation (VRF), 39, 312, 317
Verité, 432, 433
Voluntary Carbon Market Disclosures Business Regulations Act (VCMDA), 282
Voluntary Carbon Markets (VCMs), 126, 127

W, X, Y

Water stewardship, 142–143, 434
Water vapor, 114
WEF First Movers Coalition, 454
Wells Fargo employment practices, 306, 307
We Mean Business Coalition, 418, 427
World Business Council for Sustainable Development (WBCSD), 328, 413, 414, 418
World Commission on Environment and Development (WCED), 58
World Economic Forum (WEF), 370, 411, 412, 454
  challenges, 105
  Global Risks Report 2023, 101, 104
  Global Risks Report 2024, 101
  risk mitigation, 108
  top Global Risks 2023, 106, 107
  top Global Risks 2024, 102
World Food Programme (WFP), 392
World Meteorological Organization (WMO), 40, 108
World Resources Institute (WRI), 251, 252, 328, 414, 427
World Wildlife Fund (WWF), 420, 427, 429, 434

Z

Zero-carbon fuels, 465

GPSR Compliance

The European Union's (EU) General Product Safety Regulation (GPSR) is a set of rules that requires consumer products to be safe and our obligations to ensure this.

If you have any concerns about our products, you can contact us on

ProductSafety@springernature.com

In case Publisher is established outside the EU, the EU authorized representative is:

Springer Nature Customer Service Center GmbH
Europaplatz 3
69115 Heidelberg, Germany

www.ingramcontent.com/pod-product-compliance
Lightning Source LLC
LaVergne TN
LVHW010332260326
834688LV00036B/671